The Spacelab Story
Science Aboard the Shuttle

This book series presents the whole spectrum of Earth Sciences, Astronautics and Space Exploration. Practitioners will find exact science and complex engineering solutions explained scientifically correct but easy to understand. Various subseries help to differentiate between the scientific areas of Springer Praxis books and to make selected professional information accessible for you.

The *Springer-Praxis Space Exploration* series covers all aspects of human and robotic exploration, in Earth orbit and on the Moon and planets. The books tell behind-the-scenes stories of early and modern missions, both crewed and uncrewed, and cover all aspects of the space programs run by both leading and emerging spacefaring nations.

* * *

The books in this series are well illustrated with color figures and photographs. They are written in a style that space enthusiasts and historians, readers of popular magazines such as *Spaceflight* and readers of *Popular Mechanics* and *New Scientist* will find accessible.

Ben Evans

The Spacelab Story
Science Aboard the Shuttle

Published in association with
Praxis Publishing
Chichester, UK

Ben Evans
Space Writer
Atherstone, Warwickshire, UK

SPRINGER-PRAXIS BOOKS IN SPACE EXPLORATION

ISSN 2731-5401   e-ISSN 2731-541X
Springer Praxis Books
ISBN 978-3-031-53448-5   ISBN 978-3-031-53449-2 (eBook)
Space Exploration
https://doi.org/10.1007/978-3-031-53449-2

© The Editor(s) (if applicable) and The Author(s), under exclusive license to Springer Nature Switzerland AG 2024
This work is subject to copyright. All rights are solely and exclusively licensed by the Publisher, whether the whole or part of the material is concerned, specifically the rights of translation, reprinting, reuse of illustrations, recitation, broadcasting, reproduction on microfilms or in any other physical way, and transmission or information storage and retrieval, electronic adaptation, computer software, or by similar or dissimilar methodology now known or hereafter developed.
The use of general descriptive names, registered names, trademarks, service marks, etc. in this publication does not imply, even in the absence of a specific statement, that such names are exempt from the relevant protective laws and regulations and therefore free for general use.
The publisher, the authors, and the editors are safe to assume that the advice and information in this book are believed to be true and accurate at the date of publication. Neither the publisher nor the authors or the editors give a warranty, expressed or implied, with respect to the material contained herein or for any errors or omissions that may have been made. The publisher remains neutral with regard to jurisdictional claims in published maps and institutional affiliations.

This Springer imprint is published by the registered company Springer Nature Switzerland AG
The registered company address is: Gewerbestrasse 11, 6330 Cham, Switzerland

Paper in this product is recyclable.

Preface

The collaboration between the National Aeronautics and Space Administration (NASA) and the European Space Agency (ESA) to build a dedicated research laboratory which would fly in the payload bay of the Space Shuttle is without doubt one of the major success stories of the 135 flight Space Shuttle program. It provided a bridge between the Skylab Space Station of the 1970's and the International Space Station (ISS) of the 21st Century. Without its international missions (USA, Europe, Japan, Canada, and later, Russia) supporting microgravity research across multiple disciplines, it is highly unlikely the ISS would be as productive today as it has been.

"The Spacelab Story" is a success story on many levels, from programmatic management to research to international collaboration. It provided the model for the ISS, which is in turn providing a pathfinder for Cis-Lunar research on the Gateway Space Station. It is also about people with vision and common goals: engineers, managers, scientists, and crew members.

Of the many Spacelab missions described in this book, I had the opportunity and privilege to fly on three Spacelab missions: STS-61A (D-1) in 1985; STS-50 in 1992 and STS-71 in 1995. Each of them constituted a "first".

STS-61A/Deutschland One was the first Spacelab mission purchased by another country, West Germany, the first to have science controlled from outside the US (Oberfaffenhoven, Germany) and the first to support eight crewmembers. The flight was dedicated to microgravity research in materials science, human physiology, fluid physics, and biotechnology.

STS-50 or the United States Microgravity Lab-1, USML-1, was the longest Spacelab flight to date, 13 days, and was the first flight of a generation of new US research facilities destined for the ISS.

STS-71 was the first Space Shuttle mission to dock to the Russian Space Station, MIR. The shuttle mission picked up 3 crewmembers, one NASA Astronaut and two Russian cosmonauts, who had been on board MIR for 90 days. While the two vehicles were docked the Spacelab was used to conduct several physiological tests on the crew while in microgravity so that the data would not be confounded upon return to Earth gravity. It was the first to include the Russian Control Center in Moscow. STS-71 constituted the first of 9 docking missions to MIR (Phase I) used to develop a working relationship with the Russian Space Agency, RSA.

The success of "Phase I", led to "Phase II" – the International Space Station which has been in operation for more than 25 years. As part of the STS-71 operations, I had also trained as a backup to the NASA Crewmember who flew to MIR on the Soyuz 90 days prior. We trained together for 13 months with the Cosmonauts and Spacelab researchers from both the NASA and the Russian Space Agency/Institute for Biomedical Problems (IBMP) at Star City, Russia. That months long training opportunity helped to solidify the relationships between crewmembers and scientists on both sides.

These were just three missions. The full story of how Spacelab helped to shape engineering and scientific research and international collaboration in space over three decades into the 21st Century follows. Enjoy.

Dr. Bonnie J. Dunbar
NASA Astronaut 1980-2005
Mission Specialist, STS-61A/Spacelab-D1 (October-November 1985)
Payload Commander, STS-50/USML-1 (June-July 1992)
Mission Specialist, STS-71/Spacelab-Mir (June-July 1995)

Acknowledgements

This book would not have been possible without the support of a number of individuals, to whom I am enormously indebted. I must firstly thank my wife, Michelle, for her love, support and encouragement during the time it has taken to plan, research and write the manuscript. She has been uncomplaining during the weekends and holidays when I sat up late, typing on the laptop or poring through piles of books, old newspaper cuttings, magazines, interview transcripts, press kits or websites. It is to her, with my love, that *The Spacelab Story* is dedicated. I also extend my heartfelt thanks and appreciation to Dr. Bonnie J. Dunbar, one of very few NASA astronauts to have flown as many as three Spacelab missions: the joint U.S./German STS-61A/Spacelab-D1 in October-November 1985, STS-50/U.S. Microgravity Laboratory-1 (USML-1) in June-July 1992 and STS-71/Spacelab-Mir (SL-M), the first Shuttle-Mir docking mission in June-July 1995. I am grateful for Dr. Dunbar's contribution to this book.

My thanks also go to Clive Horwood of Praxis for his enthusiastic support and patience in what has been an overdue project and one which proved far more difficult to write than I imagined. Additional thanks go to Ed Hengeveld, who was gracious with his time in identifying and supplying suitable illustrations, including many 'unfamiliar' ones which will hopefully bolster the text. Others to whom I owe a debt of gratitude are Sandie Dearn and Malcolm and Helen Chawner. To those friends who have encouraged my fascination with all things 'space' over the years, many thanks: to the late Andy Salmon, to Andy Rowlands and Dave Evetts and Mike Bryce of GoSpaceWatch and to Rob and the late Jill Wood. And to my dear university friend Victoria Farmer, who passed away earlier in 2023, much too soon and far too young.

Our golden retrievers Ruby and Bella have also provided a ready source of light relief as they seized any available opportunity to drag me away from the laptop for a piece of toast, to play with them or take them for a walk.

Contents

Preface v

Acknowledgements vii

List of Illustrations xiii

1 **Unequal partners** 1
   WIND OF CHANGE 1
   A REUSABLE SPACECRAFT 2
   SEARCHING FOR PARTNERS 4
   A DOSE OF REALITY 7
   EUROPE'S OPTIONS 10
   LAST-CHANCE SALOON 14
   AN AGREEMENT AT LAST 18

2 **A difficult path to First Flights** 27
   FIRST SCIENTIFIC FLIGHT 27
   A MISSION HALVED 31
   ATMOSPHERE OF DISTRUST 34
   THE HUMAN EQUATION 44
   PATHFINDER FOR SPACELAB 48
   AMBITIOUS START, DRAMATIC END 52

Contents

3	**Verification Flight Test One**	55
	SELECTING SPACELAB'S FIRST CREW	55
	A SCIENTIFIC BONANZA	59
	PREPARING FOR SPACELAB-1	68
	RISE OF THE GREMLINS	71
	RISE OF THE COLUMBIA	73
	VOYAGE OF SCIENCE	77
4	**Verification Flight Test Two**	83
	"GOING TO SPAIN"	83
	THE TROUBLED ORIGINS OF THE IPS	87
	MULTIDISCIPLINARY PAYLOAD	91
	COLA AND SOLAR WARS	95
	A DRUNK BETWEEN THE LINES	98
	CLOCKWATCHING	102
5	**Of monkeys, mice and men**	109
	GRANDPA'S PAINTING	109
	GRAVITY GRADIENT	110
	SCREWING THE MONKEYS?	116
	UNTOUCHABLES	117
	UNDER THE COMMANDER'S NOSE	121
	SHIFTWORK AND *SEPPUKU*	124
	AROUND-THE-CLOCK SCIENCE	131
6	***Deutschland-Eins***	135
	"NOT ENOUGH BALANCE"	135
	NEW TERRITORY	137
	INSPIRATION AND HOPE	142
	SUSPENDED IN A GONDOLA	144
	"YOU WOULDN'T BELIEVE ME"	149
	DEUTSCHLAND-ZWEI	153
	RED LIGHTS AND FAILED FREEZERS	158
7	**Stargazers**	165
	"THE SUN KEPT RISING"	165
	FOUR HIGH-ENERGY 'EYES'	169
	LONG ROAD TO LAUNCH	173

	A TROUBLED MISSION	175
	ASTRO-2	181
	"INSTANT DAYLIGHT"	183
	THE LOST FLIGHTS	187
8	**Earthgazers**	199
	A FLIGHT UNFLOWN	199
	THE ATLAS METAMORPHOSIS	203
	A COMPLEX PAYLOAD	204
	PERSONALITIES AND RESPONSIBILITIES	208
	"LOOK AT THAT!"	211
	ASTRONAUT ROYALTY	216
	ENDS AND TRENDS	220
9	**X-SAR-Crossed Lovers**	225
	OF WORLDS UNSEEN	225
	MAPPING PLANET EARTH	227
	UNDER THE RADAR'S STEELY GLARE	235
	ROUND TWO	240
	LURE OF THE UNEXPECTED	242
	LASER-EYED CYCLOPS	248
	BRAVE NEW WORLD	251
10	**Mission for Japan**	261
	FAREWELL TRADITIONS	261
	"THE WAY SCIENCE IS DONE"	262
	A UNIQUE CREW	266
	"PUTTING ON A SHOW"	269
11	**Mission for Italy**	273
	"WEIRD SCIENCE"	273
	A CHANGING CREW	276
	TETHER JAM	278
	BAD SIMULATION RUN	282
	"THE TETHER IS BROKEN!"	285

12	**In the absence of *gravitas***	289
	JOURNEYS TO THE ENDS OF THE EARTH	289
	CHANGE AND TRAGEDY	291
	STAYING AWAKE FOR SCIENCE	293
	AS ONE MISSION CLOSES…	300
	"WHAT? TOGETHER?"	303
	CONSTANT SUPPLY OF CRYSTALS	307
	THE CURSE OF COMPETITION	314
	WIDE-RANGING RESEARCH	318
	GLIMMERS OF THE FUTURE	322
	TWO-FACED MISSION	330
	BROKEN ANKLE, BROKEN PLANS, BROKEN RECORDS	339
13	**The human factor**	353
	A PROBLEM WITH MONKEYS	353
	"THREE WOMEN OR THREE DOCTORS?"	357
	ANIMALIA, GALORE	359
	FIRST MEDICAL RESEARCH FLIGHT	363
	CONTROVERSIAL DISSECTIONS	369
	UNDERSTANDING THE HUMAN ORGANISM	376
	THE TRAUMA OF NEUROLAB	378
	THE PACE OF 'BLURROLAB'	385
	GAPS IN CAPABILITY	388
	A MEANINGFUL ENDEAVOUR	391

Bibliography 401

Index 417

List of Illustrations

Fig. 1.1	Described by several astronauts as a butterfly, bolted to a bullet, the asymmetrical launch architecture of the Shuttle saw it propelled uphill by the orbiter's three main engines, an External Tank (ET) and a pair of Solid Rocket Boosters (SRBs).	3
Fig. 1.2	Early conceptualisation of the Spacelab module and pallet in the Shuttle's payload bay. Note the orbiter's two-tiered cockpit, with upper 'flight deck' and lower 'middeck', the latter of which is connected to the Spacelab module by means of a pressurised tunnel.	6
Fig. 1.3	Cutaway artist's impression of the Spacelab pressurised module and pallet in the Shuttle's payload bay. Notable in this view are the ceiling-mounted Scientific Airlock (SAL, visible with deployed instruments at right) and Scientific Window Adapter Assembly (SWAA, at left).	10
Fig. 1.4	Enshrouded in white thermal insulation material, the Spacelab long module sits in Columbia's payload bay during STS-9 in late 1983. The pressurised tunnel, connecting the module to the Shuttle's middeck, is visible in the lower part of the photograph.	15
Fig. 1.5	An engineering test model of the Spacelab module undergoes fit-checks inside the Operations and Checkout Building at Florida's Kennedy Space Center (KSC).	19
Fig. 1.6	The Spacelab pressurised module (background) and U-shaped unpressurised pallet would support dozens of missions in support of science and technology, International Space Station (ISS) construction and maintenance and Hubble Space Telescope (HST) repair and servicing between 1981 and 2009.	23

List of Illustrations

Fig. 2.1	The first Office of Space and Terrestrial Applications (OSTA-1) payload, dominated by the large Shuttle Imaging Radar (SIR-A), is gently craned into Columbia's payload bay. The U-shaped Spacelab pallet, built by British Aerospace, is clearly visible.	30
Fig. 2.2	The OSTA-1 payload in orbit, captured through one of Columbia's aft flight deck windows. SIR-A is clearly visible at the left side of the frame, with FILE's pyramidal sunrise sensor at right of centre.	33
Fig. 2.3	The first shipment of Spacelab hardware is offloaded into the high bay of KSC's Operations and Checkout Building.	37
Fig. 2.4	The second Spacelab flight unit, purchased by NASA under the Follow-On Production (FOP) program, made its first mission on West Germany's Spacelab-D1 in October 1985. The module is pictured being loaded into Challenger's payload bay. Note the access hatch to the Shuttle's middeck, just left of centre, which will accommodate the connecting tunnel to the module.	41
Fig. 2.5	Bob Parker (left) and Wubbo Ockels lend a measure of scale to this interior view of the Spacelab module.	45
Fig. 2.6	The tall Thermal Canister Experiment (TCE) stands particularly prominent in this view of the OSS-1 payload during pre-launch processing.	49
Fig. 2.7	Unusual view of the OSS-1 payload during the closure of Columbia's payload bay doors; one door is already shut, the other about to be winched closed. The twin 'drums' of the Solar Flare X-ray Polarimeter (SFXP, at left) and the Plasma Diagnostics Package (PDP, with its black spherical receivers, at right), are visible in the foreground, with the gold-coloured Thermal Canister Experiment (TCE) behind.	53
Fig. 3.1	Spacelab-1 prime and backup payload specialist candidates inspect a model of the module and pallet combination. From left to right are Wubbo Ockels, Ulf Merbold, Mike Lampton, Claude Nicollier and Byron Lichtenberg.	59
Fig. 3.2	View of part of the pallet hardware, as seen through the small window in the Spacelab-1 module's end-cone. The aft bulkhead of Columbia can be seen behind.	61
Fig. 3.3	Close-up view of the tunnel adapter which connected the Spacelab module to the Shuttle's middeck. Notice the raised 'joggle' section, which was incorporated to raise the tunnel to the same level as the module's access hatch.	64
Fig. 3.4	U.S. Vice President George H.W. Bush welcomes the arrival of the first Spacelab flight unit in Florida in February 1982. From left to right are Claude Nicollier, Ulf Merbold, Bush, Wubbo Ockels and Bob Parker. Standing in the background is Owen Garriott.	67

List of Illustrations   xv

Fig. 3.5	Spacelab-1's research racks are prepared for rolling into the module's pressurized shell in the Operations and Checkout Building.	71
Fig. 3.6	The STS-9 crew sits down to breakfast in the Operations and Checkout Building on 28 November 1983. From left to right are Ulf Merbold, Bob Parker, John Young, Brewster Shaw, Byron Lichtenberg and Owen Garriott.	74
Fig. 3.7	Columbia roars aloft on 28 November 1983.	77
Fig. 3.8	Owen Garriott negotiates the 'lip' at the top of the joggle as he floats from the tunnel into the Spacelab-1 module.	80
Fig. 3.9	"Six dirty ol' men" was John Young's description of the unshowered Spacelab-1 crew as they returned to Edwards Air Force Base in California on 8 December 1983. At the foot of the steps is Young, about to shake hands with George Abbey, JSC's director of flight crew operations. Young is followed by Brewster Shaw, Bob Parker (in sunglasses), Ulf Merbold, Owen Garriott and Byron Lichtenberg.	82
Fig. 4.1	Burning three perfect main engines and two Solid Rocket Boosters (SRBs), Challenger rises from Earth on 29 July 1985. Six minutes later, calamity hit Spacelab-2.	86
Fig. 4.2	The Spacelab-2 payload is pictured during final closeout at KSC. The Instrument Pointing System (IPS), in its horizontal integration, is visible at centre, with the University of Chicago's Cosmic Ray Nuclei Experiment (CRNE) at far left.	89
Fig. 4.3	Spacelab-2 consisted of Pallet One (far right) to house the IPS, a double Pallet Two/Three with the large square baffle of the X-Ray Telescope (XRT) and the inverted cone of the Small Helium-Cooled Infrared Telescope (IRT) at centre and the Cosmic Ray Nuclei Experiment (CRNE) on its own unique support structure (far left).	92
Fig. 4.4	Loren Acton, pictured on Challenger's aft flight deck during STS-51F, was a co-investigator of the SOUP instrument.	95
Fig. 4.5	One of the few on-orbit views of the University of Chicago's Cosmic Ray Nuclei Experiment (CRNE) 'duck egg'. With the erection of the IPS, the cosmic ray detector was virtually hidden from the astronauts' view.	98
Fig. 4.6	The IPS comes alive for its first test mission on Spacelab-2.	101
Fig. 4.7	His feet anchored in place by foot-loops, John-David Bartoe works at one of the Data Display Units (DDUs) on Challenger's aft flight deck during Spacelab-2.	104
Fig. 4.8	Commander Gordon Fullerton shakes hands with George Abbey after leaving Challenger on 6 August 1985. Following the commander down the steps are Tony England, Loren Acton, Karl Henize, Roy Bridges, John-David Bartoe and shaven-headed Story Musgrave.	107

List of Illustrations

Fig. 5.1	Artist's concept of Challenger in her 'gravity gradient' attitude during the majority of STS-51B, tail directed Earthward. Note the pressurised Spacelab long module and the truss-like MPESS at the rear of the payload bay.	112
Fig. 5.2	Technicians lower themselves into the Spacelab-3 module to install the rodent and primate cages before launch. With Challenger in a vertical orientation on the pad, one technician waited in the tunnel's joggle section (top of frame), whilst a second descended into the module by means of a sling-like seat.	115
Fig. 5.3	One of STS-51B's squirrel monkeys is prepared for loading into his enclosure cage.	118
Fig. 5.4	Bill Thornton checks on one of the squirrel monkeys during his shift.	121
Fig. 5.5	Once the gold and silver shifts began operating, the two teams rarely saw each other. Here, silver-shift member Norm Thagard bales out of his coffin-like sleep station to greet gold crewmate Don Lind.	124
Fig. 5.6	Assisted by his crewmates, Taylor Wang labours to save the Drop Dynamics Module (DDM), which short-circuited and failed shortly after Challenger reached orbit. The drum-like Scientific Airlock (SAL) in the Spacelab-3 module's ceiling is visible near Wang's feet at the top of the frame.	127
Fig. 5.7	Engineers, technicians and scientists worked around-the-clock to support Spacelab-3 from the Marshall Space Flight Center (MSFC) in Huntsville, Alabama.	130
Fig. 6.1	Ulf Merbold, the first non-American astronaut to fly aboard a U.S. spacecraft, works inside the Spacelab-1 module.	136
Fig. 6.2	Wubbo Ockels trains on the Vestibular Sled in a ground-based mockup of the Spacelab-D1 module, as crewmate Reinhard Furrer looks on.	141
Fig. 6.3	Challenger launches on her last fully successful mission on 30 October 1985.	144
Fig. 6.4	Wubbo Ockels climbs into his self-designed sleeping bag.	146
Fig. 6.5	Reinhard Furrer and Bonnie Dunbar are pictured at work inside the Spacelab-D1 module during blue-shift operations.	149
Fig. 6.6	The Spacelab-D2 pressurised module is pictured in Columbia's payload bay, awaiting the installation of the tunnel adapter.	151
Fig. 6.7	The STS-55 crew, clad in clean-room garb, inspects the Spacelab-D2 facility, prior to final closeout for launch. Standing, left to right, are Ulrich Walter, Bernard Harris, Hans Schlegel, Steve Nagel and Tom Henricks. Kneeling are Jerry Ross (left) and Charlie Precourt.	155

Fig. 6.8	After a month-long delay, Columbia roars into orbit on 26 April 1993.	158
Fig. 6.9	All four Spacelab-D2 payload crew members are pictured at work during a shift changeover inside the pressurised module. Bernard Harris and Hans Schlegel work on a life science experiment, whilst Ulrich Walter winces with Baroreflex fastened about his neck and Jerry Ross (far left) tends to other duties.	162
Fig. 7.1	Playful portrait of Jon McBride's STS-61E crew, illustrating the astronomical focus of their mission, which should have been the next Shuttle flight after the Challenger tragedy. Standing, from left to right, are Ron Parise, Jon McBride, Sam Durrance, Jeff Hoffman and Dick Richards. Seated are Bob Parker (left) and Dave Leestma.	168
Fig. 7.2	Technicians prepare the ASTRO-1 payload for flight in early January 1986, only three weeks before the Challenger tragedy. The circular base and yoke of the Instrument Pointing System (IPS) is visible at centre, with the igloo in the foreground.	172
Fig. 7.3	The Hopkins Ultraviolet Telescope (HUT) is lowered gently into position aboard the ASTRO-1 payload.	176
Fig. 7.4	Technicians prepare to close Columbia's payload bay doors, ahead of STS-35. ASTRO-1 is visible on its two Spacelab pallets at the top of the frame, with the Broad Band X-Ray Telescope (BBXRT) at the bottom.	178
Fig. 7.5	The ASTRO-1 instruments stand proud after deployment.	181
Fig. 7.6	These three men were originally slated to rotate through three ASTRO missions between March 1986 and July 1987. Sam Durrance (right) and Ron Parise (centre) would fly ASTRO-1, with Durrance joining Ken Nordsieck (left) for ASTRO-2 and Parise joining Nordsieck for ASTRO-3. In the aftermath of the Challenger tragedy, Nordsieck departed the program and Durrance and Parise wound up flying both ASTRO-1 and ASTRO-2. A third mission was later cancelled.	183
Fig. 7.7	Unusual view of ASTRO-2, as seen through Endeavour's aft flight deck windows.	185
Fig. 7.8	Pictured during final closeout activities ahead of Spacelab-2, several of the IPS-based payloads from that mission were scheduled to fly the three Sunlab flights in 1987, 1989 and 1990. All three fell foul to budget cuts in the aftermath of the Challenger tragedy.	188
Fig. 7.9	STS-51F roars to orbit on 29 July 1985, carrying the IPS on its inaugural test flight. Had launch schedules and funding profiles been kinder, the pointing system may have flown many more times than the three missions it actually did.	192

List of Illustrations

Fig. 8.1	The dual Spacelab pallets of the ATLAS-1 payload are readied for installation into Atlantis' payload bay. Visible are two of the three spheres for the Space Experiments with Particle Accelerators (SEPAC) and the mission's igloo.	202
Fig. 8.2	The payload crew for ATLAS-1. Payload commander Kathy Sullivan (seated, second from left) replaced EOM-1/2's Owen Garriott. Also assigned to the mission were Mike Foale (standing, right), who went on to serve as payload commander on ATLAS-2, and payload specialists Byron Lichtenberg (standing, left) and Dirk Frimout. The latter replaced Mike Lampton, diagnosed with kidney cancer a few months before launch.	206
Fig. 8.3	Technicians install hardware into ATLAS-1's igloo, prior to integration. The twin pallets and instrument suite are visible in the background.	210
Fig. 8.4	Glorious view of ATLAS-1 in Atlantis' payload bay, backdropped by the grandeur of Earth beyond. Note the twin canisters of the Shuttle Solar Backscatter Ultraviolet (SSBUV) instrument on the payload bay wall in left foreground.	214
Fig. 8.5	Flying as a single-pallet configuration, and as a co-manifested payload, for its second and third missions, ATLAS-2 is readied for launch.	218
Fig. 8.6	The ATLAS-3 pallet is readied for installation aboard Atlantis.	221
Fig. 8.7	Backdropped by the home planet, ATLAS-3 marked the final flight of this remarkable suite of Earth-watching instruments.	224
Fig. 9.1	Technicians work on the SIR-B radar instrument, mounted atop its Spacelab pallet in KSC's Operations and Checkout Building.	228
Fig. 9.2	Challenger's Remote Manipulator System (RMS) mechanical arm, deftly controlled by Sally Ride, gently pushes one of the SIR-B antenna 'leaves' into place.	232
Fig. 9.3	Unusual perspective of the SIR-B radar, atop its Spacelab pallet, backdropped by the grandeur of Earth. The 50-foot-long (15-metre) RMS mechanical arm appears to curl, snake-like, in the foreground.	234
Fig. 9.4	SRL-1's red shift, pictured in their coffin-like sleep stations on Endeavour's middeck. From top to bottom are Sid Gutierrez, Linda Godwin and Kevin Chilton.	237
Fig. 9.5	Tom Jones looks through Endeavour's aft flight deck windows into the payload bay. Notice the large-format Linhof camera in an overhead window at far left and the Velcro patches, recommended by Godwin, to hold cameras and lenses securely in place.	241

Fig. 9.6	The SRL-2 payload is lowered onto the Spacelab pallet during STS-68 pre-flight preparations.	245
Fig. 9.7	An orbital sunrise glints over SRL-2.	248
Fig. 9.8	The tall barrel of the LITE instrument (far right) and its adjunct instrumentation affixed to the Spacelab pallet in Discovery's payload bay during STS-64.	251
Fig. 9.9	The Shuttle Radar Topography Mission (SRTM) payload is pictured during final integration in the Operations and Checkout Building. The SIR-C radar dominates the payload, with the X-SAR and mast canister visible at far right.	254
Fig. 9.10	Gerhard Thiele and another STS-99 crewmember inspect the SRTM payload during pre-flight preparations.	257
Fig. 9.11	The SRTM mast pictured fully extended over Endeavour's payload bay wall during STS-99.	259
Fig. 10.1	Mark Lee and Jan Davis, the first married couple to fly together in space, examine the interior components of the Spacelab-J module.	264
Fig. 10.2	Impressive view of the interior of the Spacelab-J module, as seen from a crewmember translating through the tunnel adapter.	266
Fig. 10.3	Mae Jemison inspects the amphibians' eggs during Spacelab-J.	268
Fig. 10.4	Commander Robert 'Hoot' Gibson emerges from the tunnel adapter and enters the brightly lit Spacelab-J module.	271
Fig. 11.1	Franco Malerba poses with the Tethered Satellite System (TSS), loaded aboard Atlantis' payload bay.	276
Fig. 11.2	Deftly manipulated by Claude Nicollier, Atlantis' RMS mechanical arm deploys EURECA. In the payload bay, the spherical TSS is clearly visible atop its Spacelab pallet.	279
Fig. 11.3	The TSS tether pictured during deployment on STS-46.	281
Fig. 11.4	A playful portrait of the STS-46 crew, seemingly entangled in the tether. Front row, left to right, are Franco Malerba, Jeff Hoffman and Loren Shriver. Back row, left to right, are Claude Nicollier, Marsha Ivins, Franklin Chang-Díaz and Andy Allen. More than half of the crew went on to fly STS-75.	284
Fig. 11.5	Franklin Chang-Díaz, Jeff Hoffman and Claude Nicollier eat a meal whilst halfway inside their coffin-like sleep stations.	286
Fig. 12.1	IML-1 experiment containers are passed through Discovery's crew hatch for stowage in middeck lockers on the day before launch. After reaching orbit, the experiments were transferred into the Spacelab module.	292

Fig. 12.2	IML-1's tunnel adapter is lowered into Discovery's payload bay, ahead of connection to the Spacelab module at one end and the middeck airlock hatch at the other.	294
Fig. 12.3	Roberta Bondar prepares for a session in the Microgravity Vestibular Investigations (MVI) chair in the IML-1 module.	296
Fig. 12.4	Dave Hilmers, wearing an instrumented helmet, participates in an MVI investigation. Hilmers is positioned in the pitch position, which Norm Thagard noted still gave the chair's occupants a curious sense that they were sitting 'sideways'.	297
Fig. 12.5	Unusual view of the IML-1 module and tunnel in Discovery's payload bay, pictured during the closure of the payload bay doors on 30 January 1992.	299
Fig. 12.6	The cryogenic oxygen and hydrogen tanks of the Extended Duration Orbiter (EDO) pallet are clearly visible at the rear of Columbia's payload bay during USML-1 pre-launch closeout operations. The insulation-covered Spacelab module sits in the middle of the frame.	302
Fig. 12.7	Dr. Bonnie J. Dunbar, USML-1's Payload Commander, prepares to store material samples from the Crystal Growth Furnace (CGF). Dr. Dunbar chaired a committee which conducted a year-long study for NASA in 1987, resulting in a proposed mission called US-1. This later was renamed USML-1 and she was assigned as its first Payload Commander.	304
Fig. 12.8	Columbia roars to orbit on 25 June 1992.	306
Fig. 12.9	Larry DeLucas at work aboard USML-1.	308
Fig. 12.10	A camera-toting Ellen Baker floats, seemingly inverted, out of the tunnel and into the brightly lit USML-1 module, a place which Bonnie Dunbar had decreed was a coffee-free (and pilot-free) zone.	311
Fig. 12.11	Carl Meade reviews procedures before conducting an experiment inside the USML-1 module. This particular double rack housed the Generic Bioprocessing Apparatus (GBA), the Solid Surface Combustion Experiment (SSCE) and hardware for the EDO Medical Program.	313
Fig. 12.12	During the pre-flight Crew and Equipment Interface Test, payload commander Rick Hieb (outstretched hand) discusses the IML-2 payload with his crewmates. Facing the camera are Don Thomas and Chiaki Mukai.	315
Fig. 12.13	STS-65 Commander Bob Cabana, wearing protective goggles, opens a science drawer aboard IML-2, as payload commander Rick Hieb looks on.	319

List of Illustrations   xxi

Fig. 12.14 Leroy Chiao (upper) and Don Thomas work at respective experiments aboard IML-2 during STS-65's blue-shift activities. ... 321

Fig. 12.15 Spacelab proved a highly versatile research facility, and a roomy one, as this IML-2 view of the module's interior reveals. ... 323

Fig. 12.16 Kathy Thornton was convinced that her lucky U.S. socks aided in USML-2's success, after six scrubbed launch attempts. ... 326

Fig. 12.17 Al Sacco is pictured at work aboard USML-2. ... 329

Fig. 12.18 Unusual view of the USML-2 module, with Columbia's port-side payload bay door positioned at 62-degrees-open to facilitate micrometeoroid protection. ... 331

Fig. 12.19 Research racks for the Life and Microgravity Spacelab (LMS) are rolled inside the pressurised module's cylindrical shell, ahead of electrical, mechanical and fluid connections. ... 333

Fig. 12.20 Canada's Bob Thirsk participates in one of LMS' many life sciences investigations, as Chuck Brady (background) conducts another. Note the multitude of foot restraints positioned across the Spacelab module's 'floor', to assist in securing crewmembers at each research facility. ... 336

Fig. 12.21 The Microgravity Science Laboratory (MSL)-1 module is readied for integration into Columbia's payload bay for STS-83. ... 340

Fig. 12.22 Don Thomas, pictured during IML-2 in July 1994, would almost miss out on flying at all in 1997, but that year he went on to log two more missions. He remains one of only a handful of astronauts to have flown three times with the Spacelab pressurised module. ... 343

Fig. 12.23 The MSL-1 facility, pictured during its short maiden voyage in April 1997. ... 346

Fig. 12.24 Technicians in the Operations and Checkout Building conduct servicing of Columbia's airlock and beneath her payload bay floor in May 1997, as preparations ramp up to refly MSL-1 the following July. ... 348

Fig. 12.25 Susan Still and Don Thomas move a piece of experiment inside the MSL-1 module during the STS-94 reflight of the laboratory mission. ... 350

Fig. 12.26 Roger Crouch works in the MSL-1 module during STS-94. ... 352

Fig. 13.1 Millie Hughes-Fulford, Bob Phillips (seated) and Drew Gaffney participate in SLS-1 training in a mockup of the pressurised Spacelab module. ... 355

Fig. 13.2 The SLS-1 module, minus its connecting tunnel, is pre-positioned in Columbia's cavernous payload bay. ... 358

List of Illustrations

Fig. 13.3	SLS-1 was one of the few single-shift Spacelab missions, dictated by the needs of its life sciences payload. Here, Rhea Seddon has secured her sleeping bag and donned an eye-patch to get some rest inside the module at the end of a long workday.	360
Fig. 13.4	Jim Bagian appears to 'fly' through the Spacelab module during a moment of levity on the busy mission.	363
Fig. 13.5	Millie Hughes-Fulford prepares to spin-up Jim Bagian on a rotating chair for a physiological study.	366
Fig. 13.6	Columbia touches down at Edwards Air Force Base in California on 14 June 1991, a landing location specifically selected to afford the SLS-1 payload crew immediate access to post-flight research facilities and personnel.	368
Fig. 13.7	Columbia launches on SLS-2 on 18 October 1993.	370
Fig. 13.8	John Blaha and Rhea Seddon work with the rotating dome experiment aboard SLS-2.	373
Fig. 13.9	Dave Wolf draws blood from Marty Fettman, the first professional veterinarian to fly in space.	375
Fig. 13.10	Payload commander Rhea Seddon prepares to 'spin-up' Marty Fettman on the rotating chair.	378
Fig. 13.11	Rhea Seddon, payload commander of SLS-2, works with a camera at the rear end of the Spacelab module. She participated to a significant degree in developing cardiovascular experiments for Neurolab.	380
Fig. 13.12	Neurolab, the final flight of the pressurised Spacelab module, is readied for launch.	383
Fig. 13.13	Dave Williams is instrumented with elaborate head gear for one of Neurolab's many research investigations.	386
Fig. 13.14	The Spacelab-Mir pressurised module sits in Atlantis' payload bay during final closeout activities. Note the Orbiter Docking System (ODS) in the forward payload bay.	389
Fig. 13.15	Gennadi Strekalov and Bonnie Dunbar are pictured at work during Spacelab-Mir's many medical activities.	393
Fig. 13.16	Norm Thagard (lower left) and Bonnie Dunbar (right) jointly became the first Americans to fly three times with the pressurised module, whilst Ellen Baker (left), as the mission's payload commander, was making her second Spacelab mission.	395
Fig. 13.17	The Spacelab-Mir pressurised module is pictured inside Atlantis payload bay as the Shuttle undocks from Mir on 4 July 1995. This view was captured by cosmonauts Anatoli Solovyov and Nikolai Budarin, station-keeping in the Soyuz TM-21 spacecraft.	397

1

Unequal partners

WIND OF CHANGE

On the day that Spacelab was born, it was pleasantly warm in Washington, D.C., but the political climate had taken a chillier turn. The skies over the United States' capital on 24 September 1973 were partly cloud-dappled, as a hot summer surrendered to a mild autumn. Eight months earlier, five burglars were sensationally convicted of wire-tapping the Democratic National Committee headquarters in the Watergate Office Building, working at the behest of corrupt officials in President Richard Nixon's re-election campaign.

As a troubled year rumbled on, Nixon became enmired in one of the greatest controversies of the last century and sought to distance himself from a mushrooming scandal. Yet the aftershocks of Watergate ruined the president's fragile reputation, exposed his culpability and forced the first resignation of a sitting U.S. head of state. Washington's early autumnal warmth that day presaged a blustery wind of change that whipped about the fine porticoes and sculpted colonnades of the White House, tightening its icy tendrils around the Watergate burglars and dozens of government officials, including two attorney generals and Nixon's chief of staff. At length, it toppled the president himself.

Elsewhere in the capital that Monday, this looming end to an embattled administration, which had weathered bitter, grinding conflict in Vietnam and acrimonious civil rights strife at home was juxtaposed against the signing of a historic Memorandum of Understanding between the United States and ten European nations to facilitate collaborative space endeavours. At the Department of State, a few blocks west of the White House, sat James Fletcher, administrator of the National Aeronautics and Space Administration (NASA), and Alexander Hocker, director general of the European Space Research Organisation (ESRO), precursor of today's European Space Agency (ESA).

Fletcher had led NASA since April 1971, following a tenure as president of the University of Utah. With his pepper-and-salt hair swept back from his forehead, a lean face framed by thick-rimmed spectacles, Fletcher shook hands with the stocky Hocker, a former Leipzig judge who had helmed ESRO for two years. It was triumphant moment for both men, as America's effort to make its reusable Space Shuttle a reality and Europe's desire to carve a niche for itself in the annals of human spaceflight finally bore fruit.

A REUSABLE SPACECRAFT

The Shuttle arose from a growing national need to reduce the cost of sending people into space. With Cold War tensions between the United States and the Soviet Union starting to thaw by the early 1970s, competition between the superpowers had lost its razor sharpness. But this changing geopolitical reality was matched by a thorny path of decline for NASA's budget, which shrank from 4.4 percent of federal expenditure in 1966 to 0.98 percent by 1975. In February 1969, the Space Task Group (STG), chaired by Vice President Spiro Agnew, identified the Shuttle as a means of putting humans into space reliably, cheaply and frequently. The concept received warm endorsement from the scientific community, pledging a "radical reduction in unit-cost of space transportation".

After a troubled genesis, in January 1972 the Shuttle's design was finalised. It comprised a delta-winged 'orbiter', 122 feet (37 metres) long with 78-foot (24-metre) wingspan, roughly scalable to a DC-9 airliner. It was powered for the first 8.5 minutes of flight by three liquid-fuelled main engines, which generated 1.2 million pounds (540,000 kilograms) of thrust. Its cavernous payload bay, 60 feet (18 metres) long and 15 feet (4.5 metres) wide, was sealed for launch and re-entry by two clamshell doors. Up to eight astronauts could live for seven to 30 days in its forward cockpit, which was divided into an upper 'flight deck' for operations and lower 'middeck' for eating, hygiene and rest (Fig. 1.1).

For launch, the orbiter sat astride a 154-foot-tall (47-metre) External Tank (ET), which housed liquid oxygen and hydrogen to feed the main engines. Flanking the 'stack' were a pair of Solid Rocket Boosters (SRBs), each rising to 149 feet (45.5 metres) and fuelled by a powder of polybutadiene acrylonitrile and ammonium perchlorate composite propellant. They generated a

Fig. 1.1: Described by several astronauts as a butterfly, bolted to a bullet, the asymmetrical launch architecture of the Shuttle saw it propelled uphill by the orbiter's three main engines, an External Tank (ET) and a pair of Solid Rocket Boosters (SRBs).

liftoff punch of 6.6 million pounds (2.9 million kilograms), igniting with the main engines on the ground in a 'parallel-burn' architecture. The boosters were jettisoned two minutes after launch, separating at an altitude of 28 miles (45 kilometres) and parachuting to an ocean splashdown for refurbishment and reuse.

Meanwhile, the orbiter/ET continued onward for six more minutes, propelled by the main engines. At 8.5 minutes after launch, the engines shut down and the ET was discarded at 70 miles (115 kilometres) to burn up in the atmosphere. After completing its mission, the orbiter returned from space to land on a prepared runway. Knifing through the atmosphere at blistering hypersonic speeds, its aluminium airframe guarded from temperatures as high as 1,650 degrees Celsius (3,000 degrees Fahrenheit) by a patchwork of 20,000 silica tiles and thermal blankets, the Shuttle swept home like an enormous unpowered glider.

In August 1972, NASA awarded the space division of North American Rockwell (later Rockwell International) in Palmdale, California, a $2.6 billion letter contract to build two orbiters. The first, 'Enterprise', performed approach and landing tests at Edwards Air Force Base in California in 1977, whilst the second, 'Columbia'—named for the United States' female personification—flew the first orbital missions. Another four orbiters (Challenger, Discovery, Atlantis and Endeavour) came online between 1983 and 1992. When the fleet retired in July 2011, this quintet of vehicles had flown 135 times, with Challenger destroyed in a launch conflagration in January 1986 and Columbia lost during re-entry in February 2003. The others now sit as oversized museum-pieces: Discovery in Washington, D.C., Atlantis in Florida, Endeavour in California.

Sixteen research missions between November 1983 and May 1998 lifted the cylindrical, school-bus-sized Spacelab 'module' into space, furnishing a pressurised 'shirtsleeve' environment for astronauts to perform experiments in the weightless environment of low-Earth orbit. Twenty-six other flights between November 1981 and May 2009 used unpressurised 'pallets' to expose scientific instruments for Earth observations, astronomy, solar physics, space plasma physics and astrophysics, as well as servicing the Hubble Space Telescope (HST) and constructing the International Space Station (ISS).

SEARCHING FOR PARTNERS

And dreams of space stations are an apt place to begin this survey. After the Apollo 11 Moon landing in July 1969, NASA had hardware built for America's first orbital station, the 'monolithic' Skylab, which supported three crews for several months in 1973-1974. But a larger outpost had crystallised into a longer-term goal. In an April 1969 request for proposals to U.S. industry, NASA outlined "a 12-man Earth-orbital space station" which "could be developed by 1975" with an operational lifetime of a decade, "subject to resupply of expendables and rotation of crews with logistics vehicles" to constitute "the initial element of a large space base". By late 1970, the Shuttle assumed pole position to assemble this station "over a period of months". According to a study that November, a small crew might inhabit the nascent outpost after its first five modules were in place and a full 12-man complement would follow. And in December, when study contracts were awarded to North American Rockwell and McDonnell Douglas, it was hoped that the station might achieve six-man capability by 1978. But in plans published in March 1971, that 'operational' date had slipped into the early 1980s. As these concepts took form, continued emphasis focused upon "assuring inherent growth potential…over a period of years".

One product which found traction in these discussions was a Research and Applications Module (RAM), nicknamed the 'Sortie Can'. It was initially part of the modular space station, before evolving into a stand-alone facility for carriage by the Shuttle. And it was the RAM which grew into the heart and bones of Spacelab.

Yet Europe's role in building Spacelab was neither clear-cut nor very likely at this juncture. A decade earlier, Europe's aspirations to play a role in human spaceflight were less lofty. Although the National Aeronautics and Space Act, enacted by Congress in July 1958, explicitly recommended "co-operation…with other nations and groups of nations", the reality proved a tougher nut to crack. NASA opened its doors in October 1958, but there were few international partners with whom the United States could work and no 'national' space agencies. ESRO was born in June 1962 with ten nations—Belgium, Denmark, France, Italy, the Netherlands, Spain, Sweden, Switzerland, the United Kingdom and West Germany—signing its convention in March 1964. Norway and Austria adopted an impartial status as observers. But ESRO's focus was upon space sciences and satellite-based communications, not human spaceflight. In March 1962, the European Launcher Development Organisation (ELDO) came into being, its charter signed by Belgium, France, Italy, the Netherlands, the United Kingdom and West Germany, with Australia as an associate member. Its remit was to build high-altitude sounding rockets, flying out of Woomera, South Australia.

Still, hope sprang eternal that Europe might foster closer ties with NASA. In May 1962, John McCloy, chair of the General Advisory Committee of the Arms Control and Disarmament Agency, appealed to delegates at an International Institute for Space Studies conference in the English town of Brighton that Europe should forego its aspirations of becoming the world's third space power. On the eve of the Cuban Missile Crisis, with relations between East and West at their lowest ebb, McCloy was aware that the Soviets tended to moderate their views whenever they sensed a united position on the part of the West.

Yet efforts to promote a distinct European space identity persisted. In May 1963, Arnold Frutkin, NASA's deputy director of international programs, addressed the European Space Flight Symposium in the West German city of Stuttgart. He recommended an "international pooling of efforts" to be "rigidly directed to solid scientific and technical objectives of mutual interest". Frutkin added that the "freest possible association of our respective programs, where they complement and support each other, is the soundest way to proceed if we are to continue to benefit in this way and if we are to reciprocate generously".

There was also a growing clamour for a pan-European equivalent to NASA: a co-ordinating body to bring the interests of member-states into equilibrium. NASA had long desired a "unitary management agency" in Europe. In February 1965, Jean Delorme, chairman of France's Aire Liquide industrial gases corporation and head of the Eurospace non-profit organisation, called for the creation of a centralising entity—"the supranational European NASA"—with sufficient authoritative clout to drive its own financial decision-making processes.

A year later, NASA extended a formal invitation from President Lyndon Johnson for European participation "in U.S. manned and unmanned space programs", although no firm commitments resulted. In many minds, a technological capability 'gap' existed between the two sides of the Atlantic; "a widening gulf," wrote Henry Lieberman in the *New York Times* in March 1967, which threatened to "leave [Europe] further and further behind as the United States assumes command of the future by its grip on the high technologies: computers, microelectronics, space, communications, nuclear energy, aircraft".

These worries echoed the sentiment of Secretary of Defense Robert McNamara, who articulated European fears that America's rapid industrialisation might "create a technological imperialism". Running hand-in-glove with that fear was the possibility that simmering malcontent in a disaffected European youth could incite renewed nationalistic fervour against the United States. Vice President Hubert Humphrey warned that "our concept of the European partnership can be eroded by the fear and concern about the power of the American capital and technology" (Fig. 1.2).

Fig. 1.2: Early conceptualisation of the Spacelab module and pallet in the Shuttle's payload bay. Note the orbiter's two-tiered cockpit, with upper 'flight deck' and lower 'middeck', the latter of which is connected to the Spacelab module by means of a pressurised tunnel.

Such a bleak landscape for partnership was not aided by Europe's weakness in higher education, as McNamara lamented in 1967. Research spending, wrote Brendan Jones in the *New York Times* that June, remained comparatively small and efforts to achieve greater technical co-operation via the Common Market and the North Atlantic Treaty Organisation (NATO) had foundered at the proposal stages. A privately funded European postgraduate institution for science and technology was one option tabled that summer as "a considerable stimulant". And European leaders began to recognise that closing the gap sat in their own domain. The fabled notion that the United States had 'created' the gap and was accountable for its closure had "entirely disappeared", said Donald Hornig, President Johnson's director of science and technology. That year's Eurospace conference in Munich exposed a damaging reality: Europe's space budget was 14 times smaller than NASA's own, a miserly 0.005 percent of the gross national products of its collaborating nations.

Gerhard Stoltenberg, West Germany's minister of scientific research, urgently called for an ESRO/ELDO merger, a funding increase and co-operation with the United States on a new launch vehicle. But a long road lay ahead. "When Europe drew pride and status from its colonies, the Americans had none; the tables are turned now," the *London Economist* opined in August 1969. "Europe faces the probability that when the planets are opened up, we Europeans will have no part in doing it. The idea, at this late stage, of a European manned space program, is nonsense. The policy that would make more sense would be to approach the United States to see if the administration will accept some foreign collaboration in the hugely expensive next years of its space program. There will be no opportunity in this generation that it would cost us more to miss."

A DOSE OF REALITY

Sadly, Richard Nixon's arrival in the White House in January 1969 brought an innate distaste for the monstrous cost of space. Upon the president's overflowing plate of priorities, exploring the heavens held relatively little sway. Ending an unpopular war in Vietnam and solving civil unrest and racial discord at home depended little on a multi-billion-dollar space program which Nixon felt contributed little to the man or woman in the street. Project Apollo, the national drive to land a man on the Moon, had been a last gasp of political adventurism to eclipse the Soviet Union. Nixon's predecessors at 1600 Pennsylvania Avenue—John Kennedy and Lyndon Johnson—aggressively drove this goal to completion with breathtaking energy to assert technological and ideological dominance over the Soviet Union. With that goal achieved, Nixon had little appetite for more.

But he did acquiesce that America needed a future in space and the Shuttle piqued his interest. When the STG submitted its report to Nixon in September 1969, it argued that a reusable spacecraft afforded substantial advantages over disposable capsules, yielding lower launch costs, higher operational flexibilities and "a broad spectrum" of missions. Its proposals for a future U.S. space roadmap demanded $5-10 billion of yearly expenditure and offered in return lunar bases, Earth-circling space stations, the Shuttle and a human voyage to Mars. All were roundly rejected by the president. NASA would have to shore up political and military support to win Nixon's ear and gain his blessing to finance the Shuttle.

The STG offered a route to earn that blessing. It recommended broadened international partnerships via parallel strands: engaging the Soviets in future space endeavours and inviting participation from U.S. allies. It advocated "the use of our space capability not only to extend the benefits of space to the rest of the world, but also to increase direct participation of the world community in both manned and unmanned space exploration and use of space". It recognised that for industrialised nations, "the form of co-operation most sought after…would be technical assistance to enable them to develop their own capabilities" and suggested the United States "should move toward a liberalisation of our policies affecting co-operation in space activities".

By the late 1960s, international enthusiasm for space had grown exponentially and NASA now had overseas players with whom it could deal. Nixon declared in March 1970 that "the adventures and applications of space missions should be shared by all people". Armed with the STG's recommendations (and a direct presidential mandate), NASA Administrator Tom Paine visited Europe, Canada, Australia and Japan in late 1969 and early 1970 to discuss post-Apollo partnerships. These visits, recalled Arnold Frutkin, were precipitated by a chance meeting with Herman Bondi, ESRO's director general in 1967-1971, after one of Paine's speeches. A mathematician and cosmologist, Bondi co-authored the 'steady state' theory of the Universe. "He was a very bright and energetic man," Frutkin said. "He was much impressed with [Paine's] talk, not in terms of its literal specifics…but in terms of there being a follow on to Apollo that would be quite elaborate and involve a space station of some sort." Frutkin suggested that Paine take his speech overseas—"a fairly elaborate dog-and-pony show"—and visits to several European capitals were arranged.

Flying foreign astronauts, explained the STG, was "the most dramatic form of foreign participation, approached in the context of substantive foreign contributions", to reinforce U.S. political leadership in a highly visible manner. Not by chance, wrote Lorenza Sebesta, Nixon "favoured the presence of astronauts from [West] Germany and Japan, the ex-enemies defeated by superior American democracy and technology". But reactions to Paine's overtures were mixed. For Australia, space was not yet a national priority and it could not invest the requisite funds for a meaningful partnership with the United States. Japan preferred to pursue autonomous technologies. "Japan was just initiating its own general-purpose space agency," wrote John Logsdon. "It took Japan some time to decide whether it wanted to respond to the U.S. invitation, particularly because its own space capabilities were at such an early stage of development. By the time that a response to the U.S. invitation was agreed on within Japan, the United States had so changed the possibilities for international participation that there was no basis for Japanese involvement." Canada proved more receptive and after "harmonious negotiations" built a multi-purpose robotic arm for the Shuttle.

Talks with the Europeans, which Paine initiated in October 1969, were underpinned by several background matters. It was *de facto* U.S. policy to discourage Europe from developing its own launch vehicle, hopeful that European satellites might fly on the Shuttle to reduce operational costs. Reservations were also harboured about the Shuttle launching foreign-made satellites for foreign-only markets. And ELDO's track record with its short-lived Europa rocket was hardly stellar: it failed on six of its 11 launches. It was essential, Paine told Nixon in November 1969, for the Europeans "to rethink their present limited space objectives to avoid their wasting resources on obsolescent developments"—a clear jab at Europa's shortfalls—"and eventually to establish more considerable prospects for future international collaboration on major space projects". Paine made clear that partnerships on the Shuttle should not be linked to Europe building its own rocket.

Another concern hinged upon White House fears that a partnership might enable European industry to gain access (at minimal cost) to advanced or sensitive U.S. technologies. Some Nixon Administration insiders regarded the president's mandate for international collaboration as involving Europeans as astronauts or scientists, not builders of physical hardware. The STG stressed that "there should be no significant technical obstacles" to sharing benefits on a global scale, but there would probably remain "economic and political issues which require recognition and effective anticipation". The need to close the technology gap and assuage European fears of U.S. imperialism sat foremost in many minds. "There will have to be a substantial raising of sights, interest and investment in space activity by the other nations…in order to establish a base for major contributions by them," the STG advised, "and creation and attractive institutional arrangements to take full advantage of new technologies and new applications for peoples in developing as well as advanced countries."

EUROPE'S OPTIONS

Early talks with Europe focused on three possibilities: (1) the RAM/Sortie Can, (2) an autonomous 'space tug' to manoeuvre payloads from low-Earth orbits into higher ones and (3) structural components for the Shuttle itself. In July 1970, at the European Space Conference in the West German capital of Bonn, Arnold Frutkin emphasised that the time for a decision on participation in the Shuttle was growing short. But a palpable sense of distrust bubbled beneath the surface. European governments, *The Economist* editorialised that August, "scare easily with each other, let alone the Americans, whom they subconsciously suspect of always having the technical emasculation of Europe in mind". Tom Paine handled the Europeans "as gently as nervous horses", but it was feared that a successor "who is less imaginative or more inclined to wear his emotions on his sleeve will undo much of this work". The result, warned *The Economist*, would "not do much harm to the American space program, but for Europe to miss this chance would be tragic" (Fig. 1.3).

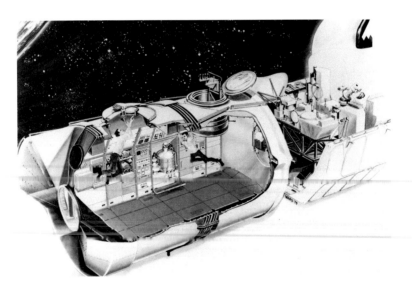

Fig. 1.3: Cutaway artist's impression of the Spacelab pressurised module and pallet in the Shuttle's payload bay. Notable in this view are the ceiling-mounted Scientific Airlock (SAL, visible with deployed instruments at right) and Scientific Window Adapter Assembly (SWAA, at left).

Still, there existed parallel benefits for the United States in having a European partner. Firstly, Nixon and Congress might respond more favourably to expensive programs like the Shuttle if they were demonstrably 'international' in scope and strengthened stable, peaceful relations with other democracies. Secondly, if such partnerships did reduce costs, it would be harder for staunchly anti-space presidents to enforce wholesale cuts or cancellations. "Prestige-fired fiscal support of multi-billion-dollar space projects is rapidly shrinking," wrote William Leavitt in *Air Force & Space Digest* in October 1970. "More cooks in today's kitchen would enrich the broth. The U.S. could bring its vast experience of the past decade, and certainly some significant funding, even in a tightened money situation. The Europeans could bring fresh enthusiasm, a good deal of skill, financial support and developmental philosophies that might be a good deal more frugal, since they have been accustomed to small budgets. Money and time could be saved all round."

The prospect of European firms subcontracting to build Shuttle parts carried some appeal, but not for long. In September 1970, the British Aircraft Corporation (BAC) and West Germany's Messerschmitt-Bölkow-Blohm (MBB) partnered with North American Rockwell on early Shuttle design studies. MBB undertook subsystem-level work on the Shuttle's attitude-control architecture, whilst BAC worked on structures, aerodynamics, flight-test instrumentation and data-handling. France and the United Kingdom expressed interest in building the Shuttle's delta-shaped wings and avionics, nose cap, radiators, tail, landing gear and payload bay doors. Sadly, these aspirations went nowhere. U.S. officials, wrote Sebesta, baulked at the difficulties which "might ensue from an intergovernmental effort to produce a relatively small number of components of a massive piece of highly complex hardware, whose timetable is pressing and in whose success the political and economic stakes are so high". One NASA assessment judged that the Europeans lacked the organisation, experience, knowledge and laboratory depth to meaningfully contribute here. The agency was concerned that Europe's aerospace sector lagged five to ten years behind U.S. industry and America risked losing more than it stood to gain. And there remained lingering fears that NASA ought not depend on other countries for critical Shuttle hardware; America's national asset, it was felt, should be untainted by foreign involvement.

Yet all was not lost. There remained a desire to build the space tug, which Sebesta called "a logical area for European participation, since it was an easily separable item with a relatively clean set of interfaces". In October 1971, representatives of ELDO and European industry met officials at NASA's Marshall Space Flight Center (MSFC) in Huntsville, Alabama, to discuss the tug. When Nixon green-lighted the Shuttle for development, the tug sat high on

Europe's list of priorities. In a January 1972 briefing, Dale Myers, NASA's associate administrator for spaceflight, revealed that European space officials had released to European industry "a fairly extensive study...leading to the definition of the tug that the Europeans might build for the Shuttle to carry". Myers added that if Europe built the tug, "we would expect it to be paid for by the Europeans". On 8-11 February, when U.S. and European representatives met in Neuilly-sur-Seine, near Paris, NASA noted that collaborative efforts might include "multilateral European responsibility for development of a major space element, such as reusable space tugs to provide access to geosynchronous and other orbits" or "Shuttle-borne orbital laboratories for research by U.S. and European scientists". It added that there should be minimal levels of technical risk, simple interfaces and a limited likelihood of frequent design changes.

Unfortunately, those hopes stalled that spring. In April 1972, as MSFC issued a request for proposals for a nine-month space tug operations and payload support study contract, the Air Force—which wanted to use the tug to launch reconnaissance satellites from the Shuttle—informed NASA of potential national security breaches if hardware originated outside the United States. It refused to guarantee support for any "foreign-built" tug if its stringent requirements could not be met. Since the Air Force's powerful political clout had gotten the Shuttle over the hurdle of gaining Nixon's blessing, its objections could not be readily dismissed. Clearly, the distrust felt by Europe toward the United States worked both ways. It is important to recall that a whole generation of U.S. officials in this era had reached adulthood after the Second World War. In their minds, America 'saved' Europe from the tyranny of Nazi Germany, then aided the continent's post-war reconstruction with the Marshall Plan and the NATO defensive alliance. "They had later witnessed Europe stand on its own feet again and become prosperous," wrote Sebesta. "They had further observed the successes of the European Communities, created in 1957, with their own strong encouragement, and the gradual transformation of Europe into a dangerous competitor on the international market by the end of the 1960s and into a recalcitrant ally in the military field." Europe's growing "assertiveness, translated into the space field" yielded in the Americans "a yet undefined but growing anxiety about future European competition in space".

In March 1972, Bill Anders, executive secretary of the National Aeronautics and Space Council, voiced his reservations to Nixon's presidential assistant, Peter Flanigan. Anders recommended that Europe should be invited to develop the RAM/Sortie Can, but not the tug. On 29 April, Secretary of State William Rogers argued against reversing America's position on this point, but the death-knell had already tolled. "There was no way that was going to happen, not from NASA's standpoint, but from the military's standpoint," said Bill Lucas, director of MSFC in 1974-1986. "That tug was to serve both NASA's interest and the military's payload interests. The military certainly would not have been willing to have a foreign entity that they had no control over to be in the loop as far as their payloads were concerned." The irony was palpable. "The U.S. government…found itself in the position of having to walk back," wrote John Logsdon, "from the European perception of the cooperative possibilities in post-Apollo activities that had been encouraged by the way the United States and Europe had proceeded to define that co-operation."

In response to Rogers' recommendation, NASA's "preferred objective" was now to "obtain European involvement to develop a specific type of sortie module" and a European-built tug was now "a distinctly second choice…and much less desirable". The Department of State did not prevail and on 1 June Henry Kissinger, assistant to the president for national security affairs, communicated Nixon's decision to remove the tug and Shuttle structural components as candidates for European involvement. Two weeks later, U.S. representatives notified their European counterparts that only the RAM/Sortie Can was an acceptable Shuttle contribution. One point raised by NASA was that "for safety reasons" the agency did not want a liquid-fuelled tug…which happened to be the only propellant the Europeans had studied in their proposal.

Having committed $20 million in study funds to the tug, the decision's arbitrary nature provoked intense displeasure among European governments and industry, creating a "tense atmosphere" at the June 1972 talks. The RAM/Sortie Can was far less challenging than the tug. "This was the part of the post-Apollo program in which Europe could have best profited from technology transfer," Sebesta wrote of the tug. In meetings with European delegates, the Department of State's Herman Pollack reported that "the conditions the United States find necessary may diminish the attractiveness of Europe in participating in the Shuttle items". An MBB-built attitude-control system, designed by West Germany, had already been rejected by NASA as "too advanced and too complex".

Europe, wrote Kenneth Gatland, had "little room for manoeuvre". On 22 June, *Today* reported that America opposed Europe's insufficient funding and dearth of technical expertise. Matters were further compounded by a U.S. unwillingness to transfer sensitive technologies and the Air Force's unswerving obstinacy in demanding full and firm control over the tug. That was a pity, for the tug covered a range of technologies of interest to Europe and its integration with the Shuttle might have fostered stronger European participation in most U.S. manned space missions in the 1980s.

But in its final summary, NASA concluded that the tug pushed propulsion technologies "well beyond what is currently envisaged in Europe". Writing in *New Scientist* in July 1972, Nicholas Valery lamented this dimming prospect of European partnership. "Having used the idea of international participation, ostensibly to ease the burden on the U.S. taxpayer, as a means of winning Congressional support for the reusable Space Shuttle," Valery opined with unalloyed bitterness, "NASA is now quietly letting it be known that it has no intention of letting Europe get its hands on anything technologically important."

LAST-CHANCE SALOON

The RAM/Sortie Can morphed virtually overnight into Europe's last-chance saloon for involvement in the Shuttle; a "take it or leave it" offer, according to Logsdon. But some within NASA regarded it with greater pragmatism. "We wanted to get European support…on the Shuttle," said Dale Myers. "Spacelab…was a separable and not fundamental requirement. We could fly without it, so if it didn't work, that would be okay. We didn't know whether the Europeans were going to carry through on the Spacelab or not, so we needed something that wasn't fundamental to the Shuttle" (Fig. 1.4).

In its earliest guise, it was intended for flights of five to seven days. 'Sortie' missions were modelled upon NASA's airborne research program, which used a four-engine Convair 990 aircraft as a high-altitude laboratory. "The goal in developing procedures for managing the Shuttle sortie missions will be to retain the relatively simple, flexible and highly responsive assets of the 990 program," NASA noted. The RAM/Sortie Can's pressurised modules "would range from 'austere' Shuttle sortie can with relatively simple laboratory equipment through more sophisticated laboratory and observation facilities, including automated free-flyers, serviced by the Shuttle". In September 1971, NASA Headquarters in Washington, D.C., directed MSFC to design a RAM/Sortie Can for potential in-house construction to reduce costs. A team in MSFC's preliminary design office laboured until January 1972, exploring the practicalities of using off the shelf hardware, devising temperature, pressure and acoustic guidelines and sizing the module at 25 feet (7.6 metres) long and 15 feet (4.5 metres) wide to fit the Shuttle's payload bay.

But with fief-like rivalries endemic among NASA establishments to seize lucrative contracts, competition raged between MSFC and the Johnson Space

Fig. 1.4: Enshrouded in white thermal insulation material, the Spacelab long module sits in Columbia's payload bay during STS-9 in late 1983. The pressurised tunnel, connecting the module to the Shuttle's middeck, is visible in the lower part of the photograph.

Center (JSC) in Houston, Texas, over who should lead the RAM/Sortie Can development. In November 1971, MSFC was identified as the lead organisation within NASA's Office of Manned Space Flight for conceptual design studies. Its work was lauded by NASA Headquarters as "the hard core of our manned payload opportunities utilising the Shuttle" and drove a directive for MSFC to move into the definition phase. By April, MSFC had established a Sortie Can Task Team, headed by Fred Vruels of NASA's program development office, with Hans Palaora as chief engineer. "A hazily defined area that opened an arena for center rivalry was the question of which center should work with customers who wanted to fly experiments on the Shuttle," wrote Andrew Dunar and Stephen Waring. "With [MSFC] assuming responsibility for payloads and payload carriers and [JSC] serving as lead center for the orbiter, someone had to satisfy user demands and minimise impact on the Shuttle."

Eberhard Rees, director of MSFC in 1970-1973, insisted that 'his' centre had well-established contacts with the scientific user community. But Chris

Kraft, who directed JSC from 1972-1982, took an opposing view. He felt the Shuttle/payload interface was fundamental and insisted that 'his' centre was aptly suited to reconcile users' requests through a co-ordinated payloads office. Consensus fell in MSFC's favour. Bill Lucas was convinced that JSC sought control over every aspect of the Shuttle. "Ultimately, they need to control the people directly and then meet the interface with the Shuttle," Lucas said. "You don't need to speak to someone in Houston to speak to your customer…the logic is that, as long as the Spacelab meets the established interface with the Shuttle, then why should the people responsible for Spacelab go through Shuttle management to get to Spacelab? That's the way it turned out to be. I like to think logic prevailed."

At first glance, MSFC seemed an odd choice, given that its history emphasised propulsion, including development of the Saturn V rocket. Its selection came via efforts from NASA Headquarters to divide tasks more equitably and its own aggressiveness in winning contracts. "When [JSC] became lead center for the Shuttle, [MSFC] was in line for compensation and Spacelab offered some solace," wrote Dunar and Waring. "But compensatory awards alone would not have been enough, had [it] not demonstrated the capacity to manage such a program." MSFC also developed Skylab, which proved advantageous when designing, developing and qualifying pressurised structures. "We had that technical capability," said Jack Lee, who managed the Sortie Can Task Team and MSFC's Spacelab program office in 1973-1980. "I think that's the reason we got it." In a March 1975 interview for *Aviation Week*, Douglas Lord, NASA's Spacelab program director in 1973-1980, regarded JSC as "the logical place to handle overall operations" and MSFC as a "leading candidate" for a Spacelab payload operations control base.

As work progressed toward a U.S.-led RAM/Sortie Can, European interest in developing the laboratory ran in parallel. In June 1972 discussions between Vruels' Sortie Can Task Team and officials from ESRO's European Space and Technology Center (ESTEC) at Noordwijk in the Netherlands, led by Dale Myers, it was stressed that if Europe intended to build the RAM/Sortie Can it should be "as nearly uniquely European as possible", with dependence on the United States kept at a minimum. NASA was pleased that ESRO seemed "eager" to undertake the project, but there remained regret on the European side that they would be unequal partners in an endeavour likely to cost over $250 million to build prototype and flight hardware. It was very much "a consolation prize", Lucas acquiesced, with less advanced technology and less impact on the Shuttle in schedule or budgetary overruns. "We are dealing with a potential supplier who is seriously considering investing $250 million of his own funds in the development of a spacecraft to be used primarily by the U.S.," Lord said. "This is not a typical joint venture, since the direct benefits are heavily in our direction." The partnership was asymmetrical, with Europe building hardware which functioned only on the Shuttle. But the Europeans had very few chips with which to bargain: in 1972, they lacked confidence in their own abilities and needed U.S. assistance to establish a manned space program. They were, wrote Logsdon, "willing to pursue co-operation on almost any terms, no matter how one-sided".

In September 1972, the Department of State advised the Europeans that it would take MSFC one year less than Europe to build its own RAM/Sortie Can. (Indeed, as late as May 1978 *Aviation Week* reported that if the facility was built in the United States, costs would have been far lower than they proved to be.) And in the absence of a formal commitment from Europe, NASA could wait until no later than 15 August 1973 before building its own system. In June 1972, three European industrial consortia—COSMOS, STAR and MESH—began preliminary (Phase A) studies on a "modular orbital system" for the Shuttle. The following November, the ESRO Council authorised Director General Alexander Hocker to move into detailed definition (Phase B) work to establish a cost-base upon which member-states could reach a final decision.

At its ministerial meeting in Brussels on 20 December 1972, ESRO agreed to build what became known not as 'RAM', or even 'Sortie Can', but 'Spacelab'. The word 'sortie' was coined by NASA to reflect a low-cost, short-term mode of operation, but the French word *sortie* is equivalent to the English word *exit*, which garnered distaste. Nor did 'Spacelab' fall easily on NASA's ears. Having recently begun Skylab, there were fears that the two names might become conflated and confused by the public. "Despite NASA's objections," wrote Lord, "once the Europeans had committed to the program, they unilaterally decided to use the name 'Spacelab', and 'Spacelab' it became." The name was transmitted to the U.S. partners during a joint NASA/ESRO Sortie Laboratory Working Group on 12 January 1973.

Two weeks earlier, European Space Conference chairman Théo Lefèvre told William Rogers that the program would be "undertaken, carried out and administered within a common European framework", adding his earnest hopefulness that it would represent "the beginning of an important new period in co-operation in space matters between the United States and Europe". The wheels of institutional change at NASA whirled in response to this brightening star of European commitment. A mission and payload integration office, led by Philip Culbertson, was established, and the Space Station Task Force was renamed the Sortie Lab Task Force, with Jack Lee appointed in May 1973 to manage it. But MSFC's in-house concept continued to preserve NASA's options if the Europeans opted out. "I had a small staff of people in program deployment and then I drew on the whole of the engineering capability of the center to put down the details on the design," Lee said. "We were pretty far along on the completion of that Phase B, so that we could either try to build it in-house or go to contracting out." Unsurprisingly, MSFC leaders observed these negotiations with keen interest; NASA Headquarters having informed them that they could expect a "substantial" role in the program if the Europeans agreed to participate.

AN AGREEMENT AT LAST

On 18 January 1973, in Paris, ESRO unanimously authorised Spacelab, with delegates from Belgium, Italy, Spain and West Germany laying down their signatures. "After definition studies of the RAM, the sortie lab final design would be agreed on by NASA and ESRO," NASA announced the next day, still using the program's older nomenclature. "ESRO would proceed to development phase, with delivery of the sortie lab to NASA scheduled for 1979." Estimated costs had swollen to $250-300 million, as worldwide inflation began to bite. West Germany pledged to fund at least 40 percent of those costs, with Italy committing 20 percent. The West Germans proposed Spacelab as "a special project" and an arrangement between ESRO member-states and the organisation was approved by the ESRO Council on 15 February. Thus armed, Hocker commenced formal negotiations with NASA and member-states to draw up the legal frameworks, which entered force in August 1973 and described Spacelab's objectives and components. One notable product was a permanent Spacelab Program Board to assume programmatic responsibilities and ensure timely interventions on technical and other matters (Fig. 1.5).

"The participation of member countries and the mode of financing," Lefèvre told Rogers, "will be settled in the near future." But some prospective partners remained unconvinced. The French agreed that Spacelab "would enable Europe to take an interest for the first time in the problems of manned flight" but felt that "none of the economic needs of the next decade would be met by the development in Europe of a sortie laboratory which can in no case be considered a substitute for a launcher program". One of the United Kingdom's conditions for adding its signature to the Spacelab contract hinged on the need for a formal ESRO/ELDO merger. And Italy remained fearful of America's vice-like control and unwillingness to cede significant technical or industrial knowledge. Only West Germany was ready, willing and able to align itself as Spacelab's main sponsor.

Fig. 1.5: An engineering test model of the Spacelab module undergoes fit-checks inside the Operations and Checkout Building at Florida's Kennedy Space Center (KSC).

Early in 1973, the Interim Spacelab Program Board, led by ESRO Vice President Massimo Trella, decided that member-states ought to possess an 'instrument of withdrawal', should the Phase B detailed definition study prove unfavourable or if original financial estimates looked likely to be exceeded. It proved a clever tactic, persuading other member-states beyond the initial four signatories to join Spacelab by offering an escape clause to withdraw before August 1973. Politically, it was easier for ministers to gain approval for a limited amount of funding, especially—as in the United Kingdom's case—it was made clear that governments could withdraw at the closure of Phase B. This new arrangement was opened for signature between 1 March and 10 August 1973.

The tactic bore fruit and nine of ESRO's ten members—Belgium, Denmark, France, Italy, the Netherlands, Spain, Switzerland, the United Kingdom and West Germany—agreed to progress into Phase B. Sweden was the only member-state which did not sign up to Spacelab, refusing to contribute to systems which might be utilised for military purposes, as the Shuttle was expected to do. (Austria voluntarily associated itself with the program later.)

By the end of Phase B, the original financial estimates were confirmed, with no anticipated cost overruns beyond an agreed 20-percent 'buffer' and none of the nine signatories withdrew. A measure of Franco-German compromise was also secured through reciprocal arrangements to participate in Spacelab and a new European rocket, Ariane. This compromise, wrote Sebesta, was reached in July 1973, after "a series of sometimes harsh discussions and horse-trading among European partners".

At the ESC council meeting that summer, the division of costs between member-states was sharpened. Due to the high percentage of expenses borne by West Germany, the prime contract to build Spacelab would go to a West German firm: either VFW-Fokker/ERNO, which spearheaded the MESH consortium, or MBB, which led COSMOS. But there existed a proviso that when choosing subcontractors, the prime contractor should consider members of the STAR consortium. And it was explicitly required that geographical distributions of subsystem-level work must "bear a close resemblance to the percentage-contributions of the participating member-states".

"In other words," wrote Sebesta, "the selection of ERNO and MBB as prime contractors did not necessarily imply the selection of the MESH and COSMOS consortia, but rather that there should be a regrouping around the prime contractors of as many as possible of the qualified industrial teams in the participating member-states." As such, the three consortia became two by February 1973. Project definition and cost studies began in March, as ESRO and NASA officials outlined a draft Intergovernmental Agreement and Memorandum of Understanding. In July, European scientists participated in a National Academy of Sciences study group in Woods Hole, Massachusetts, which emphasised Spacelab utilisation. At this stage, the Shuttle was expected to fly 50 times per annum. And NASA's 1973 plan for Shuttle use required six Spacelab units with 25 research missions per year.

On 31 July 1973, in Brussels, European ministers formally committed to the program. The Intergovernmental Agreement was opened for signature in Paris on 14 August and the Memorandum of Understanding was initialled by NASA and ESRO representatives for signing by Fletcher and Hocker at the Department of State. So it was that on 24 September 1973, NASA confirmed its readiness to manage all operational activities (including crew training and flight operations) following receipt of the first Spacelab flight unit from the Europeans in late 1978 or early 1979. "It is contemplated," noted the Agreement's Article XI, "that there will be a European member of the flight crew of the first Spacelab flight." And it added that "subsequent European flight crew opportunities would be provided in conjunction with ESRO or ESRO-member-government projects". The facility would be available "on either co-operative or cost-reimbursable basis" and NASA "would procure from ESRO any additional Spacelab units needed for U.S. programs and would not develop any further units duplicating the Spacelab's design and capabilities". That latter point became a bone of contention as the 1970s wore on.

With Fletcher and Hocker's handshake, NASA accepted the name 'Spacelab' and the Sortie Lab Task Team became the Spacelab Task Force and Task Team. On 24 October 1973, the NASA Spacelab Program Office came into being, under Douglas Lord's directorship. And in comments to the U.S. Senate Committee on Aeronautical and Space Sciences that autumn, Fletcher declared ESRO's imminent intent "to select the design concept and cost proposal of one of the two competing prime contractors", expecting that "detailed design and fabrication" would begin by June 1974.

As Spacelab came together, so ESRO and ELDO moved towards a merger. At the December 1972 ministerial meeting, it was hoped the new 'European Space Agency' would be in place "by January 1974", with member-states committing to a 'basic' work program of science, general activities and facilities (with contributions linked to each nation's gross national product) and an 'optional' program, including Spacelab. In the latter, member-states could freely decide their levels of contribution. At the ESC meeting in Brussels on 31 July 1973, the date for ESA's creation was pegged for 1 April 1974, but "divergent interests" hampered progress. The agency came into being on 31 May 1975 and included a Spacelab directorate, whose director reported to ESA's director general. "Acceptance of a European mini-NASA now promises great rewards for European industry," wrote Gatland in October 1973. "Longstanding neglect of central management in European space affairs has led to waste of limited resources and frustrated Europe's entry into potentially profitable markets. With ESA, Europe now has a chance to work at the space frontier alongside the United States and to develop other space capabilities of enormous social and economic importance."

Early in 1974, ESRO issued a request for proposals to European industry for Phase C/D, the formal Spacelab design and development contract. After evaluation, Hocker favoured the consortium led by VFW-Fokker/ERNO over MBB. It proved a tough decision, for MBB had significant expertise in aerospace production. Even after the results from ten evaluative panels were considered by ESRO's Tender Evaluation Board, the final marks between the competitors amounted to a difference of a single percentage point. Both VFW-Fokker/ERNO and MBB had, the board concluded, put forward technically sound and financially competent proposals. As such, the decision went to ESRO's adjudication committee, chaired by Hocker. The choice of

VFW-Fokker/ERNO came in May 1974. Its selection was grounded in superior technical concept, high technical preparedness and depth of design for implementation of Phase C/D, a "greater suitability of concept to user's wishes", a strong management team and "more easily repairable shortcomings".

But before contracts were signed, a potential obstacle arose. Both designs were overweight and risked undercutting the Shuttle's payload-lifting performance. Fletcher required this issue to be rectified before proceeding with the award. His reaction hit ESRO like a "bombshell", wrote Lord, and the European press were united in their mauling of NASA. High-level U.S. representatives flew to Europe to resolve the matter. After several fraught meetings, NASA and ESRO reduced Spacelab's weight and conceived 'categories' to permit differing weights on a flight-by-flight basis. Fletcher and Hocker admitted the problems were not so 'weighty' as to reject VFW-Fokker/ERNO, but the episode underlined a strand of growing frailty in NASA's relationship with Europe.

On 5 June 1974, VFW-Fokker/ERNO was awarded the six-year Phase C/D contract, worth $226 million (equivalent to $1.4 billion in 2023's dollars). West Germany led the team at 54.1 percent, followed by Italy (18 percent), France (ten percent), the United Kingdom (6.3 percent), Belgium (4.2 percent), Spain (2.8 percent), the Netherlands (2.1 percent), Denmark (1.5 percent) and Switzerland (one percent). Each member-state was required to bear a 20-percent overrun above the estimated cost ceiling. Should this 120-percent 'cap' be breached, parties could exercise their right to withdraw from the program. Under the terms of Phase C/D, VFW-Fokker/ERNO would provide a complete Spacelab unit, "fully qualified and ready for installation of experiments", for delivery to NASA by April 1979, together with two engineering mockups and three sets of ground support equipment and structural spares. The hardware would be reusable for up to 50 missions. The program passed its preliminary requirements review in November 1974, its first major milestone, which established conceptual baselines and gained approval for higher-level system specifications and plans. This was followed in 1975 by the system requirements review, a precursor for the subsystem definition and design phase, and in 1976 by the preliminary design review, a full technical analysis of the basic design approach ahead of design and manufacture of the engineering mockup. In March 1978, Spacelab passed its critical design review, which established baselines for the first flight unit. Delivery of the engineering mockups slipped to mid-1979 and the first flight unit (in a pair of shipments) to late 1979 and early 1980. This timeline would position Spacelab for its first missions in December 1980 and April 1981 (Fig. 1.6).

Fig. 1.6: The Spacelab pressurised module (background) and U-shaped unpressurised pallet would support dozens of missions in support of science and technology, International Space Station (ISS) construction and maintenance and Hubble Space Telescope (HST) repair and servicing between 1981 and 2009.

The impressive scope of those first missions was laid bare in the first Shuttle manifest, published by JSC's Space Transportation System (STS) Utilization and Planning Office in October 1977. It envisaged Spacelab-1 aboard the 11th Shuttle mission, flying from 18-25 December 1980 in a 186-mile (300-kilometre) orbit, inclined 57 degrees to the equator. The five-person crew "would probably include scientists and at least one astronaut from an ESA member country". The second Spacelab, launching on 7 April 1981, would fly for 12 days, carrying a 'train' of four instrument-laden 'pallets' and a crew of five. Orbiting 260 miles (415 kilometres) high, inclined at 57 degrees, its payload weighed 31,830 pounds (14,440 kilograms). Four more Spacelab missions were timetabled between June through December 1981, each lasting seven days and featuring five-person crews. The first would fly Spacelab alongside a deployable communications satellite, the second—operating from an altitude of 250 miles (400 kilometres)— emphasised space physics and astronomy, the third was devoted to life sciences and the fourth "might" be an ESA-dedicated flight. Notably, the life sciences mission would land with an exceptionally high 'down-mass' of 34,250 pounds (15,530 kilograms), some 20 percent greater than most flights, which planners recognised as "excessive", but necessary, "based on preliminary payload data".

Several proposed Spacelab payloads never reached fruition, yet whetted appetites for what might have been. A Colloquium on Bioprocessing in Space, held at JSC in March 1976, found that Spacelab could accommodate 300-400 experiments, "more on each flight," explained James Bredt, NASA's program manager for space processing, "than all other previous manned space missions combined". One study was the Atmospheric Cloud Physics Laboratory (ACPL), a double rack of experiments inside a Spacelab module. It would examine the "trigger actions in clouds, such as the formation of ice crystals or water droplets, electrical charges on droplets and water-droplet ice-crystal interactions to aid in the ultimate control and modification of hazardous weather conditions". It was expected that ACPL would fly 20 Shuttle missions, with MSFC noting in October 1976 that it sought to "develop techniques for accurate prediction and control of weather (increase of snowfall or rain, dissipation of fog, suppression of lightning) to improve man's environment". In February 1974, a $321,394 study contract was awarded to McDonnell Douglas by MSFC to establish the engineering constraints, requirements and design trades needed for the mission.

A formal request to industry for proposals on ACPL was issued in September 1975, with expectations that a pair of 12-month contracts would result. By November 1976, an experimental chamber to develop ACPL was nearing completion at MSFC. It included a ground test engineering model to simulate the expansion of parcels of air as they rose in the atmosphere, using carefully controlled temperature and pressure to monitor moist air as it underwent changes like those inducing the formation of cumulus clouds. Droplets formed in these simulated clouds would be photographed by cameras outside the chamber, with the microgravity conditions of ACPL allowing them to remain suspended for longer periods for longer-term analysis.

In February 1977, MSFC began a 15-month contract worth $99,663 with Desert Research Institute, part of the University of Nevada, to develop a prototype ACPL particle generator. It would form nuclei with diameters from a hundredth to a thousandth of a micron, upon which water would condense to form 'clouds'. This would enable ACPL researchers to perform cloud-physics investigations to better understand the microphysical processes in the atmosphere, with an eventual goal of predicting, altering and even controlling weather. In May, MSFC awarded a 45-month contract to the Universities Space Research Association of Houston, Texas, for ACPL support, which was then "scheduled to fly as a partial payload twice a year on Spacelab, beginning in 1980 or 1981". The mission advanced rapidly, with General Electric of Valley Forge, Pennsylvania, selected in July 1977 for a $5.6 million contract to build ACPL. A full prototype with software and spares was anticipated for delivery to MSFC by October 1979, targeting an initial Shuttle launch in January 1980. An ACPL project office was set up by MSFC in April 1978, by which time the first mission was targeted to fly Spacelab-3 in June 1981. Two years later, its launch had slipped to 1982, before it was eventually deferred in February 1980. "Technical problems in transition from ground laboratory to space environment," NASA noted, "had caused delays and increased design and construction costs…to the pointing of terminating the GE contract." It was added that the ACPL hardware would be used instead for NASA's ground-based weather and climate research.

Another payload which never made it past the early definition phase was Atmospheric, Magnetospheric and Plasmas in Space (AMPS), whose development was assigned to MSFC in January 1975. It would "study the dynamic process of the atmosphere and magnetosphere, using active and passive probing techniques". It included a laser beam to define the composition of various constituents in the atmosphere. In November 1975, contracts worth $500,000 were awarded to Martin Marietta and TRW Systems Group for AMPS definition studies, before the project fell foul to budget cuts.

It is unsurprising that in light of West Germany's lion's share contribution to Spacelab, VFW-Fokker/ERNO assumed responsibility to develop, integrate and test the system, drawing upon ten European co-contractors and 36 subcontractors. This diverse group included Italy's Aeritalia, responsible for the Spacelab module and one subsystem, the United Kingdom's Hawkey-Siddeley Dynamics (later part of British Aerospace), which fabricated the Spacelab pallets, and West Germany's AEG-Telefunken, which built the electrical power system. Firms from Belgium, Spain, the Netherlands, France and Denmark took on other projects. West Germany's Dornier built a pointing system (see Chapter Four) for pallet-mounted instruments and, in partnership with Aeritalia, designed the module's environmental control system. Several U.S. firms also offered their expertise. McDonnell Douglas, which did preliminary design work on the RAM/Sortie Can in 1971, provided consultancy services to VFW-Fokker/ERNO via a Bremen head office to manage its contractual obligations. TRW consulted on payload and avionics support, whilst Martin Marietta had earlier worked with MBB on its Phase C/D bid.

But after June 1974, concerns festered that VFW-Fokker/ERNO was a politically motivated selection. "It is…interesting to locate the geopolitical realities that it produced," wrote Sebesta, with VFW-Fokker/ERNO based in the north of Germany, at Bremen, and MBB in the south, near Munich. "The north…was the place where the Social Democrat Premier Willy Brandt had built up his political career as the mayor of Berlin, before becoming chancellor in 1969, at the head of a Social Democrat/Liberal coalition, succeeding the long-established Christian-Democrat government. The south…where MBB was located, had in contrast benefitted from governmental policies and industrial orders in the past. In fact, the development of a strong technological region in Bavaria had been always high on the agenda of the Christian-Democratic Minister of Defence, Franz Joseph Strauss."

In October 1974, Roy Gibson (who succeeded Hocker as ESRO's acting director general the previous June and became ESA's first director general from May 1975) retorted that the decision came "without any political pressure", that "geographical distribution considerations" occurred "relatively rarely" and only ever at directorial level. Gibson added that ESRO's Tender Evaluation Board based its recommendations on quality, cost and availability. The most important contracts progressed to the administrative and finance committees, which occasionally reversed the ESRO Secretariat's recommendations. "But in no case," Gibson stressed, "was the Secretariat instructed to award a contract which it considered to be technically unacceptable."

As the 1970s entered their second half, relations on both Atlantic shores had changed markedly from their position a decade earlier and a lasting partnership was born. But as with many partnerships, it proved an imperfect meeting of minds. Spacelab weathered its fair share of storms, from questions pertaining to European 'ownership' to disagreements over whether leaders ought to communicate via telephone or writing. But a door to U.S./European space collaboration had opened, if only by the merest chink. Unfortunately, on 28 November 1983, when six astronauts—including a West German physicist, the first non-American to fly on a U.S. spacecraft—reached orbit with Spacelab for the first time, the door to Europe's new laboratory almost refused to open at all.

2

A difficult path to First Flights

FIRST SCIENTIFIC FLIGHT

The first Shuttle mission to carry Spacelab hardware ended almost as soon as it began. On 12 April 1981, Columbia launched from Pad 39A at Florida's Kennedy Space Center (KSC) for STS-1, the maiden voyage of the Space Transportation System. For two days, Commander John Young and Pilot Bob Crippen circled the globe 36 times, then guided their ship back home, like a great mechanised bird of prey, to land at Edwards Air Force Base in California. STS-1 was the first of four test flights before the Shuttle entered operational service. But even as Columbia's six wheels kissed Edwards' dry lakebed runway, kicking up a rooster-tail of dust and debris in their wake, preparations for her second flight, STS-2 (and Spacelab's first taste of space) were underway. Sadly, STS-2 took a long time to reach space and concluded earlier than anyone hoped. Columbia would make history as the world's premier 'used' spaceship: the first crewed vehicle to travel into space more than once. Commanding the five-day flight was Joe Engle, with Dick Truly as pilot.

Notwithstanding the boon of getting Columbia into space on STS-1, bringing her home and flying her again, NASA's pledge of weekly Shuttle launches would never be met. The four initial test flights lacked overbearing schedule pressure, but NASA still hoped to have Columbia ready for STS-2 by September 1981. Delays quickly set in, however, due to technical maladies from STS-1 and an underestimation of the attention the Shuttle needed between flights. Despite politicised efforts to paint a picture to the contrary, the spacecraft was not like an airliner, nor would it ever be.

Columbia returned to Florida from California on 29 April 1981 for maintenance in KSC's Orbiter Processing Facility (OPF). Most pressing were her thermal protection tiles, some of which became detached from the airframe on STS-1, whilst others had received dents, dings and scrapes. Hopes to launch STS-2 on 30 September 1981 evaporated as repair workloads intensified and managers opted instead for 9 October. The mission represented a new order of magnitude above STS-1, both in complexity and risk.

STS-2 was timelined for 124 hours and ten minutes, completing 83 Earth orbits, with touchdown planned at Runway 23 at Edwards. To facilitate the extended mission—twice as long as STS-1—an additional (third) cryogenic tank was installed in the Shuttle's fuselage. Columbia would enter a 157-mile (254-kilometre) orbit, inclined 38 degrees to the equator. Her complicated flight included three days of tests of Canada's Remote Manipulator System (RMS), a 50-foot-long (15-metre) mechanical arm, and Columbia also carried a scientific platform from NASA's Office of Space and Terrestrial Applications (OSTA-1). Activated by Truly four hours after launch, it would run with limited crew interaction until the day before landing. To make room for it, a development flight instrumentation pallet at the payload bay's mid-point was shifted further aft.

The history of OSTA-1 began in 1976, when the mission was first sketched out. NASA picked six experiments from 32 proposals; a seventh—the Heflex Bioengineering Test (HBT), precursor to a payload manifested for Spacelab-1—was added later. Provided by the University of Pennsylvania, it was stowed in lockers in the Shuttle's middeck and set up in orbit. It explored the effects of weightlessness and soil composition upon the dwarf sunflower (*Helianthus annuus*) to understand relationships between plant-growth heights and moisture needs. HBT contained 72 sealed plant modules whose soil moisture varied by weight from 58 percent (below which growth would be minimal) to 80 percent (above which anaerobic conditions would prohibit further growth). On STS-2, it would determine optimal soil compositions for Spacelab-1.

Engineering analysis required the Shuttle's payload bay to be directed towards Earth in a 'Minus-Z Local Vertical' (-ZLV) attitude for 88 hours to point several scientific instruments at the home planet. Those instruments weighed 2,240 pounds (1,016 kilograms) and demanded 1,452 watts of electrical power, furnished by the Shuttle's fuel cells. They were fixed to an engineering version of the Spacelab pallet, secured in the payload bay by trunnion pins and keel fittings. Built by British Aerospace for ESA, the pallet measured ten feet (3.3 metres) long and 12.8 feet (3.9 metres) wide and weighed 2,685 pounds (1,218 kilograms). All told, OSTA-1 occupied 30 cubic feet (0.84 cubic metres) of the bay and inclusive of hardware, wiring harnesses and utilities tipped the scales at 5,375 pounds (2,438 kilograms). Operational pallet-based missions would also benefit from a temperature-controlled 'igloo' for cooling, power and data-management (see Chapter Four). Pallets would accommodate instruments which needed unobstructed fields of view. Five OSTA-1 experiments were hard-mounted onto the pallet, while the others resided in Columbia's crew cabin.

Of the pallet-mounted suite, the first Shuttle Imaging Radar (SIR-A) measured 30.8 feet (9.4 metres) long by 7.2 feet (2.2 metres) wide and weighed 400 pounds (180 kilograms). A synthetic aperture radar, its rectangular bulk overhung the pallet like an enormous dining table, filling half of the payload bay. Led by the Jet Propulsion Laboratory (JPL) in Pasadena, California, it sat atop a tubular truss which afforded it a side-looking viewing angle of 47 degrees from straight-down ('nadir'). "We were given a very short period of time to develop that OSTA-1 payload and get it integrated into the Shuttle and flown," said Gene Rice, deputy director for Earth resources in NASA's Space and Life Sciences Directorate in 1979-1981. "We had some problems with the integrity of that truss. We found out that some of the landing loads might be a little excessive, so we had to go in and do some beef-up at the last minute to make sure it would survive both launch and landing. It was quite an accomplishment…for us to get it integrated, get it up there, have it work very well and get it back in one piece so we could fly it again." During eight scheduled hours of operation, SIR-A would generate two-dimensional images of Earth's surface for geological and mineralogical research. With a 130-foot (40-metre) resolution, it could cover 31-mile-wide (50-kilometre) 'swathes' of terrain and acquire data over areas equal to the contiguous United States. It functioned in the L-band frequency to construct images, transmitting microwave signals, receiving reflected 'echoes' and recording data onto computer tapes.

Elsewhere on the pallet were four other instruments. JPL's Shuttle Multispectral Infrared Radiometer (SMIRR) determined the best spectral bands for examining terrestrial rocks to globally map mineral indicators. Ground-gathered data suggested that the infrared portion of the electromagnetic spectrum was better suited to distinguish rock types than visible light and SMIRR covered ten bands within a range from 0.5-2.5 nanometres. But SMIRR was not an imager and its radiometer readings were 'located' using a pair of cameras: one colour, another black-and-white. During STS-2, it ran during 'daytime' orbital passes, with cloud cover below 30 percent. It could not be used within ten minutes of Shuttle wastewater dumps or fuel-cell purges, as its optics were susceptible to contamination. Its remit was to assess the variability of atmospheric absorption characteristics upon data quality. Each data cycle extended from a couple of minutes to 20 minutes (Fig. 2.1).

Fig. 2.1: The first Office of Space and Terrestrial Applications (OSTA-1) payload, dominated by the large Shuttle Imaging Radar (SIR-A), is gently craned into Columbia's payload bay. The U-shaped Spacelab pallet, built by British Aerospace, is clearly visible.

The Feature Identification and Location Experiment (FILE) from NASA's Langley Research Center (LaRC) in Hampton, Virginia, characterised water, vegetation, bare ground and snow cover, clouds and ice to prioritise timings for future Earth-watching instruments and reduce unnecessary downlinked data. FILE enabled SMIRR and SIR-A to be brought online when conditions were most opportune. Activated by a pyramidal sunrise sensor when the Sun reached 30 degrees above the horizon, the output from its two television cameras—one with an optical filter, the other near-infrared—was transmitted to decision-making electronics and a Hasselblad camera acquired colour photography. Also from LaRC was Measurement of Air Pollution from Satellites (MAPS), which offered the first clear views of carbon monoxide in Earth's atmosphere. Equipped with a gas filter correlation radiometer, MAPS surveyed the lower troposphere and on STS-2 it uncovered a worrisome trend of pollution in the tropics, induced by the seasonal burning of biomass.

The Ocean Color Experiment (OCE), provided by NASA's Goddard Space Flight Center (GSFC) in Greenbelt, Maryland, mapped colour patterns of plankton and chlorophyll to identify schools of fish and assess changeable environmental and circulation conditions in the oceans. It emphasised the eastern Atlantic and eastern Pacific regions to avoid reflective interference in coastal waters. It was hoped that 25 ocean flyovers with OCE would acquire data over the 'frictional' area between the Canary Islands current and equatorial counter-current and the upwelling zone off the coast of Peru. MAPS also took measurements along the United States' eastern seaboard, near Cape Cod and Georgia.

Inside Columbia's cabin were the final experiments: HBT (described above) and the Night-Day Optical Survey of Lightning (NOSL). The latter was Truly's responsibility, activated 9.5 hours after launch. Led by the State University of New York at Albany, NOSL required the crew to photograph lightning flashes over land and water to assess new technologies to forewarn of severe storms. Engle and Truly trained to recognise cumulonimbus clouds and anvils from 'above' and were sent updated meteorological information to guide them to brewing storm sites. When targets came into view, they mounted NOSL in Columbia's overhead window, where a camera acquired film and audio/photocell signals were recorded onto a tape recorder. The shortened nature of STS-2 meant HBT and NOSL achieved partial success, but both flew on later missions.

A MISSION HALVED

Integration of OSTA-1 onto the Spacelab pallet occurred in KSC's Operations and Checkout Building in early 1981 and by July the payload was aboard Columbia in the OPF. A series of tests validated its structural compatibility with the Shuttle and Engle and Truly's ability to command it from their flight-deck stations. On 10 August, Columbia rolled from the OPF to the gigantic Vehicle Assembly Building (VAB) for stacking onto her ET and SRBs. But rollout to Pad 39A did not bring launch any nearer. Eventually, on 12

November—Truly's 44th birthday—STS-2 took flight at 10:09 a.m. Eastern Standard Time (EST), reaching orbit 8.5 minutes later. "Smooth as glass," Engle jubilantly radioed as the Shuttle slipped into space. The astronauts circularised their orbit, swung open the payload bay doors and activated OSTA-1. Then, 2.5 hours after launch, controllers were startled by a high pH indication in one of Columbia's three fuel cells. For a while, the cell sat within acceptable parameters, then registered a sharp drop of 0.8 volts. The result was an impaired ability to generate electricity and drinking water. With the risk of contaminated water, the cell was turned off. And in response to fears that water was being 'electrolysed' (potentially forming an explosive mix), it was depressurised.

Under highly conservative mission rules, all three cells had to be functional for a flight to continue. STS-2's issue was later traced to deposits of aluminium hydroxide in an aspirator, which prevented water from being properly removed from the cell. Mission managers refused to override the rules and elected to bring Columbia home at the soonest opportunity. "We think it's the prudent thing to do at this point in our test program," Chris Kraft glumly explained. "We think we can really get everything out of the mission that we had planned, with the exception of time." It was with an air of regret that astronaut Sally Ride, the Capsule Communicator (CAPCOM) in Mission Control, radioed Engle and Truly that they were coming home the next day.

"First, bad news," she began, tentatively. "Our plan is that we're running a minimum mission and you'll be coming in tomorrow."

Truly made light of the unwelcome news by feigning ratty communications. "Oh boy, I'll tell you what. You're garbled and unreadable there, Sally."

"Want me to say it again?" asked Ride.

In fact, with a squealing tone in the background, the message really was garbled, but Ride repeated her words.

"That's not so good," was all a dejected Truly could respond.

"Think of it that you got all the OSTA data and all the RMS data," she commiserated, "and you just did too good a job…we're gonna bring you in early."

A bitter laugh came in reply, then a clipped "Okay".

With their mission halved, the astronauts were forced to 'front-load' as much work as possible. This had already been built into the test flights, in anticipation of just such an eventuality. "It was the correct decision, because there was no real depth of knowledge as to why that fuel cell failed," Engle said. "There was no way of telling that it was not a generic failure, that the other two might follow, and…without fuel cells, without electricity, the vehicle is not controllable." The astronauts' concern was not that they had lost five

Fig. 2.2: The OSTA-1 payload in orbit, captured through one of Columbia's aft flight deck windows. SIR-A is clearly visible at the left side of the frame, with FILE's pyramidal sunrise sensor at right of centre.

days, but that an enormous scientific investment might yield a negligible return. In fact, OSTA-1 gathered 90 percent of its requisite data by mission's end (Fig. 2.2).

Engle credited their training as an enabler in getting so much done. "We prioritised things as much as we could," he said. "We only had the ground stations, so we didn't have continuous communication with Mission Control and [they] didn't have continuous data downlink from the vehicle, only when we'd fly over the ground stations. When our sleep cycle was approaching, we did, in fact, power down some of the systems and we did tell Mission Control goodnight, but as soon as we went Loss of Signal (LOS) from the ground station, then we got busy and…ran through as much of the other data that we could, got as much done as we could during the night. Then, the next morning, when the wakeup call came from the ground, we tried to pretend like we were sleepy and just waking up."

But burning the midnight oil did not go unnoticed. Flight Director Don Puddy later told Engle that Mission Control knew they were awake. "We could see," he said, "you were drawing more power than you should've been if you were asleep."

Engle and Truly's work ethic paid dividends in OSTA-1 output, with 64 of 94 data-takes completed. SIR-A collected its eight hours of radar data, imaging North America, southern Asia and Europe, Australia and Oceania, North

Africa and northern parts of South America. It allowed determinations of surface 'roughness' and identified faults, drainage patterns and stratification, as well as ancient watercourses beneath Egypt's arid sands. The radar spotted the sunken Nazi battleship, *Admiral Graf Spee*, scuttled off the Uruguayan coast in December 1939. OSTA-1 was finally deactivated at 9:26 a.m. EST on 14 November and the pallet was powered-down at 11:35 a.m.

After two days in space, the astronauts prepared for their fiery return to Earth. Touching down at Edwards at 1:23 p.m. Pacific Standard Time (PST), STS-2 snatched triumph from the jaws of defeat.

"Good, solid bird, all the way," a happy Engle radioed.

"Welcome home," replied CAPCOM Rick Hauck from Mission Control.

"Good ol' Eddy day," gushed Engle, who knew Edwards well as an Air Force test pilot.

"Yessir."

ATMOSPHERE OF DISTRUST

Short and disappointing though STS-2 had been, it offered an enticing glimpse of Spacelab's potential. But the troubled mission mirrored a stony path that the program had navigated from its inception in the 1970s. Leading Spacelab were NASA in the United States and ESA, a supranational entity of a dozen European member-states. Within this pan-European framework, three directorates were established, including one for Spacelab. And whilst program implementation previously rested with the ESRO Council, the new ESA directorates reported to the ESA director general, facilitating more direct action and less dependency upon the inclinations of member-states.

Program implementation fell to two program directors, one from the United States and one from Europe, with Douglas Lord leading NASA's Spacelab Program Office from 1973-1980 and Jean-Pierre Causse heading ESRO's Spacelab Program Office until April 1974. Causse was replaced by Heinz Stoewer, who held the dual role of Spacelab program director until March 1975 and project manager through March 1977. The ESA directorship was filled by Bernard Deloffre until June 1976. After Deloffre came an 'interregnum', with ESA Director General Roy Gibson and Technical Inspector Massimo Trella covering the role. In November 1976, a new ESA program director arrived in the form of Michel Bignier, former head of the Centre National d'Études Spatiales (CNES, the French national space agency). He led Europe's side of the program until 1980.

Kenny Kleinknecht, NASA's deputy associate administrator for spaceflight for European operations in 1977-1979, remembered Bignier's courteousness. "I didn't understand any French," he said. "I thought they'll all be talking their own language and I won't know what was going on. But I was received with welcome arms." Bignier knew four languages and tried his utmost to speak English whenever Kleinknecht was around. If ever he struggled to find the words—"I'm sorry, Kenny, I can't express what I want to say in English"— Bignier always ensured that someone promptly translated what he had said from French into English for Kleinknecht.

Beneath the directorships, project managers from NASA and ESA oversaw day-to-day technical co-ordination functions. With MSFC serving as NASA's lead center for Spacelab, its inaugural project manager was Jack Lee, with Heinz Stoewer as his European counterpart. Stoewer was replaced by Robert Pfeiffer until July 1983. NASA/ESA activities were co-ordinated by a Joint Spacelab Working Group (JSLWG, nicknamed 'jizzlewig'), which met monthly under the directors' co-chairmanship. Despite management splits along geopolitical/industrial lines—with the European directorship frequently held by a French national, despite West Germany's financial leadership—co-operation was achieved via consultative joint groups.

But collaboration was not always straightforward. "Whenever a co-operation is invested with strong political significance," wrote Lorenza Sebesta, "the credibility of both partners is equally at stake and the risk of losing it is a great cement for the enterprise." As Spacelab moved into adolescence, the U.S. position preferred that dissent ought to be communicated verbally, via telephone or face-to-face, rather than the European manner of letters and written documents. Conversely, mechanisms for presenting reports through graphs and tables were resented as an implied criticism for European engineers used to operating under greater autonomy. Written communication, noted Sebesta, tended to polarise positions, forcing the other partner to retaliate with similar formality. "To force a showdown can be counterproductive," Sebesta added, "if the negotiating partner is in a weak position relative to other political actors at home."

At one June 1978 meeting between Roy Gibson and NASA Administrator Robert Frosch, these dangers came damagingly to the fore. "It quickly became apparent that their main concern was concentrated on the fact that I had written, rather than telephoned," Gibson reflected. Frosch was embarrassed at having a letter on record, a matter exacerbated by the difficulty of how to respond in writing. NASA's view was that telephone calls and private, non-recorded conversations might influence international negotiations, without leaving a tangible trace. "My attitude was to say that the contents of the letter represented the minimum compatible with the complaints of member states," said Gibson, "and that these had reached the point they could not be explained by telephone calls." Communication difficulties expanded into other areas. "We were dealing with ten different cultures," said MSFC's Bill Lucas. The problems were both cultural and institutional. Jack Lee fervently believed that Spacelab may have been completed sooner, were it not for troubles securing agreements among European members. "I suspect that we waited on them more than they waited on us," he said. On the U.S. side, Congress tended not to interfere, but ESRO was "more of a parliamentary process, so quite often we would have to wait for a year".

Day-to-day relationships between U.S. and West German personnel induced other challenges, some bordering on farce. "In all their offices, they had drink," said NASA engineer Ernie Reyes. "They had liquor, they had coffee, they had soda pops, they had beer, they had wine, they had all that good stuff, but they never went to excess." One day, as teams mulled over a process workflow chart, a German engineer announced he was thirsty. Thinking he would grab a Coke from the vending machine, Reyes was floored when the engineer poured alcohol into a paper cup.

"Whoa, they don't allow liquor out here," Reyes cautioned.

"Why not?"

"Because Americans don't know how to control liquor," Reyes replied. "They have liquor, then they want to fight or take your women or something."

"But there's no women here."

"No, you don't understand. We can't drink and work. It's just not done in America. We'll drink sodas and we'll drink coffee. After work, when we're not here, we'll go to Cocoa Beach or Titusville or my house, then we will drink."

Beneath the surface, there lingered Europe's unforgotten fear that ESA was not being treated as a partner. In legality, the United States and the Europeans *were* partners, sovereign states having sanctioned the Intergovernmental Agreement and diplomats on both sides of the Atlantic recognised Spacelab as an international program. "It was very much, by necessity, a partnership relationship," said Lucas. "Europeans were very sensitive about that. They were supplying most of the money, so you couldn't think of it as a contractor." Several NASA leaders sympathised with ESA's position. "They developed and delivered it to us for free," said Kleinknecht. "They were ready long before the Shuttle was. They could have delivered [Spacelab] at least two years before it was needed."

But NASA still found itself subconsciously treating ESA as a contractor. MSFC had to "act like we had a contractor, but not let them know that," said Lucas. "In other words, we had to give them a lot of guidance, but we had to

Fig. 2.3: The first shipment of Spacelab hardware is offloaded into the high bay of KSC's Operations and Checkout Building.

do it in a discreet way, rather than like you would work with a contractor here…it's just much less direct than the contract relationship." Lee understood ESA's concerns and avoided dictating NASA specifications, seeking instead to provide criteria to judge "whether what we were going to fly was acceptable", then allowing the Europeans to decide how to proceed. "I saw it better to let them have the flexibility of working against performance specification," Lee said, "instead of me having to follow along with all the detailed specs" (Fig. 2.3).

In furnishing such support, NASA trod a razor-thin tightrope with European engineers who regarded their American 'friends' with distrust, even suspicion. It became disappointingly apparent that the United States would rebuff access to new technologies not directly linked to Spacelab, effectively hamstringing Europe's hopes of gaining advanced "program management and systems engineering experience". Speaking in 1985, CNES Director General Frederic d'Allest spoke of "the bitter experience of…co-operation in the Spacelab program" and berated "the declared policy of limiting transfers of technology and technical information to the minimum needed to ensure compatibility of peripheral European elements", which demonstrated "unambiguously the limits of co-operation…in a strategic sector".

Mel Brooks, NASA's assistant chief for systems and payload operations at JSC in 1972-1975, co-ordinated the Shuttle/Spacelab interfaces. "When we first got there, they obviously looked upon us as spies and not there to help them," said Brooks. "When I saw things that really drastically needed changing, I had two choices." One was to call his boss in the United States. "That didn't work…coming from that direction." The alternative was to coax one of the European managers into conversation and cajole them to implement the change. "And if I could convince him, he would carry the ball forward and it would go," he said. "So we managed to get some things done that way." Other engineers were less tactful; some tried hard not to offend, but a few—after European hubris carried them to the ends of their tethers—dispensed with tact altogether.

Dovetailed into this environment were European fears that NASA might 'copy' Spacelab or create systems to supplant it. In August 1972, U.S. and European teams took steps to assuage those worries. It was agreed that the laboratory was an essential part of the Shuttle and would not be developed in the United States, so long as Europe committed to building it. But NASA was adamant that this did not guarantee Europe 'preferential' treatment in using the Shuttle; rather, European nations would enjoy a "priority right" to Spacelab's use and an entitlement to appoint crewmembers to its missions.

It was also pledged that "substantial duplication" of Spacelab should not occur in the United States, with NASA promising to refrain from "separate and independent development of any Spacelab substantially duplicating the design and capabilities of the first Spacelab…unless ESRO fails to produce such Spacelab". But the application of this clause caused great dispute, with blurred lines between duplication and 'substantial' duplication difficult to disentangle. Whereas the United States regarded Spacelab as just another international payload, for Europe it was the first step in a hoped-for long-term program of co-operation. In furtherance of this goal, the Europeans insisted on having provisions inserted into the Intergovernmental Agreement "for consideration of the timely expansion and extension of this co-operation as…mutual interests warrant". As early as 1970, ESRO Director General Herman Bondi called for collaboration to be focused "towards an area of activities…rather than *ad hoc* projects".

But as the decade progressed, European concerns festered that the Americans were indeed duplicating Spacelab. In March 1979, Frosch and Gibson established a Duplication Avoidance Working Group (DAWG) to flesh out greater technical scope in areas left vague by the Intergovernmental Agreement, including what "substantial duplication" really meant in practice. At the first DAWG meeting the following July, teams led by Michel Bignier and NASA Deputy Associate Administrator Philip Culbertson discussed several U.S. conceptual designs—Rack Integration Aids, a Pallet of Opportunity and a Department of Defense Pallet—which constituted very close copies. One system that drew particular interest was the Sortie Support System, which in ESA's opinion substantially duplicated their Spacelab pallet. NASA agreed partly with this European characterisation and sought clarification from the Department of Defense. It was found that whilst the Intergovernmental Agreement reserved for Europe a 'first refusal' to build the hardware, U.S. procurement laws (enshrined in the DoD Appropriation Act of 1973) prevented awards of research and development contracts outside the United States if satisfactory U.S. sources were available at lower cost. The Department of Defense and the Air Force considered this problem solvable by selecting a U.S. prime contractor, procuring from European sources.

In December 1979, new guidance was approved by the JSLWG, under which NASA would communicate to ESA any new requirements which might lead to substantial duplications. If the JSLWG determined that duplication had taken place, NASA should refrain from its development and procure from European sources, unless it could not be made available in accordance with agreed schedules and at reasonable prices. If this was not possible, NASA should provide opportunities to make proposals. In such cases, European and U.S. proposals would be treated on an equal footing. Frosch added that any NASA studies and plans for Spacelab-like systems should be made known to NASA Headquarters in sufficient time that an effective decision on European involvement could be made.

Europe also desired 'ownership' of the facility, with unfettered use of its first flight unit for themselves and their partners. But NASA's view was that such a move might be perceived as 'shared ownership' of the Shuttle. Here, the language of the Intergovernmental Agreement comprehensively favoured the United States, which could make "unrestricted use of the first Spacelab unit, free of costs" and retain "full control of the first Spacelab unit", including any modifications it desired. Spacelab remained ESA's property, insofar as the Europeans were fully liable "for damage occurring in connection with a Space Shuttle launch, flight or descent" on the first mission and shared liability for later missions. Nor did the fact that the Europeans had built Spacelab afford improved access for European astronauts after Spacelab-1. A European request for "a formal commitment to have unconditional access and use of the Shuttle at least when it is used for a European space mission" was flatly rejected. Instead, the United States would "make the Space Shuttle available for

Spacelab missions…of the European partners and their nationals on either a co-operative or cost-reimbursable basis".

In fairness, NASA retained overall responsibility for interfaces, quality control and safety, but Europe's lack of ownership—and a perceived absence of tangible returns on its investment—would return to bite. Spacelab-1 would be "jointly planned on a co-operative basis", equating to a non-cost for Europe as far as Shuttle launch services were concerned. But follow-on missions remained hazy. NASA's elusiveness was understandable, as the reality of building the Shuttle hit home. As development costs soared, prices were continually revised upwards. In 1977, the first Shuttle Reimbursement Policy for commercial and foreign customers established a fixed flat rate for the first three years of operational service (1983-1985), based on best estimates of the numbers of missions available. The standard price of a 'full' Shuttle mission was $18 million (in 1975 dollars), but with cost increases in operations, reductions in anticipated flight rates and effects of inflation, by 1982 the prices for 1986-1988 amounted to $71 million.

With these burgeoning costs came a startling likelihood that Spacelab would slip entirely under U.S. control and European member-states—whose own national space budgets had dwindled in the decade's second half—would be unable to afford to use their own facility. In May 1978, Wolfgang Finke, director of West Germany's department of space and transportation, told the House Committee on Science and Technology that Shuttle costs imposed "a real brake on our more ambitious plans to utilise the new system". And Hubert Curien, director general of CNES in 1976-1984, added that the "anticipated cost of Spacelab experiments now reaches such heights…that their funding raises very serious problems for national budgets and constitutes a very severe limitation".

Hopes of 50 Shuttle missions per year evaporated like raindrops on a hot summer's day. The lure of 25 annual Spacelab missions led ESA to believe that NASA would order four to six flight units, a sizable return on a European investment which had ballooned from $250 million in 1973 to $369.6 million by March 1974 and almost $500 million by the end of 1975. Worldwide economics and politics played a role here. Oil-price increases struck the West with a vengeance from 1973 and inflation attained its highest levels since the Second World War, directly driving cost increases in that first year alone. And in 1974-1980, worldwide inflation soared to 9.2 percent in the United States, 11.1 percent in France, 15.9 percent in the United Kingdom and 16.8 percent in Italy.

Member-states initially pledged 120 percent of their financial commitments until September 1979. But as early as November 1976, ESA's cost projections already exceeded 100 percent and by the July 1978 ESA Council meeting it was obvious that Spacelab would outgrow 120 percent. "We have always been very much afraid of being forced to exceed 120 percent," said Bignier, lamenting that "a certain number of systems are now behind schedule and...it will be difficult to catch up completely." West Germany requested increased national contributions, arguing that the program "had not been originally sufficiently precisely designed" and "major design modifications had to be made, while appreciable technical difficulties subsisted". They concluded that future overruns should be avoided and stressed a need to adopt "an extremely firm attitude towards NASA's demands...at the risk of...some deterioration of the good relations with NASA" (Fig. 2.4).

Following protracted deliberations, ESA proposed increasing the funding cap to 140 percent as a "large number of interface modifications...and the

Fig. 2.4: The second Spacelab flight unit, purchased by NASA under the Follow-On Production (FOP) program, made its first mission on West Germany's Spacelab-D1 in October 1985. The module is pictured being loaded into Challenger's payload bay. Note the access hatch to the Shuttle's middeck, just left of centre, which will accommodate the connecting tunnel to the module.

delivery to NASA of more hardware than initially foreseen…greatly contributed to this increase in expenditures". In March 1980, member-states voted to finance Spacelab under this new scale, ensuring its completion at levels of 120-140 percent. West Germany's share increased to 64.4 percent, that of France climbed from ten to 12.7 percent and the United Kingdom's contribution rose from 6.3 to 7.6 percent. The remainder was divided among the other members. Only Italy, hitherto Spacelab's second largest sponsor, at 18 percent, refused to endorse the new cost ceiling and its share was reduced. Italy had shouldered one of the highest inflation-rate burdens of any member-state and suffered poor returns for its industry. Exchange rates were also unfavourable for the Italian lira.

In testimony before the House Committee on Space Sciences and Applications in November 1975, NASA Associate Administrator for Manned Space Flight John Yardley noted that U.S. costs were largely unaffected as the Intergovernmental Agreement precluded the Europeans from recouping additional development fees. Yet the Americans remained empathetic to Europe's woes. "Spacelab ended up costing quite a bit more than the Europeans originally thought, partially because the Shuttle kept changing," said Stanley Reinartz of MSFC's special projects office. "If you're trying to do two things in parallel, it can run up the bill, particularly if you're trying to do one thing in this country and one thing in another."

ESA was in store for another harsh dose of reality. NASA had pledged to procure "at least one" additional Spacelab flight unit, through a Follow-On Production (FOP) program. That procurement would occur no later than two years before the delivery of the first flight unit, provided it was "available to the agreed specifications and schedule and at reasonable prices to be agreed". But successive NASA administrators argued that the agency should not pursue a second Spacelab until the first proved its worth. (Indeed, an article in the Intergovernmental Agreement highlighted "the desirability of gaining operational experience with the first flight unit before ordering additional units".) The Europeans were eager to avoid costly gaps in industrial production and unsuitable overlaps between the deliveries of the first and second Spacelabs. In October 1978, *Aviation Week* articulated Europe's concerns that U.S. delays "in ordering the second flight unit might disrupt skilled manpower employment and management performance". For European industry, an early start on FOP was of utmost import. Although a start was made in 1976, not until the decade's end were negotiations with NASA concluded.

As costs climbed, the issue of how Europe might use the Shuttle returned to the fore. Ideas for a 'barter' arrangement were floated, aimed at offsetting the cost of several Shuttle missions and averting the risk of ESA getting 'priced-out' of its own hardware. In 1977, ESA proposed that NASA should provide four 'free' Shuttle launches in exchange for the second Spacelab flight unit. Two flights would be for ESA and two for West Germany. Although agreed in principle in September 1978, it met stiff opposition from several ESA members, not least West Germany itself, which argued that committing to four missions was premature. European plans to use the Shuttle ought to be definite, their delegation pointed out, before barters could take place. West Germany knew that if it could not use the Shuttle's full capabilities on its missions, it would be obligated to 'fill-up' the vehicle or absorb the losses. Matters were not aided by the voracious rise of NASA's Shuttle pricing policy and increasing launch costs in West German deutschmarks against a devalued U.S. dollar.

In any case, ESA's Spacelab Program Board rejected the barter plan in November 1978. Although ESA favoured it, *Defense/Space Business Daily* noted that European governments refused to commit themselves before the Shuttle had proven itself. In July 1979, ESA contracted with NASA for long-lead-time materials for a second Spacelab, anticipating hardware to be delivered "progressively" in 1982-1983. And in January 1980, NASA exercised its contractual requirement to purchase one additional flight unit, at a cost of $183.96 million. This second Spacelab—a duplicate of the first, plus operational spares—was slated "to be delivered in 1984".

Then NASA struck an iceberg of its own. The agency's share of the U.S. budget shrank from 0.98 percent of federal expenditure in 1975 to 0.82 percent in 1981 and $30 million in cuts during the first year of President Ronald Reagan's administration imposed one-year delays on several early Spacelab missions. NASA had already postponed Spacelab-1 from its placeholder target of 1980 (on STS-7, the Shuttle's first operational mission), due to a desire to have a pair of geostationary satellites available to support its communications and data-relay needs. This proved ironically fortuitous for the Europeans, who would have been hard pressed to get the first flight unit delivered and processed before 1980. But additional delays proved highly unwelcome. "Over the past four years, the Spacelab-1 launch has slipped three years," lamented James Harrington, who became NASA's Spacelab program director in 1980. "Additionally, the manifest of Spacelab flights has been reduced from four or five flights per year to the current two flights per year through 1986."

As the 1970s ended, there was cause for dissatisfaction on both sides: the Europeans resented paying over the odds for a facility they might hardly use, whilst NASA disliked ESA's unwillingness to accept risks or responsibility for development difficulties. In May 1979, NASA's European representative, James Morrison, complained to Bignier of "a clear inconsistency in ESA's strongly held position of not wanting to be treated like a contractor by NASA…but as a partner in a development program…and then, on the other hand, wanting to set a date…after which all programmatic risk is assumed by NASA". In Morrison's view, the latter position did not "admit the former". Risk formed a natural tenet of the space business. "And he who does not wish to share the risk," warned Morrison, "cannot really be called a partner."

THE HUMAN EQUATION

Efforts on the ground, meanwhile, inserted real humans into the equation. In July 1973, a Concept Verification Test (CVT) at MSFC simulated high-data-rate experiments and test data-compression methods. Early the following year, a general-purpose laboratory resembling the Spacelab module was added to the CVT and a preliminary requirements review was concluded in May 1974. Seven CVT simulations between January 1974 and July 1976 supported ionospheric research, cloud physics, metal alloy preparation, high-energy astrophysics and life sciences. During one, in December 1974, four female MSFC scientists—metallurgist Mary Helen Johnston, physicist Ann Whitaker, astronautical engineer Carolyn Griner and aerospace engineer Doris Chandler—spent five days operating 11 experiments and established a benchmark for resolving on-board anomalies. (Two experiments were saved by the crew's expertise and corrective actions.) Another simulation in November 1975 included a television downlink capability and two-way voice communications with principal investigators.

Meanwhile, NASA's Ames Research Center (ARC) in Mountain View, California, established an airborne science program using the Convair 990 aircraft to evaluate Spacelab experiments. Four instrument-laden Learjets flew out of ARC in 1972-1974, with NASA astronaut Karl Henize and MSFC's Lee Weaver as 'operators' on the final flight, marking the first 'man-in-the-loop' Spacelab test. Experiments included a broadband infrared photometer and Henize and Weaver measured infrared radiation emissions from various astronomical sources. They underwent 140 hours of training in May-September 1974 and the hardware aboard the Learjet mirrored a 'real' Spacelab layout, including miniaturised workbenches and restricted storage volume.

Preparations expanded with airborne simulations in a larger aircraft and ground-based trials in a full-scale mockup of the Spacelab module. In March 1974, the Europeans proposed a joint Airborne Science/Spacelab Experiment System Simulation (ASSESS) to determine design parameters and operational concepts and conduct 'real' research. The first mission (ASSESS-I) began on

2 June 1975 with five Convair 990 flights over six days, emphasising infrared astronomy, upper atmospheric physics and ultraviolet planetary observations. Flights began from Moffett Field, followed a triangular ground track over Santa Barbara and onward to Payette in Idaho and El Paso in Texas, then returned to California. The crew comprised three NASA personnel (including astronaut Bob Parker) and two Europeans, who spent their off-duty hours between flights in quarters adjacent to the aircraft parking lot. Since the Spacelab module could only support three crewmen for a six-hour 'shift', ASSESS-I furnished an opportunity to wring out Spacelab's human logistics. To assure close parallels between simulations and the real thing, crewmembers were moved from their quarters to the aircraft via a 'lift van' and isolated from the flight crew for the entire mission (Fig. 2.5).

Although a second round of flights was not originally planned, in August 1976 NASA announced that ASSESS-II would fly a different route from the Pacific Ocean to the northern and southern borders of the United States, as far east as the Dakotas. NASA and ESA selected prime and backup 'payload specialists' for an atmospheric physics mission using a synthetic aperture radar and microwave limb sounder. In December 1976, Henize was assigned to lead the mission, backed up by Parker. ASSESS-II began on 16 May 1977 and completed nine daily flights, each lasting six hours. The five scientists—two Europeans and two from NASA, led by Henize—were confined to the

Fig. 2.5: Bob Parker (left) and Wubbo Ockels lend a measure of scale to this interior view of the Spacelab module.

Convair 990 and their off-duty living quarters. Dedicated to Earth resources, atmospheric monitoring and infrared astronomy, the crew was highly experienced. Henize and Parker were astronomers, whilst the primary U.S. payload specialists were JPL's Robert Menzies, principal investigator for ASSESS-II's laser absorption spectrometer, and David Biliu of the synthetic aperture radar team, backed up by Lee Weaver. The primary European payload specialists were Switzerland's Claude Nicollier and the United Kingdom's Michael Taylor, backed up by West Germany's Juergen Fein and Klaus Kramp.

Elsewhere, at JSC, three Spacelab Mission Development (SMD) simulations were conducted in a full-sized mockup of the pressurised module in the Life Sciences Payload Facility in JSC's Building 37. On 1 October 1974, physician-astronaut Story Musgrave and Dennis Morrison of JSC's Bioscience Payloads Office, started a week-long test of operational procedures and 12 biomedical experiments. During their workdays, they occupied the Spacelab mockup, then moved to a mobile home for their off-duty hours. Musgrave took the role of 'mission specialist', a career astronaut who would co-ordinate payload operations and liaise with the Shuttle's commander and pilot to resolve hardware anomalies. Morrison adopted the role of a non-career payload specialist, which NASA envisaged being personnel selected by academia or industry for specific experiments. Test directors, data managers and a science manager monitored SMD-I from consoles outside the Spacelab mockup, just as Mission Control and a payload operations control room would oversee real flights. Musgrave stressed a need for physical conditioning on longer Spacelab missions and recommended that commanders and pilots take an active role in the scientific research.

In January 1976, Musgrave returned for the five-day SMD-II, accompanied by nuclear chemist Robert Clark from NASA's Planetary and Earth Sciences Division and cardiopulmonary physiologist Charles Sawin of the Biomedical Research Division. This time, the Spacelab was adjoined by a Shuttle crew cabin mockup (including middeck and aft flight deck stations), which enabled the crew to simulate off-duty hours not unlike a 'real' mission. Musgrave, Clark and Sawin supported 20 biomedical experiments and 14 operational tests to evaluate personal hygiene, general housekeeping, cleaning and maintenance activities in space. The crew compartment setup enabled them to work with a pallet based experiment—in SMD-II's case, a cosmic ray detector—commanded from displays on the aft flight deck. For SMD-II, it sat behind the module to represent the placement of pallet-mounted experiments. Musgrave, Clark and Sawin were awake for 18 hours per day and did 40 percent more research than planned.

The final test (SMD-III) came on 17-23 May 1977. Led by physician-astronaut Bill Thornton, it carried 26 experiments, including over a hundred rats, four monkeys, two frogs, more than 40 mice and thousands of flies. (The crew even designed a tongue-in-cheek patch, with a large primate riding the Shuttle to underscore the difficulty of keeping animals in holding cages during a week-long mission.) Thornton was joined by biochemist Carter Alexander of JSC and biophysicist Bill Williams of ARC. But Thornton harboured concerns that they risked becoming little more than "puppet[s] on an electronic string", manipulated by flight controllers. He felt that mission and payload specialists ought to be an extension of principal investigators: fully responsible for an experiment's conduct and success.

The selection of Alexander and Williams followed a process which Thornton hoped would yield guidelines for picking Spacelab payload specialists. But it was mired with organisational rivalry, particularly as JSC had historically exercised full control over astronaut selection and training and did not relish relinquishing this monopoly to other NASA establishments. "A tentative selection of PS was being made between JSC and [ARC] and unofficial joint training was being accomplished to an acceptable degree," wrote Dave Shayler and Colin Burgess. "News that the PS selection might not be ratified and that [ARC] was looking to train its own selection…did not go down well at JSC. There was even a rumour that the JSC PS selected might be replaced by a second person of unknown technical experience." Thornton was concerned that last-minute payload specialist selections could impair mission success and impose greater weight on the mission specialist to juggle a new crewmember as well as the experiments.

The payload specialists shared duties in prime or backup capacity on SMD-III's experiments, providing dual coverage and redundancy and enabling Thornton to adopt a 'global' troubleshooting outlook. Although the mission suffered communications glitches between principal investigators and the crew, Thornton regarded SMD-III as essential preparation for a real Spacelab flight. "I shudder to think what might have been if we had not done it," he said. "There were some first-rate trainers, but the procedures in getting some of the experiments defined for the simulation were hard enough, without trying to evaluate them for spaceflight conditions."

PATHFINDER FOR SPACELAB

With STS-2, NASA proved that Europe's Spacelab pallet could handle Earth-watching instruments. But before the first dedicated Spacelab could occur, STS-3 in March 1982 looked outward into space. Designated the 'Pathfinder Mission', STS-3 laid the groundwork for flights with sensitive astronomical, solar physics and space plasma physics sensors. It would study 'outgassing' from wastewater dumps and Shuttle thruster firings, as well as the molecular-level 'cleanliness' of the payload bay. Waste products were known to deposit debris 'films' on spacecraft surfaces, threatening delicate optics. If the Shuttle was ever to study the heavens, understanding these impacts was vitally important. Commanded by Jack Lousma, with Gordon Fullerton as pilot, the mission benefitted from a fourth cryogenic tank to support seven days in orbit. And though STS-3 surpassed all planned objectives it taught an enduring lesson: to always expect the unexpected.

Not only did Lousma and Fullerton quadruple the time spent in space by Columbia's first crews, but they also tested the mettle of the RMS arm by handling 'real' payloads for the first time. The Spacelab pallet and the Shuttle's middeck carried nine experiments from NASA's Office of Space Science (OSS-1). By the time STS-3 flew, the office had been renamed the Office of Space Sciences and Applications, but the OSS-1 nomenclature stuck. According to *Flight International* in March 1981, OSS-1 was earmarked for STS-4 but was advanced by one flight when the Department of Defense exercised an option to carry a classified payload on the fourth mission. And on the murky morning of 22 March 1982, Lousma and Fullerton embarked on the most demanding Shuttle mission so far (Fig. 2.6).

"A real barn-burner" was Lousma's description of the launch, as Columbia sprang from the pad at 11 a.m. Eastern Daylight Time (EDT), bound for a 150-mile (240-kilometre) orbit, inclined 38 degrees to the equator. The ride's exhilaration was worth it, save for the nauseous malaise of space sickness. Both astronauts took anti-sickness medications, consumed their requisite daily calories and found as they acclimated to the strange weightless environment they functioned more effectively. "I kinda felt fifty-fifty," Fullerton recalled. "You're pretty happy to just float around and relax, rather than keeping on charging…and into the second day, this is really fun and great and you feel a hundred percent."

Glancing out of Columbia's aft flight deck windows, their Spacelab pallet sat the midpoint of the payload bay. Fullerton powered-up the pallet three hours after launch and activated its experiments 90 minutes later. Inclusive of

Fig. 2.6: The tall Thermal Canister Experiment (TCE) stands particularly prominent in this view of the OSS-1 payload during pre-launch processing.

cables and utilities, OSS-1 totalled 6,890 pounds (3,125 kilograms). Its most visible component was the nine-foot-tall (2.7-meter) box of GSFC's Thermal Canister Experiment (TCE). Weighing 805 pounds (365 kilograms), it investigated simpler thermal concepts to protect scientific instruments from temperature extremes in space. Its aluminium walls were temperature-equalised by longitudinal heat pipes, which collected thermal energy dissipated by electrical heaters simulating scientific instruments in operation. Heat was transferred to the heat pipes, mounted to external radiators and radiated into space. During STS-3, it operated at several set 'temperature points' and demonstrated thermal controllability when Columbia's payload bay was directed towards the Sun or deep space.

Atop the TCE box was the Microabrasion Foil Experiment (MFE), provided by the United Kingdom's University of Kent, the first Shuttle payload built outside the United States. It included 50 aluminium foil sheets of varying densities, bonded to a Kapton substrate and measured numbers, chemical compositions and densities of micrometeorite impacts, with lighter particles gouging shallow 'craters' in the foil and heavier ones penetrating the aluminium to provide clues about their relative densities.

Elsewhere, the Contamination Monitor Package (CMP), led by GSFC and funded by the Air Force Space Division, measured molecular and gaseous contaminants around the Shuttle. Its data helped mission planners understand how contaminants from by wastewater dumps, thruster firings, venting of relief valves and outgassing might hinder future instruments. CMP's four temperature-controlled quartz crystal microbalance sensors viewed the interior and exterior of Columbia's payload bay. Two passive 'witness' mirrors, supplied by the Naval Research Laboratory, were coated with magnesium fluoride, a material used in ultraviolet optics. Their reflective qualities were tested before and after the mission to gauge the effect of outgassed contaminants. CMP revealed "very little" molecular contamination during colder phases of flight, but "increased slightly" as the airframe warmed up.

The University of Florida's Shuttle-Spacelab Induced Atmosphere (SIA) utilised a photometer to understand how evaporating volatiles and dust particulates generated localised 'clouds' around spacecraft. These were known to impair scientific instruments, degrading optical surfaces through scattered sunlight. SIA's observations during orbital daytime emphasised contaminants from the Shuttle itself, whilst the action of interplanetary dust was studied during orbital darkness.

A pair of Sun-scanning instruments began gathering data only hours after launch. The Solar Flare X-ray Polarimeter (SFXP) investigated the polarisation of X-rays emitted by solar flares. It observed a large 'X-flare'—the most energetic type of solar event, capable of triggering worldwide radio blackouts and radiation storms—on the last night of STS-3. Built by the Astrophysics Laboratory at Columbia University in New York, it worked with the Naval Research Laboratory's Solar Ultraviolet Spectral Irradiance Monitor (SUSIM), whose twin spectrometers made 20 hours of ultraviolet observations to assess the changeability of Earth's atmosphere and climate over the 11-year solar activity 'cycle'.

To evaluate the Shuttle's performance under differing thermal loads, Lousma and Fullerton oriented Columbia for ten hours with her payload bay facing the Sun, then 30 hours tail-to-Sun and 80 hours nose-to-Sun, before returning to bay-to-Sun for a further 26 hours. Most of OSS-1's experiments (and particularly the solar instruments) gathered their principal data during these bay-to-Sun periods.

Most intriguing was the University of Iowa's Plasma Diagnostics Package (PDP), a drum-like canister of electromagnetic and particle sensors to 'sniff' the Shuttle's environment, monitoring fields, waves and plasmas. It was known that the Shuttle produced radiation 'noise' via its power distribution system, transmitted and pulsed electric currents and low-level magnetic and electric fields as it moved through its ionised plasma environs. PDP worked during the nose-to-Sun and tail-to-Sun attitudes, its four receivers observing very low frequencies across 16 channels, mid-frequencies in eight channels and very high/ultra-high frequencies in four channels.

On the second day of STS-3, Fullerton uncradled the RMS from the portside sill of the payload bay. One issue which cropped up during this four-hour checkout was the failure of the arm's wrist-mounted television camera, which denied him the visibility to grapple and manoeuvre an interim environmental control monitor to demonstrate RMS handling qualities. With the wrist camera out of action (and with mission managers fearful that another camera might fail), Fullerton could not safely berth the monitor into the bay and it was deferred to another Shuttle flight.

Instead, the smaller PDP was employed for these 'loaded' RMS tests. Pre-flight plans already called for the 350-pound (160-kilogram) PDP to be unberthed from the OSS-1 pallet and manoeuvred around the bay for a pair of eight-hour data-gathering sessions: firstly on 24 March, 50 hours into the flight, and again on the 26th, at 94.5 hours. It would also conduct 36 hours of data collection whilst mounted onto the pallet, measuring electrical and magnetic fields within 45 feet (14 metres) of the Shuttle, as well as ion and electron densities, energies and spatial distributions and electromagnetic waves over a broad frequency range. Lousma and Fullerton conducted three PDP deployments over three days, yielding insights into the strange plasma 'wake' generated by the Shuttle as it passed, boat-like, through the ionosphere at low orbital altitudes.

Utah State University's Vehicle Charging and Potential Experiment (VCAP) rounded out the OSS-1 suite, its charge sensors and current probes sitting at the pallet's corners. It examined the electrical characteristics of the Shuttle, including interactions with natural plasmas in Earth's ionosphere, disturbances induced by the active emission of electrons, the extent to which electric charges accumulated on insulating surfaces and how 'return currents' could be established through a limited area of conducting materials to neutralise active electron emissions. VCAP aided the design of particle accelerators for later Spacelab missions. Emissions from its thousand-volt fast-pulse electron generator were adjustable in repetitive rates or deliverable in steady streams. Those emissions, which ran from 500 nanoseconds to several minutes, were observed and recorded aboard the PDP.

With eight OSS-1 experiments on the Spacelab pallet, a ninth—the Plant Growth Unit (PGU), supplied by the University of Houston—resided in a file-cabinet-sized container in Columbia's middeck. It housed growth lamps, temperature sensors, batteries, fans and data-storage equipment for 96 plants in six terrarium-like chambers. It assessed if 'lignification' was a response to gravity or a genetic process with little environmental interference. Lignin is a structured polymer, enabling plants to maintain vertical posture. The experiment assessed whether lignin reduction in microgravity lowered plants' structural strength. Lousma and Fullerton relayed temperature data to mission controllers twice daily. STS-3 flew Chinese mung bean, oat and slash pine seedlings, all capable of growing in closed chambers with relatively low lighting. Pine is a 'gymnosperm', capable of synthesising large amounts of lignin, and it was thought that its growth was directly affected by gravity. Unlike the mung bean and oat seedlings (germinated only hours before launch), the pine samples were germinated several days earlier.

The seedlings supported three investigations. Several mung beans experienced orientation problems, although the oats suffered no ill effects. The results pointed to a reduction of lignin in space-grown plants, but the difference was statistically insignificant. The mung beans and oats were used for chromosomal studies, confirming their root cells were affected by microgravity. And the plants' gravity-sensing tissues (including root caps) were examined to gauge adaptation behaviours.

AMBITIOUS START, DRAMATIC END

Plans called for STS-3 to last 171 hours and 24 minutes, before returning to Edwards. The astronauts' homebound flight path would cross the California coastline, north of Morro Bay, descend over the San Joaquin Valley, pass near Bakersfield, Tehachapi and Mojave to land—possibly under crosswind conditions—on Runway 23 at 10:24 a.m. PST on 29 March.

But all that changed a week before launch.

Unseasonal rain showers left Edwards' runways sodden and on the 18th NASA called up the backup landing site: a great mountain-ringed blotch of salt and gypsum in New Mexico's Tularosa Valley, aptly named 'White Sands'. Selected as a contingency Shuttle landing strip in 1979, its runways were widened, lengthened and compacted until their gypsum surface achieved a consistency as strong as granite. Two trains were chartered from the Santa Fe Railroad to transfer heavy cargo over 800 miles (1,300 kilometres) from Edwards to White Sands. Non-mobile equipment was transported on flatbed trucks and a makeshift barracks and press site was assembled within three days (Fig. 2.7).

Fig. 2.7: Unusual view of the OSS-1 payload during the closure of Columbia's payload bay doors; one door is already shut, the other about to be winched closed. The twin 'drums' of the Solar Flare X-ray Polarimeter (SFXP, at left) and the Plasma Diagnostics Package (PDP, with its black spherical receivers, at right), are visible in the foreground, with the gold-coloured Thermal Canister Experiment (TCE) behind.

However, despite 90-percent year-round perfect weather, on 29 March 1982—the very day Lousma and Fullerton were due to land—White Sands was battered by its worst wind and sandstorm in a quarter-century. Gypsum drifts gathered to waist-height against buildings. One NASA manager saw his car almost totally sandblasted of its paint. And Rick Nygren, head of NASA's vehicle integration test team, found a layer of dust covering his car…on the *inside*, having pushed through the seals of his rolled-up windows. "It got into everything," Nygren said of the dust's ubiquitousness. "Everything."

Blissfully unaware of the poor weather, early on 29 March Lousma and Fullerton deactivated OSS-1 four hours before landing. But conditions at White Sands were unacceptable, with visibility less than two miles (3.2 kilometres). CAPCOM Steve Nagel told the crew it was "not a good day" and they would stay in orbit another 24 hours.

"We had a good drill," quipped Lousma.

The winds subsided after sundown. Ten road graders were deployed onto the badly eroded runway that night, headlights ablaze, to re-compact the gypsum and prepare for a landing the following morning. Nygren remembered seeing convoys of recovery vehicles heading out to the runway, their tall snorkels and purge masts barely visible above the low-level gloom, "like dinosaurs creeping across the lakebed".

Early on the 30th, Lousma and Fullerton re-entered the atmosphere above the South Pacific, east of Samoa, crossed Baja California and the Mexican state of Sonora, then headed for White Sands. Touchdown came at 9:04 a.m. Mountain Daylight Time (MDT) on Runway 17. As Columbia sped down the runway, her forward gear in the process of coming down, the nose unexpectedly pitched upward, due to a flaw in the flight software, producing what Fullerton called "a kind of wheelie".

STS-2 and STS-3 buoyed NASA and ESA with confidence that Europe's laboratory was ready to fly as an integrated system. Two pallet-based flights had demonstrated Spacelab's capacity to accommodate Earth-watching and space-watching instruments with great precision. But the program had barely begun and its Golden Age still sat, mirage-like, on a far-off horizon. Ahead lay Spacelab-1, a triumph of international collaboration whose legacy endures to this day.

3

Verification Flight Test One

SELECTING SPACELAB'S FIRST CREW

Four days past Thanksgiving, on 28 November 1983, six men in bright blue flight suits strode from KSC's Operations and Checkout Building into a bewildering blaze of media flashbulbs. Leading the sextet was John Young, chief of NASA's astronaut corps, veteran Moonwalker and the first human to leave Earth a sixth time. His pilot was Brewster Shaw, an excited rookie's grin cleaving his face from ear to ear.

These two military pilots—Young a Korean War aviator, Shaw a decorated Vietnam War veteran—were followed by four civilian scientists. Mission specialists were electrical engineer and former U.S. Navy electronics officer Owen Garriott and astronomer Bob Parker. And this particular flight marked the debut of a new subset of crewmember, the payload specialist: a non-professional astronaut chosen by academia or industry for specific experiments. Rounding out Young's crew were Byron Lichtenberg, a former Air Force fighter pilot and biomedical engineer from Massachusetts Institute of Technology (MIT), and Ulf Merbold, a West German physicist from the Max-Planck Institute in Stuttgart. Merbold, who fled East Germany as a teenager, just before the rise of the Berlin Wall, became the first non-American to fly on a U.S. spacecraft. STS-9, the ninth voyage of the Shuttle and the sixth by Columbia, was also the first to fly six crewmembers. "For the first time in spaceflight," Young joked, "the doctors outnumber the pilots, four to two."

On 1 August 1978, six men—four payload specialist candidates from Europe and the United States and the two NASA mission specialists—were identified for Spacelab-1. The lead mission specialist was Garriott, a Skylab veteran and deputy director of NASA's Science and Applications Directorate since August 1974. That role saw him work extensively on Spacelab's early evolution. Parker was a former astronomy professor from the University of Wisconsin at Madison. He had not flown in space yet accrued a wealth of expertise on the ground, including ASSESS-I and chief of the astronaut corps for the Science and Applications Directorate since August 1974.

As outlined in Chapter Two, the mission specialist had crystallised into a co-ordinator for Spacelab scientific activities and a representative of the crew at high-level meetings on the ground. In November 1974, NASA noted that a high degree of usefulness existed in having astronauts who were also professional scientists to oversee experiments and address payload issues with principal investigators. In time, this leadership role would evolve into the 'payload commander', a common staple of later Spacelab flights. "We're the eyes and ears and hands of the guys on the ground," Parker said. "If they want us to turn that switch, and do this thing this way, we'll do it for them."

When Garriott and Parker joined Spacelab-1, their launch had shifted inexorably to the right. In May 1977, NASA outlined up to six Shuttle test flights, with Spacelab-1 "probably" on the seventh mission in May 1980, "unless another flight test was needed". A year later, with Garriott and Parker assigned, launch had moved into "the early 1980s"—Parker privately anticipated flying in 1981—for a seven-day mission. But as the Shuttle's maiden voyage slipped, so Spacelab-1 slipped in tight lockstep. By May 1980, launch was aimed for late 1982; by July 1980, it had moved to May 1983; and by April 1982, it was scheduled for September 1983. "To some extent, training expands to fill the vacuum," said Parker. "To other extent, we basically had a year when we weren't doing a whole lot of training for it. It was kind of a year's hiatus. It didn't mean we didn't do anything, but we just really weren't terribly active during that year."

Selecting four payload specialist candidates—primes and backups from the United States and Europe—fell to an Investigators Working Group (IWG) of 35 principal investigators, one for each Spacelab-1 experiment, chaired by Mission Scientist Rick Chappell of MSFC. In September 1977, ESA chose 53 candidates from 12 European nations for 'its' seat, with an expectation that six finalists would be selected for testing and evaluation by NASA in January-April 1978, then reduced to three by mid-1978. One would be picked shortly before launch, with the other pair serving as backups.

Meanwhile, the United States was busy choosing its own candidates. In November 1977, 18 U.S. scientists (including NASA's Bill Thornton) entered consideration. And on 22 December six U.S. and four European candidates were chosen as Spacelab-1 finalists. In addition to Lichtenberg, they included

JSC flight surgeon Craig Fischer, physicist Mike Lampton, meteorologist Robert Menzies, astronomer Richard Terrile and physicist Ann Whitaker. Competing for the ESA seat with Merbold were Italian physicist Franco Malerba, Swiss astrophysicist Claude Nicollier and Dutch physicist Wubbo Ockels.

On 1 June 1978, Lichtenberg and Lampton were formally selected by the Payload Specialist Selection Group (PSSG), a subgroup of the IWG, as the U.S. candidates for Spacelab-1, following a two-day meeting at MSFC. "Thus, the researchers who designed and built the experiments and who are to analyse the results," wrote Walter Froehlich, "helped select from their peers the two in-flight specialists who are to be in charge of carrying out research in orbit." It honoured an oft-stated NASA pledge that on the Shuttle, scientists would get to operate their own hardware in space. At the same time, Merbold, Ockels and Nicollier were selected for Europe's payload specialist place. According to Michael Cassutt in *The Astronaut Maker*, ESA hoped all three might fly Spacelab-1, alongside a NASA commander, pilot and mission specialist. (Certainly, physician-astronaut Joe Kerwin had long pushed for at least one NASA mission specialist to lead the payload crew on this highly complex scientific voyage.) But as the Shuttle evolved, it became apparent that a 'basic' crew needed two pilots and two mission specialists. The result was that ESA got not three payload specialist seats on Spacelab-1, but just one.

As well as a difficult pill to swallow for ESA, this decision proved tough, politically, for France—Spacelab's second highest financial sponsor—as no French candidate made the cut. "The French president was very upset with that result, because France is a main contributor to ESA and half of the French space budget goes to ESA," said French astronaut Jean-Loup Chrétien. "It was the first time Europe was selecting astronauts and the express [wish was] for scientists…they were looking for pure scientists." When the selections were made, pure scientists were picked, but the successful applicants also had flying experience. "All these guys had it," Chrétien said, "but the guys that France had selected were pure scientists, so they were eliminated."

Training began in October 1978, with orientation tours at various U.S. and European sites, ahead of briefings from January 1979. That June, the two mission specialists and five payload specialist candidates conducted a three-day crew station review in the Spacelab mockup at MSFC. It sought to "evaluate experiment designs, together with written procedures…[to] ensure proper crew-equipment interface, especially the location of switches, controls and displays". And in July 1980, the United States and Europe agreed to train Nicollier and Ockels as fully fledged mission specialists, with ESA reimbursing the costs. This decision was induced by the Spacelab-1 delays and "in recognition of ESA's contribution to the STS in funding Spacelab development". However, NASA Administrator Robert Frosch made clear to ESA Director General Erik Quistgaard that Spacelab-1 took priority over the mission specialist training program. The pair concluded their training in September 1981. Ockels resumed his Spacelab-1 payload specialist duties in January 1982, but Nicollier remained on the mission specialist 'track'.

"These were essentially payload-operating people, but they were also going to operate a lot of the Spacelab equipment," remembered Mel Brooks. "I argued like crazy with the Europeans that they should define the requirements according to the mission specialists, because that's the only way you'll ever get them to be considered as astronauts…if you get them into the mission specialist training and get their wings. Otherwise, they will be forever passengers and scientists." Brooks got Nicollier and Ockels into mission specialist training, but not Merbold, an eventuality he ascribed to George Abbey, JSC's director of flight crew operations. According to Cassutt, ESA's willingness to pay for the mission specialist training left Abbey with little choice but to agree. Ockels did not stay on the mission specialist track for long. "When I had to make the [payload specialist] selection for…the first Spacelab flight, we had to bring one of them back," Brooks said. "Wubbo and Claude decided that Claude should stay, because he was a pilot and more acceptable to the NASA people". And Merbold, wrote Melvin Croft and John Youskauskas in *Come Fly With Us*, "could not meet the medical requirements to become a mission specialist" (Fig. 3.1).

Spacelab-1 training expanded on 20 April 1982, with Young and Shaw named as commander and pilot. And the following 20 September, Merbold and Lichtenberg were picked for the European and U.S. payload specialist seats. "Winners," noted MSFC, "were chosen in secret ballot by a panel of 36 U.S. and European scientists."

"I was thrilled to have a flight assignment," said Shaw. "I was more thrilled to be able to fly with John Young." But there was uncertainty from the scientists. "He's motivated differently," Garriott said of Young. "Standard sort of prototypical test pilot." But Young "really jumped in and assisted with the conduct of the science…we all really enjoyed having him on board". A lesser man than Shaw might have felt a pang of intimidation flying with America's premier astronaut, but the pair knew each other's capabilities. "John knew I could fly airplanes," Shaw said, "and I figured that wasn't an issue."

Fig. 3.1: Spacelab-1 prime and backup payload specialist candidates inspect a model of the module and pallet combination. From left to right are Wubbo Ockels, Ulf Merbold, Mike Lampton, Claude Nicollier and Byron Lichtenberg.

A SCIENTIFIC BONANZA

As the crew gelled, so too did an ambitious mission. A collaborative venture between NASA and ESA, Spacelab-1 comprised a pressurised 'long' module and single pallet with five scientific focuses: atmospheric physics/Earth observation, space plasma physics, materials science/technology, astronomy/solar physics and life sciences. In March 1976, NASA issued a request for proposals and the following July identified an 80-strong academia/industry team to examine the tendered responses, after which the Space Science Steering Committee selected a group of investigations for Spacelab-1. In February 1977, after evaluating 2,000 proposals, NASA and ESA chose 222 scientists from the United States and 14 other nations (including India, Canada, France, Belgium, Austria and Norway) to participate in the mission. By August 1978, when Garriott *et. al.* were named, the mission featured "about 40 experiments". This figure tightened by May 1980 to 37 experiments—13 from NASA, 24 from ESA—with each agency allocated 3,000 pounds (1,400 kilograms) of hardware. Another experiment was added by September 1982, bringing the total to 38, supporting 71 investigations. Managing the U.S. share of the mission was the Solar Terrestrial Division of the Office of Space Science at NASA Headquarters, with ESA's share overseen by the Spacelab Payload Integration and Co-ordination in Europe (SPICE), based at Porz-Wahn, southeast of Cologne in West Germany.

"The whole mission was oriented towards making clear that the Spacelab…was useful for the whole range of scientific disciplines," said Garriott. "My interest is interdisciplinary work, so not only a lot of biomedical work, but also astronomy, fluid physics, materials processing, atmospheric sciences: all were represented with differing experiments. I found that quite interesting and I think it does relate to the fact that you need an interdisciplinary background to try to conduct experiments. You need somebody who has got some degree of competency in all of the variety of areas."

Garriott, Parker, Lichtenberg and Merbold spent a lot of time in West Germany, where most experiments were based. Spacelab-1 represented Europe's first foray in human spaceflight, a fact not lost on the ESA member-states who had spent a decade making it happen. "Owen and I spent a lot of time going to Europe," Parker said. "First European mission, so we had to go to every European country that had a piece of it in ESA." They travelled to Denmark and France, too, but West Germany was the nexus of everything, "so we spent most of our time in Germany".

Four experiments emphasised atmospheric physics. The Imaging Spectrometric Observatory (ISO), provided by Utah State University, had five spectrometers to investigate oxygen, nitrogen and sodium abundances in the mesosphere and measure atmospheric airglow spectra across extreme ultraviolet to infrared wavelengths. ISO also worked with Spacelab-1's space plasma physics suite to examine the atmosphere's vertical structure. The Grille Spectrometer (which used a grille for part of its optical system and a mirror for the other) came from Belgium's Institut d'Aeronomie Spatiale de Belgique and France's Office National d'Etudes et de Recherches Aerospatiales and sat on the pallet. It studied atmospheric trace constituents (including carbon dioxide, water vapour and ozone) involved in photochemical processes at altitudes between 9-93 miles (15-150 kilometres). And a pair of experiments, both supplied by France's Service d'Aeronomie du Centre National de la Recherche Scientifique, photographed cloud-like 'waves' in the oxygen-hydrogen emissive layer of the middle atmosphere at 50 miles (85 kilometres) and examined sources of Lyman-alpha emissions using a spectrophotometer (Fig. 3.2).

Fig. 3.2: View of part of the pallet hardware, as seen through the small window in the Spacelab-1 module's end-cone. The aft bulkhead of Columbia can be seen behind.

Two Earth observation experiments, provided by Germany and the Netherlands, included the large-film Metric Camera to evaluate high-resolution imaging in space. Mounted by the crew into an optical-quality window in the Spacelab module's roof, it was operated remotely by ground controllers, with the crew periodically changing its black-and-white and infrared film magazines and filters. And the Microwave Remote Sensing Experiment (MRSE) furnished all-weather radar imagery of Earth's surface in support of applications from agriculture to fishing and shipping to crop monitoring. It comprised an antenna on the pallet and systems inside the Spacelab module.

Six pallet-based experiments explored space plasma physics. Of these, the most complex was Space Experiments with Particle Accelerators (SEPAC), developed by Japan's Institute of Space and Astronautical Sciences (ISAS) and the University of Tokyo. Its 'electron gun' investigated ionospheric dynamics, firing pulses of gas and high-intensity electrons to understand atmospheric aurorae, Earth's magnetic field and charged-particle environment and plasma effects on the Shuttle. Despite the failure of its electron beam assembly in high-power mode, SEPAC achieved almost all its allocated tasks. Another gun, provided by France's Centre National de la Recherche Scientifique, was the Phenomena Induced by Charged Particle Beams (PICPAB), which included an 'active' unit on the pallet and a 'passive' recorder in the module. Like SEPAC, it produced artificial aurorae and although some data was lost when one of its gas bottles failed, PICPAB's overall success topped 60 percent.

Lockheed Palo Alto Research Laboratories supplied the Atmospheric Emissions Photometric Imaging (AEPI) sensor to observe faint emissions from natural and artificially induced phenomena using a low-level camera/photometer, whilst two German experiments from the Max-Planck Institute and the University of Kiel measured heavy cosmic ray nuclei and low-energy electron 'echoes' and a study from the Austrian Academy of Sciences examined magnetic fields in the Shuttle's environs.

Spacelab-1 also featured three astronomical investigations. The University of California at Berkeley's Far Ultraviolet Space Telescope (FAUST), situated on the pallet, was a wide-field-of-view instrument tasked with observing faint ultraviolet emissions from distant sources to understand the lifecycles of stars and galaxies. During STS-9, fogged film ruined most of FAUST's images and a post-flight investigation recommended that its next mission should record incoming photons electronically. But it achieved 95 percent of its planned objectives, including the first far ultraviolet image of the Cygnus Loop, a relatively 'local'—at 2,400 light-years away—supernova remnant. Both the space plasma physics and astronomy experiments benefitted from a Scientific Airlock (SAL) in the Spacelab module's roof, which could expose up to 220 pounds (100 kilograms) of hardware to vacuum. At alternate intervals during STS-9, the SAL housed PICPAB's 'passive' detector and the Very Wide Field Camera (VWFC) from France's Laboratoire d'Astronomie Spatiale in Marseilles, which imaged ten astronomical sources, yielding ultraviolet views of a 'bridge' of hot gas between the Large and Small Magellanic Clouds. The third astronomy experiment, a spectrometer from ESTEC in the Netherlands, examined X-ray sources using a gas scintillation proportional counter.

Three other pallet-mounted experiments were devoted to solar physics. JPL's Active Cavity Radiometer (ACR) measured the Sun's total 'irradiance' (a phenomenon termed the 'solar constant') using three detectors spanning far ultraviolet to far infrared wavelengths. The Solar Spectrum (SOLSPEC) from the Service d'aéronomie of France's Centre National de la Recherche Scientifique, tracked the Sun's output across the electromagnetic range. And the Measurement of Solar Constant (SOLCON), supplied by Belgium's Institut Royal Météorologique de Belgique, used a high-resolution radiance sensor to trace absolute values of the solar constant.

With these experiments sitting externally atop the pallet, 15 life sciences investigations provided by researchers in the United States, Germany, the United Kingdom, Italy and Switzerland resided inside the module for hands-on interaction from the crew. Spacelab-1 flew a 'long' module of two halves. Its 'core' segment housed data-processing equipment, a workbench, air-cooled research racks lining its port and starboard walls and a 0.9-foot-wide

(30-centimetre) optical-quality Scientific Window Adapter Assembly (SWAA). And its 'experiment' segment supplied room for additional research racks and the deployable SAL to expose payloads to space.

Experiment racks were a 'standard' 48.2 centimetres (19 inches) wide, roughly refrigerator-sized, and assembled outside the module, checked out as a unit, then slid into the cylindrical shell for integration with subsystems and the primary structure. Between the two walls of racks, a central aisle offered additional floor and ceiling room for experiments. "The racks were pretty much standard," said Gene Rice. "You either had a drawer in a rack or you had a whole rack or…a double rack, depending on the magnitude or size of the experiment. We would help [customers] through the process of designing their experiment, integrating it into a Spacelab rack, doing the testing that they needed to do. They would have to show that they met the safety requirements to put it into the Spacelab and to fly it."

Although the core segment could fly alone as a 'short' module—usually with pallets—this exceptionally heavy configuration was never used and all module-based missions employed the longer format. With each segment measuring 13.1 feet (4.2 metres) in length, the long module extended to 23 feet (7.5 metres) and its diameter of 13 feet (4.1 metres) snugly fitted the width of the Shuttle's payload bay. It provided a pressurised volume of 2,600 cubic feet (75 cubic metres). The system was covered with white passive thermal insulation to guard against the temperature extremes of low-Earth orbit and secured by three longeron fittings on the payload bay walls and one keel fitting in the floor. Its truncated end-cones measured 2.6 feet (0.8 metres) deep, with a 'large' end of 13.5 feet (4.1 metres) in external diameter and a 'small' end of 4.3 feet (1.3 metres). Each cone had three 'cutouts', 1.4 feet (0.4 metres) wide: one at the top for a viewport and two at the bottom to support feedthrough plates for utilities.

Connecting the module to the Shuttle's cockpit was a flexible tunnel, available in two lengths: 8.7 feet (2.7 metres) for short-module missions and 18.9 feet (5.8 metres) for long-module missions. The tunnel had an internal, unobstructed diameter of 3.3 feet (1.1 metres). On short-module missions, the Spacelab sat 'forward' in the payload bay (hence the shorter tunnel), but for long-module missions centre-of-gravity constraints dictated it reside further aft. Work on the tunnel started in April 1975, when an early conceptual design from Goodyear, was evaluated; it folded out like an accordion from a stowed length of 1.9 feet (0.6 metres) to a full extent of 14.1 feet (4.3 metres). Rigidised by steel rings, the Goodyear proposal was made from layered aluminium foil, Capran film and nylon, coated with a spongy outer shield to guard against micrometeoroid impacts. These tests validated the strength,

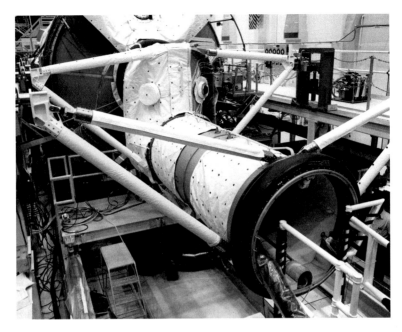

Fig. 3.3: Close-up view of the tunnel adapter which connected the Spacelab module to the Shuttle's middeck. Notice the raised 'joggle' section, which was incorporated to raise the tunnel to the same level as the module's access hatch.

airtightness and behaviour of tunnels in space-like environments. Ultimately, McDonnell Douglas was picked to build the tunnel. In April 1977, JSC signed a $3.1 million contract with Rockwell International for tunnel design changes and a subsequent $5.2 million contract covered eight engineering adjustments (Fig. 3.3).

Since the access hatch into the Spacelab module sat 3.3 feet (1.1 metres) 'above' the level of the Shuttle's cockpit airlock hatch, a 'joggle' section was built into the tunnel to resolve this vertical offset. "The joggle also permitted a small amount of longitudinal movement," explained Shayler and Burgess, "during the ascent and descent phases of the flight, helped absorb movements caused by the differential expansion rates of the orbiter and Spacelab module and minimised any overstressing that might have occurred in a 'straight' tunnel/adapter configuration."

But this posed another problem. If Young and Garriott needed to go outside on an Extravehicular Activity (EVA) perhaps to manually close the Shuttle's payload bay doors in an emergency or retract the SAL if it malfunctioned—the tunnel connected directly to the cockpit, bypassing the airlock. "That changed all our profiles and we had to have another way to do EVA once this tunnel adapter made it to the airlock outer hatch," remembered tunnel adapter project manager Hank Rotter. An EVA hatch, ducting and pipework was therefore built into the tunnel's roof to allow spacewalkers to get outside if needed. In such cases, the module would be vacated of crew and its hatch closed—as would the hatch into the Shuttle's cockpit—allowing Young and Garriott to depressurise the whole tunnel. "We essentially doubled the volume of the airlock by having these two modules…mated together," said Rotter. "We did those pressure profiles, so we could know how long it would take to depress and repress."

Spacelab-1's life sciences haul included dosimeters to measure the extent to which neutrons, protons and highly charged particles penetrated the Spacelab module, an advanced 'biostack' of biological materials to assess cosmic radiation effects and 300 microorganisms in four containers for direct exposure to the space vacuum environment.

An MIT-furnished set of vestibular experiments sought to examine the role of body and eye movement and vestibular control in triggering space sickness. It included body restraints, cameras, tape recorders and other tools for visual stimulation and recording of crewmembers' responses. The experiments proved insightful by linking head movements to space sickness; indeed, three of Spacelab-1's four scientists suffered varying degrees of the malaise, although, as Douglas Lord commented, many tests sought to drive them to the brink of sickness, "so a high percentage of such problems in this mission would not have seemed unusual". Other experiments investigated spinal reflexes and the effects of weightlessness upon the inner ear, specifically the 'otolith', an organ which helps to maintain humans' upright posture. Instrumented backpacks holding miniature accelerometers and electrocardiographs assessed heart and respiratory rates and voluntary limb motions during physical activity, whilst belt-worn recorders and head-worn electrodes measured sleep patterns and physiological responses to the rigours of ascent and re-entry.

Blood samples were taken before, during and after the mission to assess immune-system responses, changeability in antibody levels, reductions in circulating red and white blood-cell masses and hormones. Lichtenberg and Merbold measured each other's central venous pressure using sterile needle strain gauges to record bodily functions known to undergo change in space. Owing to Young and Shaw's responsibility to land the Shuttle, they were immune to blood draws. "They weren't supposed to take our blood," Shaw said, "and I don't think anybody took any of my blood." Young joked that whenever he entered the module, he was at risk from needle-toting scientists. "Everytime I came down there, they wanted to draw my blood," Young drawled, "so I had to leave."

Closing out Spacelab-1's life sciences suite was the University of Pennsylvania's *Helianthus annuus* experiment, whose dwarf sunflower seedlings flew in a smaller format on STS-2. The State University of New York sponsored an experiment to grow fungus in nutrient-filled tubes, and in complete darkness, to determine if Earth-like day/night circadian rhythms persisted in space-grown plants; it found that they did. And a particularly fond memory for Shaw was the Mass Discrimination During Weightlessness Experiment—provided by psychologist Helen Ross of Scotland's University of Stirling. This comprised a 20-minute series of tests before, during and after the mission, using 24 small steel balls of equivalent size, but differing mass.

"Helen's Balls," exclaimed Shaw, clearly savouring the memory. "She had a bunch of little yellow balls that had different mass, different weight. Since there's no weight, there's only mass in zero-gravity, we had to try and differentiate between the mass of these balls. You would take a ball in your hand and you would shake it and you would feel the mass of it by the inertia and the momentum of the ball as you would start and stop the motion. They were numbered as to which was the most massive to the least massive."

Spacelab-1's materials science haul came from West Germany, Austria, the United Kingdom, Italy, France, Sweden, Belgium, the Netherlands, Spain and Denmark. Spearheading the payload was West Germany's refrigerator-sized Materials Science Double Rack and four furnaces: an isothermal heating facility for 14 solidification, diffusion, metal and composite casting experiments, a gradient heating facility for five crystal growth and unidirectional solidification experiments, a mirror heating facility for four crystallisation experiments featuring silicon and cadmium telluride and a fluid physics module for seven studies on damping, spreading, convection and stability of liquids in weightlessness.

Mel Brooks, who managed the double rack for NASA, recalled a troubled evolution, with issues pertaining to toxicity of its samples, limited clearances and high operating temperatures of 1,600 degrees Celsius (3,000 degrees Fahrenheit). "The more I met with the [MSFC] people, I could see that they were already planning some backup, in case this thing doesn't fly," he said. "I had to pull it out of the fire and go through all those gates to get it accepted by NASA and the safety people and all the design reviews. It had everything that would attract the attention of the safety guys and it was a real challenge to get that developed [and] qualified to our own satisfaction that it would work." Years later, Brooks derived "tons of pleasure" as it performed admirably on Spacelab-1.

But as Bill Thornton found on SMD-III, the astronauts considered themselves little more than laboratory technicians, riding an ever-revolving

Fig. 3.4: U.S. Vice President George H.W. Bush welcomes the arrival of the first Spacelab flight unit in Florida in February 1982. From left to right are Claude Nicollier, Ulf Merbold, Bush, Wubbo Ockels and Bob Parker. Standing in the background is Owen Garriott.

hamster wheel. "We put materials sealed in cartridges into furnaces, heated, melted, solidified the materials, pushed buttons and started computer programs," recalled a non-plussed Merbold. But Spacelab-1 comprehensively underscored the need for a 'man-in-the-loop'. The double rack was a great success, although both the isothermal and mirror heating facilities suffered partial power failures. ("Probably some air in the water loop," mused Merbold.) Fortunately, the mirror heating facility was later recovered, thanks to maintenance inputs by the crew (Fig. 3.4).

The Spacelab-1 module and pallet totalled 33,252 pounds (15,265 kilograms), the heaviest payload lifted by the Shuttle at that time. Science operations were co-ordinated from the 4,000-square-foot (370-square-metre) Payload Operations Control Center (POCC) at JSC. For the first time, the POCC enabled principal investigators to talk directly to the astronauts, rather than relaying messages via a CAPCOM in Mission Control. Led by a payload operations director, its team included alternate payload specialists Lampton and Ockels to troubleshoot issues. Instructions from the POCC were sent to the Shuttle's on-board text and graphics machine or directly to the experiments. Astronaut Bryan O'Connor was lead CAPCOM in Mission Control during STS-9. "It was part of my responsibility to make sure that the CAPCOMs and the equivalent, the folks that were in the POCC talking to the science crew, were co-ordinated," he said. "We practiced what happens if there's an emergency or we have to use that communications loop to solve an orbiter problem and the protocols that are used for that: handing off back and forth between them, writing the rules for how we're going to do that…a lot of planning and co-ordination."

Generally speaking, the process worked well, although on occasion the crew found themselves overloaded with requests. Parker once lost his patience when the POCC asked him to adjust a medical procedure, recharge a battery, restart a furnace *and* check an experiment on the pallet, all at the same time. "If you guys would recognise that there are two people up here trying to get all your stuff done," he snapped, "I think you might be quiet until we got one or the other of them done!"

PREPARING FOR SPACELAB-1

Spacelab-1's hardware began arriving at KSC in October 1981, when a double-rack-sized life sciences mini-lab was delivered to the Operations and Checkout Building. Modification of this 320,000-square-foot (29,700-square-metre) building for Spacelab started in April 1976, with contracts to Pan American Technical Services for changes to its gaseous nitrogen, helium, high-pressure air, water and air conditioning infrastructure and installation of ESA-provided ground support equipment. Early the following year, NASA contracted with McDonnell Douglas to assemble Spacelab in the building. And in December 1978, NASA announced its intent to physically integrate all Spacelab experiments at KSC.

The mission's showpiece, the long module and pallet, arrived in Florida on 11 December and an official unveiling ceremony in February 1982 was attended by Vice President George H.W. Bush. "More than 300 invited guests from Europe and the U.S. gathered in the high bay area," wrote Douglas Lord, "where they could see in the background both the engineering and flight-unit hardware of the Spacelab work stands." Bush lauded Spacelab as "the fruit of a lot of hard work" and praised Europe's role in "the largest co-operative space project ever". He added that "if today can be considered

Spacelab's birthday, then there are a great many proud parents celebrating…let us continue to be partners".

Elsewhere, Spacelab-1's research facilities continued to arrive. The high-voltage power supply for SEPAC marked the first major piece of experiment hardware installed onto the pallet. "The experiments were brought in by their various scientific teams," remembered Spacelab-1 Mission Manager Harry Craft of MSFC. "We would let them check the experiments out initially in an off-line capability and then we'd bring them into a room and just make sure the instrument had met the transportation environment and still worked, [then] they'd turn it over to us."

By the late summer of 1982, all experiments were installed and a mission sequence test in November verified their compatibility with on-board systems. Seventy-nine continuous hours of the week-long flight were simulated, with ground-support equipment taking the role of Shuttle systems to demonstrate high-data-rate recording and playback functions. In May 1983, the payload was hooked up to cargo interface test equipment, which duplicated the Shuttle's systems in high-fidelity mode to validate their integrated performance, and in July a closed-loop test was conducted with the POCC. Finally, on 16 August 1983 Spacelab-1 was moved to the OPF and loaded aboard Columbia. The tunnel was installed and compatibility testing was conducted remotely with the POCC. But major design deficiencies cropped up in the command and data-management system, following tests in September 1982. Although these problems were remedied, some U.S. engineers derided Spacelab-1's computers as "marginal, if not obsolete technology".

STS-9 was Columbia's sixth space voyage. Following her fifth flight in November 1982, she was stood down for several months of 'Spacelab Only' modifications. In addition to Young and Shaw's seats, a third (collapsible) seat was added for Parker, who served as flight engineer during ascent and re-entry, assisting the pilots with any anomalies. Three more collapsible seats resided on the middeck for Garriott, Lichtenberg and Merbold. "Seat floor beef-up at attach point of mission specialist and scientist operational seats on crew compartment flight and middeck floor to support 20G crash-load requirements," noted NASA's summary of the Spacelab Only modifications. Parker's dual role as flight engineer enabled him to build a tight relationship with Young and Shaw. "At the same time as we were training on the experiments, I was training, particularly the last year, with Brewster and John on ascents and entries," Parker said. "For a good last six months or so, I'd be training on Mondays with them doing ascents and entries, fly to [MSFC for] experiment training, fly back and do ascents and entries on Wednesday, back and forth. I got to…keep in touch with them, maybe a lot better than the others."

As well as extra seats, the Spacelab Only work saw the text and graphics machine fitted, improvements to the Shuttle's brakes and tyres and structural/electrical upgrades. The payload bay floor was strengthened to handle the module and pallet, payload consoles were added to the aft flight deck and a fifth cryogenic tank was installed. These changes supported a mission which had expanded from seven to nine days. *Flight International* reported in October 1982 that this helped "relax crew workload", but ESA had been pushing for the longer flight duration since at least September 1980.

STS-9 added one-third greater complexity to many Mission Control functions over earlier flights, specifically power distribution, life support, cooling and cryogenic consumables. "Management of cryogens for fuel cells," NASA noted, "will be a more significant duty on this flight, in part because of the power levels to be experienced, but more because consumption must be monitored and budgeted over a longer-duration flight." Several console occupants found their duties multiplied. The mechanical systems officer, for example, also monitored the Spacelab module's windows and pressures and venting during the SAL deployment. And a brand-new console, the command and data-management systems officer, was responsible for Spacelab-1's two main computers.

Three bunk-like sleep stations were fixed to the middeck's starboard wall, containing sleeping bags and covers for eyes and ears to facilitate restful off-duty hours on a mission intended to operate around the clock for the first time in human spaceflight. Each sleep station included personal storage, a light, ventilation and a retractable privacy door. Plans called for the 'red' team of Young, Parker and Merbold to start their first sleep period five hours into the flight, with the 'blues' of Shaw, Garriott and Lichtenberg continuing Spacelab-1 activation. The blues would then go to bed 14 hours after launch (Fig. 3.5).

Typical daily activities saw Young's team manning the overnight duty from 9:30 p.m. to 9:30 a.m. EST and Shaw's team running the daytime shift for the opposing dozen hours. Each team's time entailed eight to ten hours of payload operations, with the remainder devoted to breakfast, lunch or dinner, shift handovers, exercise, daily planning with ground controllers and pre-sleep/post-sleep hygiene. As the mission neared its end, the shifts staggered slightly to ensure that all six men were awake, with Young's reds finishing their last sleep period 12 hours before landing and Shaw's blues awakening four hours before landing. "It was just a giant step ahead," Parker said. "In the old days, [if] we wanted to observe something…we had to wake the crew up two hours early. Now there was always a crew available, so you could do that. And whoever was up, did it."

Fig. 3.5: Spacelab-1's research racks are prepared for rolling into the module's pressurized shell in the Operations and Checkout Building.

On 23 September 1983, Columbia rolled to the VAB for stacking onto her ET and SRBs, then moved to Pad 39A five days later. After a decade of waiting, Spacelab-1 was ready to fly, with liftoff targeted for 28 October.

But fate had other cards to play.

RISE OF THE GREMLINS

On 7 September 1983, an important test—vital to Spacelab-1's success—was conducted. Controllers in the POCC remotely commanded hardware aboard the module and pallet, which included in the loop both the ground-bound Columbia and a new NASA communications satellite in geostationary orbit. Yet that very satellite conspired to delay STS-9 until the end of 1983, threatening the mission's scientific harvest with ruin.

An essential pre-requisite for Spacelab was NASA's Tracking and Data Relay Satellite (TDRS), intended to provide near-continuous voice and data communications between Mission Control and the POCC in Houston with astronauts in space. Positioned in geostationary orbit, at an altitude of 22,300 miles (35,700 kilometres), TDRS replaced a cumbersome worldwide network of ground stations and ships used to track human space missions. As Shuttle development accelerated in the mid-1970s, it was recognised that two satellite relays would afford astronauts reliable voice and data communications for 85 percent of each orbit.

NASA selected TRW as prime contractor to build TDRS in January 1977, with plans for the first satellite to fly in the summer of 1980, well ahead of Spacelab-1. In November 1978, the space agency announced that Spacelab-1 should fly "as soon as feasible" after the first two satellites were in orbit, noting that "an operational TDRS is required by the [Shuttle] in order to initiate the…Spacelab program". But as the Shuttle's launch schedule slipped, so too did the first TDRS, which by May 1980 had moved from March to September 1982. For Spacelab-1's data downlink needs (estimated at 50 megabits per second), NASA desired two functional satellites, the first of which (TDRS-A) was deployed by Shuttle Challenger's STS-6 crew in April 1983. Unfortunately, its Boeing-built Inertial Upper Stage (IUS) booster failed to lift it to its geostationary perch and left it loitering in a low orbit. Months of firings of the satellite's tiny thrusters eventually raised TDRS-A to its correct orbit but wasted two-thirds of its station-keeping propellant. Even when operational, it did not fare well: after its communications payload came alive in July 1983, one of its Ku-band single-access diplexers failed, followed by a Ku-band travelling-wave tube amplifier. As such, TDRS-A was not fully functional until December 1984, a year after STS-9.

"Spacelab-1 needs two TDRS in orbit to ensure a complete record of the results of its experiments," *Flight International* reported in May 1983. "One estimate suggests that a single TDRS will allow 60-70 percent of Spacelab-1 data to be achieved." Even after reaching orbit, it demanded months of checkout before it could support any Shuttle mission, much less one as complex as STS-9. "Should Spacelab-1 be flown with only a single relay satellite in place," Lord rhetorically asked, "or should it be delayed until the two-satellite system was ready?" Many scientists felt a meaningful Spacelab-1 was unattainable without dual-TDRS support. And although the ailing TDRS-A enabled some data-traffic provision, its role was limited. Having just one TDRS in place, said Lord, meant "a number of fingers would be crossed". But even the second TDRS was no closer to launch. Following the almost-failure of its predecessor, the IUS was grounded for a year of repairs and grounded along with it were its future TDRS passengers.

STS-9 correspondingly slipped from 30 September to 28 October 1983. But more trouble was afoot. Days after Columbia reached the pad, delay struck again. During the STS-8 launch in August, an SRB sustained excessive corrosion in its nozzle 'throat', leaving it perilously close to rupture. The fault was traced to a 'bad' batch of resin used in the boosters and STS-9 was postponed for repairs. This required a rollback to the VAB on 19 October and disassembly of the boosters to allow engineers to access the nozzles. During the enforced, month-long down-time, Spacelab-1 was serviced and camera films and batteries for several experiments were replaced.

Launch was rescheduled for 28 November but flying so late in the year threatened seven astronomy, space plasma physics, atmospheric and Earth observation experiments. Some required maximum orbital darkness to work, whilst the Earth-observing instruments demanded good visibility and viewing opportunities as autumn turned to winter fell by 60 percent. Others needed darkness at northerly and southerly latitudes. One experiment called for a new Moon on the fifth day of the mission. And the Grille Spectrometer would achieve barely 16 percent of its objectives, thanks to unfavourable viewing conditions in December as opposed to September. "In general," *Flight International* opined, "the scientists associated with these experiments would prefer to delay Spacelab-1 until observing conditions are more favourable." But a later launch was unacceptable to ESA, not only due to potential losses from other experiments, but also the cost of maintaining the hardware at flight-ready levels: an estimated $300,000 for each month of added delay.

Columbia returned to the VAB on 3 November for hoisting astride her ET and SRBs. She returned to the pad on the 8th, targeting a 14-minute 'launch window' which opened at 11 a.m. EST on the 28th. Subsequent launch opportunities opened and closed at the same time each successive day until 5 December, after which the daily window duration narrowed to 12 minutes, on account of a requirement for daylight conditions at Columbia's transoceanic abort landing sites.

RISE OF THE COLUMBIA

Launch morning on the last Monday of November 1983 dawned fine and reasonably dry, despite concerns about thunderstorms over Cape Canaveral. An emergency landing site in Spain, which Young and Shaw would use in the eventuality of a launch abort, was also iffy in terms of weather.

"The causeways were lined with the usual assortment of campers, cars, signs, flags, vendors, public address speakers, portajohns and sunbathers," wrote Lord. "The VIP stands were filled with enthusiastic supporters of the Spacelab program and a scattering of dignitaries and luminaries from the entertainment, political and international arenas. A thriving business was underway in Spacelab and STS-9 mission mementoes, first-day covers, hats and T-shirts. The huge countdown clock in front of the viewing area moved

Fig. 3.6: The STS-9 crew sits down to breakfast in the Operations and Checkout Building on 28 November 1983. From left to right are Ulf Merbold, Bob Parker, John Young, Brewster Shaw, Byron Lichtenberg and Owen Garriott.

ever so slowly and paused at the planned holds for what seemed an eternity. Photographers manoeuvred for the best spots and telephoto lenses looked like small howitzers aimed at the distant Shuttle launch complex. The public address announcer droned on with a running monologue of the countdown, but most people concentrated on looking around to see who they could recognise. Members of the Spacelab team not needed in the Launch Control Center...exchanged greetings and wished each other good luck" (Fig. 3.6).

Early that morning, Young's crew awakened, showered and breakfasted in their quarters in the Operations and Checkout Building, then headed to the pad. The six men ascended the elevator to Columbia's middeck hatch, where technicians assisted them into their seats: Young and Shaw at the front of the flight deck, with its wraparound windows, Parker behind them. Downstairs on the darkened middeck sat Garriott, Lichtenberg and Merbold. The countdown proceeded crisply and at T-31 seconds, command and control was handed over from the ground launch sequencer to Columbia's four computers, a pivotal moment called 'autosequence start'. With ground computers now taking a backup role, the Shuttle's electronic brain monitored hundreds of functions as the final seconds of the countdown sapped away.

"Coming up on the 30-second mark…and we are Go for autosequence start," intoned the NASA launch commentator, his voice echoing across the KSC bleachers, by now brimming with excitement. "The SRB hydraulic power units have started; these move the solid motor nozzles to steer the vehicle…T-20 seconds…18, 17, 16, 15, 14, 13, 12…ten…"

In those final seconds, the gigantic 'rainbirds' of the sound suppression system began drenching the pad and flame trench with water, to protect the Shuttle from acoustical energy and rocket exhaust. At ten seconds, sparkler-like igniters swirled to quell unburnt hydrogen gas lurking beneath Columbia's three main engines.

"…We have a Go for Main Engine Start…eight, seven, six…"

The engines roared alive at 120-millisecond intervals, their low growl intensifying into a thunderous crescendo. A gout of translucent orange flame gave way to three dancing Mach-diamonds, as supersonic exhaust gases surged from the engine-bells. A vast cloud of steam obscured the Shuttle from view, billowing high into the clear morning air. Strapped into their seats, the astronauts braced—"You feel that noise in the cockpit, quite clearly," Young said later—as they realised they were about to ride a wild animal into space. Within three seconds, the engines attained nominal performance and Columbia's computers gimballed all three to liftoff configuration.

"We *have* Main Engine Start," came the announcer's call, his voice notching up an octave in pent-up excitement. "Three, two, one and…"

His next words were drowned by the staccato crackle of the SRBs, which belched fire at precisely T-0. Eight frangible nuts anchoring the boosters to the pad were detonated, the final umbilicals disconnected and the main engines were commanded to full throttle.

"…Solid motor ignition…and *liftoff*…liftoff of Columbia and the first flight of the European Space Agency's Spacelab…the Shuttle has cleared the tower!"

For the astronauts, the vibrations and acceleration unmistakably reminded them that something enormously life-changing was happening. "Things are shakin'," Young remembered. "Mostly, your knees!"

Ten seconds after liftoff, the computers rolled the stack onto its back, at an exceptionally rapid rate of 15 degrees per second, to manoeuvre Columbia onto the proper azimuth for a northerly uphill climb and insertion into a 150-mile-high (240-kilometre) orbit. Ranging as far north as Scotland and as far south as Tierra del Fuego, the 57.5-degree-inclined orbit was the highest ever attained by a Shuttle crew and Columbia became the first U.S. human spacecraft to overfly Russia in daylight. The roll manoeuvre lessened aerodynamic stresses on the vehicle in the lower atmosphere, affording the crew a better orientation for communications and navigation. This "real high roll rate", said Young, was necessary to achieve the high-inclination orbit demanded by Spacelab-1.

"Soon, the reverberations from the Shuttle main engines and its boosters reached the viewing stands and overwhelmed the cheers from those looking on," wrote Lord. "The Shuttle quickly rolled around its axis and started to pitch over as it passed through the layer of clouds. Camera shutters clicked rapidly, old friends hugged each other with delight and tears coursed the cheeks of many space-hardened veterans. There is nothing quite like those few moments after liftoff, when everyone is of a single mind, trying to help push the launch vehicle into orbit."

A minute into the flight, the wind-noise outside Columbia's heavily reinforced cockpit intensified into a scream-like trill as the stack passed through peak aerodynamic turbulence. Here, atmospheric forces imparted their most severe stresses on the airframe, known in engineering parlance as 'Max Q'. Passing supersonic speed, visible shockwaves formed around the tips of the SRBs and Columbia's nose. To avoid exceeding structural limitations on the airframe, the computers throttled the main engines back to 67 percent of rated performance. Seventy seconds into the flight, with Max Q safely behind them, the engines returned to full power.

The incessant guttural snarl of the SRBs became more sporadic and decreased to virtually nothing as the time approached, two minutes into ascent, for their jettison. Young, Shaw and Parker witnessed a yellow-orange flash of light stream across Columbia's nose, as the boosters' separation motors pushed them away. Their departure was accompanied by a harsh grating sound and a fair amount of sooty gunk deposited on the Shuttle's windows, although both SRBs performed nominally (Fig. 3.7).

With the boosters gone, the six men oddly felt that they had ceased accelerating, a sensation that they were falling back to the water. But by now, Columbia was above much of the 'sensible' atmosphere and Young and Shaw found it easier to flip switches.

The vehicle continued onward, her main engines shutting down 8.5 minutes after liftoff. By this stage, the Shuttle was moving at an orbital velocity of 17,640 miles per hour (28,400 kilometres per hour). Suddenly, the equivalent to three times the force of terrestrial gravity—like a gorilla sitting on their chests—was gone, instantly replaced by absolute serenity, ethereal silence…and weightlessness. Nineteen seconds later, the ET was discarded to burn up in the atmosphere over a sparsely inhabited stretch of the Indian Ocean.

Young and Garriott knew the feelings, sights and sounds of space well, but for the four rookies came euphoria, as the vestiges of weightlessness

Fig. 3.7: Columbia roars aloft on 28 November 1983.

manifested themselves in the form of washers, filings, screws and bits of wire liberated from every nook and cranny in the cabin, all floating comically in mid-air. An hour after liftoff, the two Orbital Manoeuvring System (OMS) engines in Columbia's tail circularised their orbit.

Spacelab-1 was officially underway.

VOYAGE OF SCIENCE

The six men divided into their shifts to operate Spacelab-1 around the clock. Although 24-hour operations were not needed on all Spacelab missions, its demonstration on STS-9 maximised crew flexibility. Typically, the teams met twice daily for a handover, discussing experiments and sharing meals, with half of the crew wolfing down dinner as the others took breakfast. Meal-making was aided by an airliner-style galley, built by General Electric under a September 1978 contract with NASA. It provided hot and cold water, an oven, serving trays, a personal hygiene station (including handwashing area, light and shaving mirror) and water heater. No refrigerators or freezers were aboard the Shuttle and foods were typically rehydratable, thermostabilised, irradiated or in natural form. Sleeping, at least for Ulf Merbold, was time wasted. "My strategy was that I might sleep a lot for the rest of my life and I did not want to waste any minute to sleep more than necessary, so that's what I did," he said. "I think the average hours of sleep during the entire mission was between four and five hours [per shift]."

Spacelab-1 was the first of two Verification Flight Tests of the integrated system; Spacelab-2 would evaluate a 'train' of unpressurised pallets, an igloo and an array of scientific instruments on a mission planned for March 1985. A total of 264 sensors verified that the passive thermal control system could keep temperatures inside the module and on the pallet within requisite limits, whilst also preventing condensation and heat leakage. Thermal, acoustic and structural responses were carefully monitored and the astronauts found the module offered a pleasant working environment, with the small window in its end-cone allowing them to photograph experiments on the pallet.

But despite Spacelab-1's eventual success, the mission hit a snag only hours after reaching orbit, as the crew struggled to access the tunnel adapter. "We couldn't get the hatch open," said Shaw. It was a momentary scare, particularly looking into the faces of Garriott and Parker—"these guys," Shaw sympathised, "are seeing their lives pass in front of their faces, if we can't get in there and do the stuff they've been training to do"—but by the evening of the 28th Europe's laboratory was open for business. "Almost like home," Garriott said as he turned on the lights, 2.5 hours after launch. In fact, several experiments had already begun, Lichtenberg and Merbold having worn biomedical head sensors to monitor their eye motions during ascent.

"It was fun to watch Owen back in the module, because you could tell right from the beginning he'd been in space before," reflected Shaw. "He knew exactly how to handle himself, how to keep himself still, how to move without banging all around the place. And the rest of us…were bouncing off the walls until we figured out how to operate. But Owen, it was just like…he was here yesterday. The human body is remarkable in its ability to remember adapting to a previous thing."

Typically, the pilots remained on the flight deck and spoke to their crewmates in the module via intercom. "Since there was only one of us awake on the flight deck…you didn't want to leave the vehicle unattended very much, because this is still STS-9, fairly early in the program," said Shaw. "We hadn't worked out all the bugs and everything and neither John or I felt too comfortable leaving the flight deck unattended, so we spent most of our time there."

Orbital manoeuvres were periodically needed for different thermal attitudes, but the pilots primarily monitored systems performance. "After a few days of that, boy, it got pretty boring, quite frankly," Shaw said. "We spent a lot of time looking out the window and taking pictures…but there was nobody to talk to, because the other guys were…in the back end in the Spacelab, working away."

As such, Shaw spent much time gazing out of the window. Earth observations were important to both pilots. "One of our first experiments had to do with a rotating dome, which would be taking a picture of your eye," Garriott said. "That camera…broke very early in the flight and…all of this film and camera was sitting there without any particular use, so John Young helped to find a use for that. He took the camera and all these rolls of film up to the flight deck. I think he spent a major fraction of the flight…taking pictures out that left-hand window with the nice format camera." STS-9's post-flight report acknowledged that over 7,000 frames were acquired. Young happily accommodated other crewmates who ventured to the flight deck to take photographs (Fig. 3.8).

The only real problem was a temperature glitch in a remote acquisition unit, which serviced NASA's experiments on the pallet. Analysis revealed the temperature of Columbia's freon coolant loop was partly to blame and the issue was resolved. Unfortunately, as part of its resolution, a 'patch' inserted into the software caused the module's computer to crash and temporarily affected data-collection efforts from the pallet experiments.

"We didn't get to participate to a great length in the science that was going on," Shaw said. He and Young regarded themselves as truck drivers to get Columbia into orbit and home. "In the meantime, the other guys did all the work." That work was extended on 3 December, when NASA and ESA lengthened STS-9 to ten days, contingent on satisfactory landing weather on the dry lakebed Runway 17 at Edwards Air Force Base, which was needed to handle the Shuttle with the heavyweight Spacelab aboard. Reserves of critical supplies, including fuel cell reactants, and a power consumption rate 1.2 kilowatts lower than predicted easily permitted an extension—from 216 hours and 145 orbits to 240 hours with 160 orbits—and a longer flight would benefit Spacelab-1's scientific yield. (After the mission, it was found that the experiments did not operate as much as planned and heater duty cycles were less than predicted during pre-flight 'cold' tests, producing a lower power consumption rate.) Even at this halfway point, Mission Scientist Rick Chappell lauded STS-9 as a "very successful merger of manned spaceflight and space science".

Late on 7 December, the crew shut down Spacelab-1, stored and secured samples and deactivated the module and pallet, ahead of landing at 8:01 a.m.

Fig. 3.8: Owen Garriott negotiates the 'lip' at the top of the joggle as he floats from the tunnel into the Spacelab-1 module.

PST on the 8th. But five hours before landing, as Young and Shaw configured Columbia for re-entry, a computer failed. "This is the first computer failure we had on the program," Shaw recalled. The pilots' eyes widened with alarm. But worse was to come. Six minutes later, a second computer stopped. Young and Shaw brought it back online, but their efforts to restart the first one proved fruitless and it was powered down. The mood in the cockpit was tense. "My knees started shaking," Young said of the first failure. "When the next computer failed, I turned to jelly."

Mission Control waved off the landing for eight hours. (It later turned out that several tiny slivers of solder, floating freely inside the computers, had induced the failures.) Then, an inertial measurement unit—a critical piece of navigation hardware—failed, its death-throes accompanied by a harsh banging sound. As Columbia lingered in orbit, with Spacelab-1 deactivated, the crew maintained themselves in a state of preparedness for re-entry, whenever that might come. Eventually, they began their homebound return on the afternoon of the 8th, touching down at Edwards at 3:47 p.m. PST after ten days and 167 orbits, the longest Shuttle mission so far. Yet even the landing did not go well. Four minutes before touchdown, one of Columbia's three Auxiliary Power Units (APUs) exhibited higher than normal temperatures. The second failed computer, which Young and Shaw had earlier brought back to life, conked out, seconds after the Shuttle's wheels hit the runway. And six minutes after that, one of the APUs experienced an 'underspeed' condition and shut down, as did another shortly thereafter. The crew did not know it, but one of the APUs caught fire, due to a hydrazine leak from a cracked injector tube, its flickering flame clearly visible as Columbia raced down the runway. "There was more excitement on that flight than we should ever really have again," said STS-9 propulsion systems officer and future Shuttle program manager Wayne Hale. "We learned a number of lessons in how to prepare the vehicle."

After touchdown, the six men—"six dirty ol' men," Young joked, after ten days without a shower—disembarked to meet a "very happy" scientific community for post-flight tests. Eleven life sciences experiments met with full success, the rest achieved 50-90 percent of their planned objectives. The space plasma physics suite scored 80-100 percent, with a failure of SEPAC's electron beam assembly in its high-power mode the sole anomaly of note. Atmospheric physics and Earth observations hovered between 75-100 percent successful, thanks to unfavourable viewing parameters. Key Spacelab-1 accomplishments included the first silicon melt and crystal growth in space, confirmation that vision helps to orient the human body in weightlessness and evidence that fungus does indeed maintain Earth-like circadian rhythms, even in complete darkness and a virtual absence of terrestrial gravity (Fig. 3.9).

On 19 January 1984, a report landed on the desk of NASA Administrator James Beggs, praising a successful mission. It certainly did not harm President Ronald Reagan's decision later that spring to invite Europe to build a future space station—nor as a lip-service 'partner', as had been the case early in Spacelab's evolution, but rather as a full and real partner. In fact, as early as February 1982, Beggs and ESA Director General Erik Quistgaard had discussed the potential for space station collaboration; "Europe," wrote John Logsdon, "was thus given the opportunity to be involved in the station program almost from its inception." Pages of scientific journals throughout 1984 overflowed with Spacelab-1 results. And in February 1985, ESA determined that all but two of the mission's materials science experiments were successful.

Fig. 3.9: "Six dirty ol' men" was John Young's description of the unshowered Spacelab-1 crew as they returned to Edwards Air Force Base in California on 8 December 1983. At the foot of the steps is Young, about to shake hands with George Abbey, JSC's director of flight crew operations. Young is followed by Brewster Shaw, Bob Parker (in sunglasses), Ulf Merbold, Owen Garriott and Byron Lichtenberg.

That same month, the National Society of Professional Engineers recognised MSFC's role in Spacelab-1 as one of the United States' ten outstanding accomplishments of the previous year.

With STS-9, NASA and ESA triumphantly staged the first of two Verification Flight Tests of Spacelab, validating the module and pallet. Ahead lay Spacelab-2, which would deploy a 'train' of pallets and an instrument pointing system. But if Spacelab-1 ended traumatically, Spacelab-2 experienced its fair share of trauma both before—and during—launch. In fact, the second Verification Flight Test came within a whisker of not making it into space at all.

4

Verification Flight Test Two

"GOING TO SPAIN"

Sixty-seven miles (108 kilometres) above the home planet, racing to orbit at a dozen times the speed of sound—six times faster than a bullet—everything in the world dropped out six minutes into the Spacelab-2 mission. It was the afternoon of 29 July 1985 and Shuttle Challenger had just launched with a seven-man crew. Theirs was the second Spacelab Verification Flight Test to evaluate a 'train' of three unpressurised pallets, laden with solar physics, space plasma physics and astrophysics instruments.

Seated on the flight deck, Commander Gordon Fullerton and Pilot Roy Bridges watched their displays as Challenger roared uphill. Behind them, flight engineer Story Musgrave leafed through the contingency procedures book, for the STS-51F launch was not going well. What the three men saw and heard was alarming in the extreme.

One of the Shuttle's three main engines had just shut down, halfway to space, raising the likelihood of an aborted mission and an emergency landing. "The ground made the call *Limits to Inhibit*, which is, for us, an extremely serious omen…[it] means the ground is seeing problems that are going to shut you down," Musgrave said. "I'm looking through the procedures book and thinking we're going to land at our transoceanic abort site in Spain. I'm rehearsing all the steps and my hands are moving through the book. And I'm thinking *We're going to Spain. Things are bad.*"

The emergency unfolded five minutes and 45 seconds after Challenger's 4 p.m. EDT liftoff. The SRBs had been jettisoned three minutes earlier and the orbiter/ET stack was hurtling to orbit at 9,200 mph (15,000 km/h), the crystalline limb of Earth and the darkness of space affording a spectacular view through Challenger's flight deck windows. Suddenly, temperature data for the upper main engine's high-pressure turbopump indicated it was operating above its maximum 'redline', prompting the computers to shut it down. At this stage, with three minutes of powered flight remaining and one engine down, Fullerton's crew were too high and too fast to return to KSC. That meant either a Transoceanic Abort Landing (TAL) at Zaragoza Air Base in northeastern Spain—close to the nominal ground track for STS-51F's targeted orbital inclination of 49.5 degrees—or a tricky Abort to Orbit (ATO) manoeuvre, whereby Challenger would burn her tail-mounted OMS engines to augment the two remaining main engines and limp into a low, stable orbit.

Leading the Mission Control team in Houston that afternoon was Ascent Flight Director Cleon Lacefield, with astronaut Dick Richards at the CAPCOM's console. "We had worked on that flight for six months and the CAPCOM…can't say anything without the flight director being okay with it," remembered Richards. "We had developed this sort of silent communication there and he'd just turn to me and nod his head…it worked very well, because it was very quick. That paid off in spades on that particular flight, the only time in ascent where we really had a time-critical emergency that the Mission Control Center could actually do something about."

Booster Officer Jenny Howard suspected that bad sensors (rather than bad engines) lay at the root of the problem. Now, Challenger's right-hand engine appeared to be on the verge of failure. If the data was wrong, it reared the frightening prospect that the computers might unwittingly shut down a healthy engine.

"Flight, we think an erratic set of sensors shut down the first engine," Howard told Lacefield, "and I'm watching, in the right engine, one of the two sensors is already disqualified and the other sensor is starting to move around." With one engine already down, TAL or ATO remained the safest abort options, but if a second engine also went south, Fullerton and Bridges might be forced to land the Shuttle as far downrange as Souda Bay Naval Air Station in Crete. Averting a second engine shutdown required Mission Control to urgently bypass the bad sensors.

"Flight," Howard began, tentatively, "I'm thinking about taking the limits to inhibit."

Lacefield nodded.

"Limits to inhibit, Gordo," radioed Richards.

"Roger, limits to inhibit," Fullerton acknowledged, his breathing laboured under three times the force of terrestrial gravity.

The inhibit command instructed Challenger's computers to ignore the over-temperature signals, preventing them from shutting down the right-hand engine. It was a dicey judgement call. If Howard had not made it, the computers might shut down the healthy right-hand engine. But if fault really did lie with the engines, the propulsion community preferred to safely shut them down, rather than risk one exploding.

It proved a timely intervention. Two minutes later, the right-hand engine's sensor indeed exhibited off-nominal temperatures. Howard's inhibit call ensured the engine did not shut down and Challenger flew safely onward. "We were in an extremist situation for the second engine failure of where we were going to end up," Richards said. "Cleon understood all this. That was the right thing to do."

Meanwhile, on the edge of space, Musgrave's fingers worked through the procedures book for a diversion to Zaragoza. Seated at his right shoulder on the flight deck, astronaut Karl Henize watched apprehensively. "He didn't know what was going on," Musgrave said. "He looks over and sees the top of my page: SPAIN. He's looking, poor Karl, and I'm going down the checklist that I'll be reading to the guys in the front seats when we abort. Karl's looking over at me and I sense a really severe stare."

Henize dared ask the inevitable question. "Where we going, Story?"

"Spain, Karl." Then, he retracted it. "We're close, but not yet."

Not yet, indeed, for the 'window' to reach Zaragoza had closed a half-minute earlier. Challenger was sufficiently high and fast to avert an emergency landing in Spain and make the next available option, aborting directly to orbit.

"He's ATO, Flight," said Flight Dynamics Officer Brian Perry.

Lacefield nodded his concurrence to Richards.

"Abort ATO, abort ATO," Richards radioed. Years later, he hoped he would remain the only CAPCOM ever to call 'abort' to a crew during ascent. Richards inadvertently panicked Fullerton's wife, Marie, who was watching the launch in Florida. When she heard the word *abort*—not once, but twice—poor Marie feared the worst.

At 4:06:06 p.m. EDT, six minutes and six seconds after launch, Fullerton fired the OMS engines for 106 seconds to commit the Shuttle to a low, but safe orbit. The two healthy main engines burned an extra 49 seconds, shutting down 9.5 minutes into the flight. "We never did get the call for the transoceanic emergency landing," Musgrave said, "but we ended up making it to orbit and finishing the mission."

STS-51F's orbital path, with an apogee of 143 miles (230 kilometres) and a perigee of 108 miles (174 kilometres), required three OMS burns to attain and was far lower than the 240 miles (390 kilometres) planned. It also impaired several scientific objectives. The ATO was the only in-flight shutdown ever experienced by the Shuttle and raised many eyebrows, not least

Fig. 4.1: Burning three perfect main engines and two Solid Rocket Boosters (SRBs), Challenger rises from Earth on 29 July 1985. Six minutes later, calamity hit Spacelab-2.

because all engine parameters appeared normal during countdown, ignition and liftoff. Subsequent analysis verified Howard's suspicions: the issue lay not with the engines, but with faulty sensors which incorrectly pointed to an overheating situation (Fig. 4.1).

As they settled into orbit and divided themselves into two shifts for around-the-clock operations—a 'red' team led by Bridges, with Mission Specialist Henize and Payload Specialist Loren Acton, a 'blue' team led by Musgrave, partnered with Mission Specialist Tony England and Payload Specialist John David Bartoe, and with Fullerton anchoring his duties across both—the seven astronauts barely had chance to reflect not just upon an eventful day, but a crisis-ridden month. Originally scheduled to fly at 4:30 p.m. EDT on 12 July, they had been thwarted by a shutdown of Challenger's engines on the launch pad, just prior to liftoff.

"At T-7 seconds, the main engines start with a rumble from far below," recalled Bridges of the events of 12 July. "I watch the chamber pressure indicators come to life and surge towards 100 percent." Suddenly, he spotted something awry. "The left engine indicator seems to be lagging behind. Before I can say a word, it falls to zero, followed by the other engines. With less than three seconds before our planned liftoff, we have an abort." Over the intercom came groans from the crew: Musgrave and Henize on the flight deck, England, Acton and Bartoe on the middeck. Fullerton beadily eyed his pilot across the cockpit. Bridges, whose responsibilities included the engines, was convinced that Fullerton thought he had screwed up. He showed his commander both hands, palms upturned. "Gordo, I didn't touch a thing," Bridges insisted. "It was an automatic shutdown!"

That shutdown was triggered by the sluggish closure of the left-hand engine's chamber coolant valve and precipitated a 17-day wait for another launch try. But this delay caused other problems. The launch window on 12 July ran for two hours, calculated to satisfy lighting requirements for Spacelab-2's astronomy and space plasma physics experiments. For the next five or six days, the window opened at the same time—4:30 p.m. EDT—before a launch became unfavourable. The mission's infrared astronomical observations demanded total orbital darkness. "The optimum launch time is four to five days before a new Moon, when the sky is darkest," noted the STS-51F press kit, "but a launch can occur a few days beyond that, without seriously affecting investigations." Even the 29 July launch, originally scheduled for 2:23 p.m. EDT, was postponed by 90 minutes, thanks to an erroneous command to one of Challenger's computers.

THE TROUBLED ORIGINS OF THE IPS

Karl Henize's reaction to an impaired flight was understandable, as a mission to which he devoted eight years of his life seemingly slipped through his fingers. At 58 years old, the former astronomy professor was the oldest human to travel into space; he knew that STS-51F would be his only Shuttle flight and he intended to make it count.

Spacelab-2 was physically unlike Spacelab-1, with no pressurised module. Instead, Challenger carried three unpressurised pallets and a specialised support structure, holding an eclectic group of 13 experiments—11 provided by the United States and two by the United Kingdom—which encompassed seven scientific disciplines: solar physics, space plasma physics, high-energy physics, atmospheric physics, technology, infrared astronomy and life sciences. Two experiments occupied middeck lockers, another was conducted from the ground using the Shuttle as a target and the remainder demanded an unobstructed perspective of the sky. That demand was best met by the wide-field-of-view capabilities of the pallets, rather than a module, which some scientists disparagingly considered an "expensive nuisance". But Spacelab-2 had already been leapfrogged by the module-only Spacelab-3 mission (see Chapter Five), due to delays which hit its most critical piece of hardware, the problem-plagued Instrument Pointing System (IPS).

Fullerton called Spacelab-2 "the gangbuster payload", on account of the sheer amount of work to bring it to fruition. In March 1975, it was identified as a mission of three interconnected pallets housing 15,000-17,000 pounds (7,000-8,000 kilograms) of hardware, with two (or possibly three) payload specialists supporting up to a hundred hours of observations over a week-long flight. "This second mission would be very ambitious," wrote Douglas Lord, "with experiment resources considerably beyond those provided for the first mission." NASA issued a request for proposals from the scientific community in August 1976. Twelve months later, 59 scientists—47 from the United States, the rest from the United Kingdom—were selected for Spacelab-2, then scheduled for 1981.

Key to its success was the IPS, which Lord described as more challenging to develop than any other Spacelab component, in terms of technical complexity, organisational and schedule risk and ability to readily outgrow its established cost parameters. When ESRO selected the industrial consortium led by VFW-Fokker/ERNO to build Spacelab in June 1974, West Germany's Dornier firm was contracted for IPS design and definition. It was intended as a three-axis system, capable of pointing scientific instruments with an accuracy close to a single arc-second. The IPS would measure up to 13.5 feet (4.1 metres) long and seven feet (2.1 metres) in diameter, weighing 4,400 pounds (2,000 kilograms). The sheer size of its yoke prompted Dornier to propose an 'end-mounted' engineering approach, its three gimbals situated on a Spacelab pallet and supporting a circular framework onto which instruments were mated. A 'payload clamp' supported the IPS/instruments in a horizontal configuration during ground operations, ascent and re-entry and control electronics elevated the 'stack' upright in space to attain requisite levels of pointing accuracy (Fig. 4.2).

Authorisation to proceed with its design was granted in December 1974. But costs spiralled and other options were considered to reduce expenses by holding Dornier to less stringent specifications. If ESA was nervous, so too were the Americans, internal NASA assessments having already raised concerns that the IPS design failed to satisfy users' pointing requirements. By September 1975, it was rejected on the grounds of unacceptable breaches of cost limits and severe schedule risks. Three months later, ESA issued another

Fig. 4.2: The Spacelab-2 payload is pictured during final closeout at KSC. The Instrument Pointing System (IPS), in its horizontal integration, is visible at centre, with the University of Chicago's Cosmic Ray Nuclei Experiment (CRNE) at far left.

request to industry for a new concept. Early the following year, a joint bid was tendered by Dornier (this time in partnership with MBB) and a second from VFW-Fokker/ERNO. Dornier/MBB's proposal was selected, with contracts initialled for signature in June 1977. But the storm-clouds of cost, schedule and technical difficulty refused to dissipate and tensions grew between the United States and Europe until, early in 1977, some ESA leaders proposed deleting the IPS in its entirety and building something new. However, strong backing from West Germany and France—the program's financial heavyweights—ensured that the IPS was not 'descoped' and retained its place within the Spacelab budget, despite fierce opposition from five ESA memberstate delegations.

This anti-IPS mood was aptly captured by Michel Bignier in a letter to Lord. "I have accepted the constraints imposed by them [the delegations] on the program," he wrote, "in the spirit of the TV program, *Mission Impossible*, in which the actors always succeed in carrying out their impossible task." By

June 1977, ESA (with some reluctance) agreed to press on and Dornier/MBB envisaged delivery of the first IPS by mid-1980. Yet the problems persisted. Evidence of a susceptibility of components to stress-related corrosion and uncertainty over the reliability of the IPS software were flagged in December 1977. When the ESA member-states objected to funding the IPS, Dornier was forced to turn in an almost complete redesign, which imposed additional delay. Originally targeted in the mid-1970s to fly in 1981, by January 1982 Spacelab-2 had slipped to 1984 at the earliest. A year later, that date moved to September 1984. And a year after that, launch moved to March 1985. By the time the mission specialists were announced in February 1983, it was clear that Spacelab-3—a module-only flight—would precede Spacelab-2 into orbit.

Keen to visibly signal its confidence that ESA and Dornier could resolve these problems, in May 1980 NASA purchased a second IPS for $20 million "for delivery by the end of 1983". The quantity of planned Spacelab missions requiring IPS capabilities, it noted, drove this decision. As circumstances transpired, the first IPS did not arrive at KSC until November 1984. "The last few months of checkout," wrote Lord, "were fraught with debates about the state of readiness of both the hardware and software and the adequacy of documentation and operating instructions." NASA found itself torn between pushing for the completion of IPS qualification testing in Europe and an early delivery of the hardware to the United States to begin Spacelab-2 integration. For its part, Dornier offered the Americans "iron-clad assurances" that open actions and tests would be completed before launch and both NASA and ESA duly accepted the first IPS.

In its final form, the 2,600-pound (1,180-kilogram) system could handle instruments weighing up to 15,430 pounds (7,000 kilograms), pointing them to within two arc-seconds and targeting them to 1.2 arc-seconds. An optical sensor package comprising a boresighted fixed-head star tracker and a pair of skewed fixed-head star trackers would execute attitude corrections and configuration for solar, stellar and Earth-viewing applications. For Spacelab-2, one end of the IPS was mounted directly to the forward pallet and the other to an 'integration ring', which held the three-axis gimbal. It could move its instruments backwards and forwards, and from side to side, and even 'roll' them in a 22-degree arc around its vertical position.

These motions were commanded from the Spacelab subsystem computer and a pair of Data Display Units (DDUs) on the Shuttle's aft flight deck: a starboard-side console for a mission specialist to operate the IPS and a port-side console for a payload specialist to operate the instruments. The IPS could be operated in manual or automatic modes and was capable of spending long periods focusing on single objects or conducting slow-scan mapping operations. Its reaction times were sharper than the Shuttle's attitude control system: the orbiter could achieve a pointing precision of a tenth of a degree at best, whereas the IPS could hit accuracies of one-thirty-six-hundredth of a degree. That precision was analogous to an instrument on the steps of Washington's Capitol Building keeping perfect focus on a dime atop the Lincoln Memorial, 2.2 miles (3.5 kilometres) away. Even crewmembers, equipment or Shuttle engine firings—whose motions tended to resonate through the vehicle in a disruptive harmonic—were compensated by built-in accelerometers which kept the IPS locked onto its targets.

MULTIDISCIPLINARY PAYLOAD

Early guidelines for Spacelab-2 identified a pallet-only configuration, featuring a pair of two-pallet 'trains', described by mission planners as a '2+2' approach. Mounted vertically on a crossbeam at the forward end of the first train was the 'igloo', a temperature-controlled cylinder housing subsystems for the instruments and providing electrical power, cooling and command and data acquisition services. Standing 7.9 feet (2.4 metres) tall and 3.6 feet (1.1 metres) wide, with an interior volume of 53 cubic feet (1.5 cubic metres) and weighing 1,450 pounds (660 kilograms), the igloo—which one engineer likened to "a five-gallon oil drum"—was built of aluminium alloy forged 'rings', pressurised to 14.7 psi, and held systems which might otherwise have been accommodated in Spacelab's 'core' module. Several pallet-only missions in the 1990s benefitted from the igloo's unique functionalities.

Planning changed in late 1977, when the Cosmic Ray Nuclei Experiment (CRNE) was added. The University of Chicago won a $3.4 million contract in October 1978 for "cosmic ray investigation provisions" on Spacelab-2, its CRNE resembling a giant duck's egg in appearance. The CRNE sat at the rear of the payload bay and was directly exposed to space, its detectors recording incoming cosmic ray nuclei, from boron to nickel, counting their prevalence and energies, ascertaining their provenance and transmitting data to the ground. It would record about 24 million particle 'events', of which around 30,000 were in a largely unexplored energy range, spanning hundreds of billions to trillions of electron-volts (Fig. 4.3).

The Spacelab-2 layout was redesigned as a single train of three pallets at the midpoint of the payload bay and a 'specialised support structure' at the back for the CRNE. And it was this specialised support structure which earned the ire of Bignier, temporarily re-igniting European fears (see Chapter Two) that NASA was "substantially duplicating" Spacelab systems to supplant what

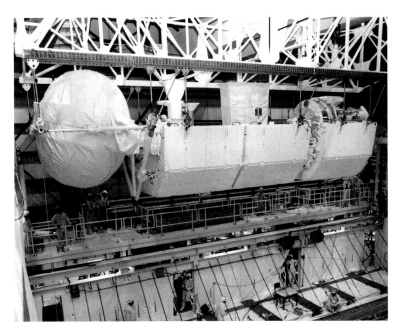

Fig. 4.3: Spacelab-2 consisted of Pallet One (far right) to house the IPS, a double Pallet Two/Three with the large square baffle of the X-Ray Telescope (XRT) and the inverted cone of the Small Helium-Cooled Infrared Telescope (IRT) at centre and the Cosmic Ray Nuclei Experiment (CRNE) on its own unique support structure (far left).

pallets could already do. But the CRNE was too big—some eight feet (2.4 metres) in diameter and weighing 4,400 pounds (2,000 kilograms)—to be straightforwardly stationed on a pallet and needed a unique support structure.

Like a game of musical pallets, the payload thus morphed to its final configuration in 1980, when Jesse Moore, deputy director of the Solar Terrestrial Division in the Office of Space Science at NASA Headquarters, proposed switching the three-pallet train and igloo to a single pallet with an igloo and a separate two-pallet train. Although the result still amounted to three pallets, only two were physically linked. Moore's rationale was that the Spacelab-2 instruments attained 12,264 pounds (5,500 kilograms), above the 11,000-pound (5,000-kilogram) threshold for a three-pallet train. The structure was held in place by five attachment fittings—four along the payload bay walls and a keel 'pin' in the floor—with aluminium ducts and trays on its port and starboard sides to route cables and utilities. Thermal control was enabled by multi-layered insulation and Spacelab's own freon coolant loop, which collected excessive heat via a set of 'cold plates' and rejected it through the Shuttle's heat exchanger. Spacelab-2's status as a Verification Flight Test saw the pallets and igloo outfitted with sensors to assess their performance, including thermal, acoustic and structural loads.

The three pallets arrived at KSC in 1982 for checkout. A year later, support equipment was installed, the igloo was set in place and by early 1985 Spacelab-2's instruments were integrated. Elsewhere, the crew stations on Challenger's aft flight deck underwent modification: in addition to the DDUs, a backup high-data-rate recorder was added and a contingency panel to jettison the IPS in case it malfunctioned or could not be lowered to permit payload bay door closure. A closed-loop test in May 1985 allowed all experiments to be remotely operated by POCC controllers and the payload's compatibility with Shuttle systems was validated. These tests included TDRS-A, now fully functional and capable of supporting high-rate data transfer. On 8 June, the payload—weighing 33,145 pounds (15,030 kilograms)—was loaded aboard Challenger.

The forward pallet and igloo held the IPS, whose four instruments would investigate the ever-changing Sun and its influence upon the home planet. Of the four IPS instruments, three emphasised solar physics, the other atmospheric physics. The Solar Optical Universal Polarimeter (SOUP), provided by Lockheed Solar Observatory in Palo Alto, California, used a telescope/camera to observe the Sun's magnetic behaviour across different wavelengths and polarisations in visible light. The Naval Research Laboratory's High-Resolution Telescope and Spectrograph (HRTS) studied emissions from the outer solar atmosphere and the United Kingdom's Coronal Helium Abundance Spacelab Experiment (CHASE), jointly developed by Mullard Space Science Laboratory of University College London and Rutherford Appleton Laboratory, used a telescope/spectrometer to detect hydrogen and helium emission lines and assess relative abundances. Finally, the Naval Research Laboratory's SUSIM atmospheric physics experiment flew again after STS-3. During Spacelab-2, it was tuned to a narrow range of the ultraviolet spectrum, activating itself automatically whenever the IPS was turned towards the Sun.

The second and third pallets were integrated into a chain. Pallet Two was dominated by the X-Ray Telescope (XRT), provided by the United Kingdom's University of Birmingham. Its two co-aligned telescopes employed a 'coded-mask' technique to create X-ray images at energies between 2.5-25 keV. The two masks contained differently sized 'holes' for different angular resolutions to enable the higher-resolution telescope to make detailed studies of brighter celestial sources and its lower-resolution counterpart to examine fainter regions of more diffuse emission. In *Come Fly With Us*, Croft and Youskauskas noted that the proximal closeness of the IPS on Pallet One and XRT on Pallet Two left them vulnerable to a collision risk. "NASA had developed collision-avoidance software to prevent a potentially embarrassing situation," they wrote. "The software had to be activated six times during the flight, each time requiring ten to 30 minutes to recover." Fullerton recounted that at least three solar observing 'passes' were lost performing collision-avoidance recovery protocols.

And Pallet Three included the Small Helium-Cooled Infrared Telescope (IRT) from the Smithsonian Astrophysical Observatory in Cambridge, Massachusetts. It included a 250-litre (55-gallon) container of liquid helium to cool its optics and was one of the few instruments not adversely affected by Challenger's low orbit. By the end of the flight, it acquired over 75 hours of data, including observations of eight galactic clusters and images of the 11,000-year-old Vela supernova remnant.

Also on Pallet Three was Spacelab-2's only technology investigation, the JPL-furnished Superfluid Helium Experiment (SFHE), an insulated dewar and cryostat filled with 'superfluid helium'—helium cooled almost to absolute zero—to test a potential cryogen for future scientific instruments. Superfluid helium moves freely through pores so small that they block normal liquids and conducts heat a thousand times more efficiently than copper. Before Spacelab-2, many of its subtleties were unknown because gravitational effects distorted the superfluid state in ground-based experiments. Of particular interest were capillary waves and their sloshing motions and temperature variations. SFHE functioned whilst the Shuttle occupied a 'gravity-gradient' attitude, her tail facing Earth, to minimise disturbances. Results suggested it could be managed efficiently in space, used a porous plug cryostat.

The final payload on Pallet Three was the University of Iowa's Plasma Diagnostics Package (PDP), previously tested on STS-3. For this second mission, it was to be deployed 51 hours and 50 minutes after launch to fly freely for seven hours, before being recaptured and locked back onto Pallet Three. Working in tandem with PDP was its old friend from OSS-1, Stanford University's Vehicle Charging and Potential Experiment (VCAP), whose electron-gun emissions were recorded by three plasma probes on Pallets One and Two. A number of VCAP studies also involved PDP as it manoeuvred through their artificially generated electron beams.

The final two experiments in Challenger's middeck were devoted to life sciences. The University of Wisconsin at Madison provided a study of the crew's vitamin D metabolite levels to understand causes of bone demineralisation (loss of density) and mineral imbalances in microgravity. Astronauts returned from earlier missions exhibiting a loss of lower-body mass (especially in the calves), decreased muscle strength and negative calcium balance. This process was not unlike the onset of osteoporosis and muscle wastage, observed in bedrest patients. Before, during and after Spacelab-2, three vitamin D metabolites were measured in blood specimens from four crewmen. Although the levels of two metabolites remained unchanged, the third underwent an intriguing pattern, showing a rise in the level of blood samples gathered early in the flight, dropping at the mission's mid-point, then returning to normal after landing.

The second experiment was a reflight of the University of Houston's plant lignification study from OSS-1 and Spacelab-1. It featured another crop of Chinese mung bean, oat and pine seedlings. Preliminary observations revealed that the mung beans and oats behaved normally and the pine seedlings grew well in space. Some reduced growth (on the order of 15-20 percent) was observed in the mung beans and both they and the oats grew 'above' the supporting medium, indicative of some disorientation.

COLA AND SOLAR WARS

Thirteen principal investigators for Spacelab-2 formed an IWG, under the chairmanship of Mission Scientist Eugene Urban of MSFC. One of its visible responsibilities was picking payload specialist candidates and in April 1978, the *Marshall Star* reported that eight physicists (six Americans and two Britons) had been provisionally chosen, to be reduced to four candidates for two Spacelab-2 seats (Fig. 4.4).

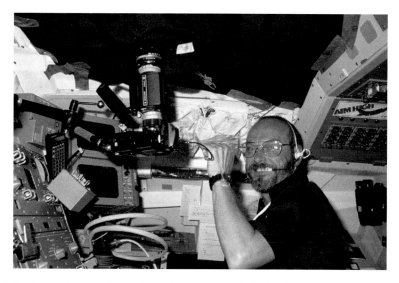

Fig. 4.4: Loren Acton, pictured on Challenger's aft flight deck during STS-51F, was a co-investigator of the SOUP instrument.

Nominally, the mission was baselined for seven to 12 days. Candidates included physicists John-David Bartoe and Dianne Prinz of the Naval Research Laboratory in Washington, D.C., Jack Harvey of Arizona's Kitt Peak National Observatory, Paul Peterson of Ball Research Corporation in Colorado, George Simon of the Air Force Geophysics Laboratory in New Mexico and Loren Acton from California's Lockheed Palo Alto Research Laboratory. Candidates from the United Kingdom were physicist Bruce Patchett of Rutherford Appleton Laboratory and astronomer Keith Strong of Mullard Space Science Laboratory. On 9 August 1978, Acton, Simon, Prinz and Bartoe were selected. They began training the following April with orientation tours—Chicago, Iowa City, Palo Alto, Pasadena, Washington, D.C., and Cambridge, Massachusetts, in the United States, and Abingdon, Dorking and Birmingham in the United Kingdom—to visit Spacelab-2's principal investigators. All four were closely involved with the experiments; Acton and Simon were co-investigators for SOUP, Prinz for SUSIM and Bartoe for SUSIM and HRTS.

On 5 June 1984, Acton and Bartoe were selected as prime payload specialists, with Prinz and Simon as alternates. By that time, the rest of the seven-man crew was also in place. On 18 February 1983, NASA named astronomer Karl Henize and geophysicist Tony England as mission specialists for a flight initially labelled 'STS-24', but later renamed 'STS-51F', a clumsy nomenclature used by NASA before the Challenger accident. The pair were 'career' NASA astronauts, chosen in August 1967, but England resigned in August 1972 to join the U.S. Geological Survey, then rejoined the space agency in June 1979. England considered himself "kind of an oddball" and never felt he truly fitted in with the older astronauts or the newer ones. His return to NASA came with a self-imposed condition: to stick around long enough to get a space mission, then return to academia. Henize, conversely, participated in the 1974 Learjet flights and ASSESS-II. "A joy to work with" and "total enthusiasm" were phrases Acton used to describe Henize. He recalled that Henize regularly prised him from his work to gaze out of Challenger's windows, drinking in the grandeur of Earth or deriving pleasure from weightlessness.

On 17 November 1983, the final three Spacelab-2 crewmen were named. Commanding the mission was Gordon Fullerton, joined by pilot Dave Griggs and flight engineer Story Musgrave. By now, launch had slipped from September 1984 to March 1985. Griggs was already assigned to another

Shuttle flight, originally scheduled for August 1984, but postponed into early 1985. As the gap between his two flights narrowed to a point that he could not train for them both, in October 1984 NASA replaced Griggs with Roy Bridges.

"It was about as multidisciplinary as you could imagine," Acton said of Spacelab-2's seven scientific fields and 13 experiments. "One of the things we learned was that we tried to accommodate and carry out a great variety of experiments." But in the world's eyes, STS-51F is remembered today as 'The Coke and Pepsi Flight'. Decades after the mission, Acton glumly related that his visits to talk to schoolchildren were dominated by questions not about solar physics or astronomy, but about carbonated drinks. "Coca Cola had gotten permission to do an experiment in space to see if they could dispense carbonated beverages in weightlessness," he recalled. "They got approval to build this special can, put significant money into it and were all set to fly it on one of the early Shuttle missions."

At the time, the 'Cola Wars' were bitterly playing out between Coca Cola and Pepsi on an international stage. A high-level Pepsi executive demanded that Coke should not fly the first carbonated beverage in space and it was withdrawn until both colas could fly together. "It turned out that our mission ended up getting the privilege of carrying the first soda pop in space," said Acton, cynically.

The astronauts were required to taste them and set the date/time functions of their cameras to 'prove' which cola was consumed first. (It turned out to be Coke, but the astronauts were unimpressed with either drink.) Photographs showed Bridges' red shift drinking Pepsi and Musgrave's blues sipping Coke, but the experiment's distracting effect proved an extreme irritation. Matters came to a head the morning before launch. "There is always a briefing, during which all the last-minute things that need to be talked about get talked about," Acton remembered. "We were about halfway through that briefing on the latest data concerning the Sun, when who should walk in, but the chief counsel of NASA, who began to brief us once again on the Coke and Pepsi protocols."

Acton hit the roof. "We've been getting ready for this mission for seven years," he thundered. "It contains a great deal of science. We have a very short time to talk about the final operational things that we need to know. We don't have time to talk about this stupid carbonated beverage dispenser test. Please leave!"

Without another word, the chief counsel turned on his heel and walked out.

A DRUNK BETWEEN THE LINES

Validating the IPS was critical on Spacelab-2, for the pointing system's next scheduled flight in March 1986 would carry ASTRO-1, a set of ultraviolet telescopes to observe Halley's Comet. Since this celestial wanderer frequented the inner Solar System once every 75 years, it had never been seen from space. "If the IPS doesn't work," an ASTRO-1 official told *Flight International*, "the whole mission is down the drain." During Challenger's first day in orbit, the crew unstowed the IPS from Pallet One, raised it upright at a pace of about a single degree per second and aimed it at a number of solar targets to assess its capabilities (Fig. 4.5).

But success took time to achieve, SOUP in particular proving irksome. "About eight hours after its activation, it shut itself off and would not accept the turn-on command," said Acton. "The crew did everything it could but

Fig. 4.5: One of the few on-orbit views of the University of Chicago's Cosmic Ray Nuclei Experiment (CRNE) 'duck egg'. With the erection of the IPS, the cosmic ray detector was virtually hidden from the astronauts' view.

ended up having to forget it." Not until 4 August, six days after launch, did SOUP awaken and in the last day of the mission recorded several hours of sunspot activity and active solar regions, including 6,400 photographic frames hailed as "unique". Its results proved far more consistent in frame-by-frame quality than were achievable using high-altitude rockets. But Acton did not see his instrument come alive, having suffered debilitating space sickness. It was left to Bartoe to relay the happy news. "I got sick as a dog," said Acton. "Thirty seconds after the main engines shut off, I felt like my stomach and my innards were moving up against my lungs. I was sick for four days and learned very quickly that you cannot unfold your barf bag as fast as you barf. When Bartoe came to tell me SOUP was alive, I was feeling so bad I didn't even get up to go look."

Post-flight analysis of the data revealed bubble-like convective cells, called 'granules', crawling across the Sun's surface in near-continuous motion at 2,500 mph (4,000 km/h). These granules covered almost the whole solar disk, except where sunspots were prevalent, and SOUP revealed them to 'float', like colossal corks, astride larger convective cells, four times wider than Earth.

Notwithstanding SOUP's difficulties, Spacelab-2 was activated and online by 15 hours into the mission. CHASE examined ultraviolet emissions from hydrogen and ionised helium, both in the Sun's visible disk and corona, revealing that active-region material formed 'bridges' between hotter and cooler regions. But its potential was compromised by STS-51F's low orbit since there was sufficient 'free' helium at the lower altitude to interfere with its measurements. Elsewhere, HRTS acquired eight hours of video and still photography in ultraviolet light, imaging fine-scale structures in the visible chromosphere, corona and the transitional zone between them. It recorded 19,000 exposures of sunspots and high-speed gaseous jets, known as 'spicules'. Eighteen observations were performed (some many times over) across 23 orbits, using virtually all available film and acquiring 600 full-frame and 18,000 short-frame spectrographic exposures. "It has the ability to zoom in on very small features on the surface of the Sun," said Bartoe. "The primary goal is to try to understand how the Sun makes the 'solar wind'. Some interesting things happen right on the surface of the Sun. For instance, the temperature goes up very dramatically as you go just above the surface. We're trying to look at that region right there, where that sudden transition of temperature takes place. Most of the light emitted there is in the ultraviolet."

But reflected sunlight induced higher than expected temperatures in the payload bay and IPS pointing problems complicated the observations. On several occasions, HRTS was temporarily shut down to prevent its on-board computer and film from exceeding temperature limitations. And by the second orbit of its operations, its spectrograph began to lose sensitivity, which impaired its overall yield.

With Bartoe watching from Challenger's aft flight deck, and Dianne Prinz watching from the POCC in Houston as one of the alternate payload specialists, the mission demanded close co-ordination between the ground and space. "The position of alternate payload specialist…is at the top of the pyramid over all the experimenters who are trying to get information up to the crew," Bartoe said. "Dianne had to listen to five or six telephone conversations at a time, listen to the crew and then try to sort it all out. In my opinion, this is the toughest job, much more difficult than flying. We only had seven people and two telephone lines up there!"

"It was really fatiguing trying to keep track of everything at once," Prinz admitted. "Each experimenter was very interested in getting dedicated information about his experiment, but I'd try to prioritise what we could ask of the crew without overtaxing them. I was the interface for all the experiments—not just the solar ones—and I'd be called into back rooms to iron out problems that occurred. It was that kind of understanding of what the intent of the mission was that made the crew such an outstanding group. We had a fantastic relationship with each other; very, very close."

Adjoining SOUP, CHASE and HRTS on the pointing system was SUSIM, which measured solar irradiance across 120-400 nanometres, with an accuracy better than ten percent. It monitored long-term variability in solar ultraviolet radiation, which (although only a tiny percentage of the Sun's total output) is the main energy source entering Earth's upper atmosphere. But solar ultraviolet radiation caused satellite-borne instruments to degrade and lose sensitivity over time. The result was that long-term solar change could not be effectively distinguished from change within the instruments themselves, leading to misinterpretations of long-term changes. By flying SUSIM routinely on week-long Shuttle missions, then returning the instrument to Earth for recalibration between flights, solar scientists could compensate for the deterioration of sensors on long-term satellites. SUSIM's absolute sensitivity was calibrated at the National Bureau of Standards' Synchrotron Ultraviolet Radiation Facility in Gaithersburg, Maryland. Results from Spacelab-2 indicated a 30-percent sensitivity loss from the solar spectrometer and a 20-percent degradation in its calibration spectrometer; both also exhibited a declining loss at longer wavelengths. Although this magnitude was relatively low, it was significant when compared to anticipated changes in solar ultraviolet output (Fig. 4.6).

All four IPS-mounted instruments handsomely proved their worth, but their prospects looked grim on the evening of 29 July 1985, when Flight Director Lee Briscoe told journalists that the pointing system's optical sensor

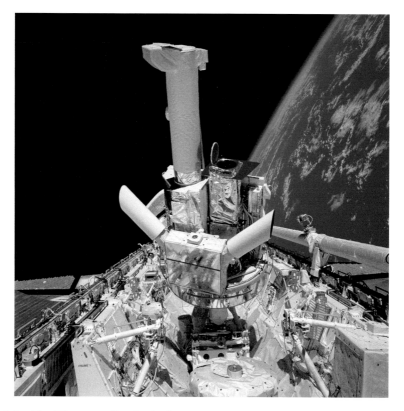

Fig. 4.6: The IPS comes alive for its first test mission on Spacelab-2.

package could not track solar targets smoothly. "It appears that we are able to find the Sun, find the stars, get into what we call a 'rough-track' mode," he said. "But we never appear to get into a fine-track and actually finish the total tracking." The optical sensor package tended to 'wander' erratically. "It's like a drunk that can hold it between the ditches," remarked one observer, "but can't stay between the white lines!" Henize managed to fine-tune the tracking system—and, for a time, CHASE and HRTS pulled double duty as pointing sensors—but it refused to stay 'locked' onto targets during Shuttle manoeuvres. The *Houston Chronicle* waspishly reported on the 30th that, far from tracking a dime at two miles, the IPS "would do good to hit the broad side of a barn". But Mission Manager Roy Lester expressed confidence that "it will work very well before the mission is over". During troubleshooting, a number of software 'patches' were uplinked and most observations were completed. Economical use of consumables enabled NASA to extend STS-51F by an extra day—from a planned 167 hours and 109 orbits to 190 hours and 127 orbits—to gather additional data.

In the absence of a Spacelab module, the astronauts laboured from Challenger's aft flight deck. "I had a joystick control in my hand, like on a video game, that permitted us to move the solar pointing telescopes around to point at particular features on the Sun," recalled Bartoe. "We would have a conference call, a solar conference, just before sunrise on each orbit. This gave us a chance to talk to the investigators, so we'd know what we were trying to do on that orbit. We also received about 20 feet of typed messages over the teleprinter, every orbit." The crew even had to replace the teleprinter paper five days into the mission, due to the extensive replan after the ATO and Challenger's low orbit.

And that replan proved a constant source of worry for Fullerton. He knew that Spacelab-2 was complicated even if it ran perfectly; much more so if it hit anomalies. "Is this going to squarewave the whole flight plan and mess everything up?" he mused after the ATO. "It did to some extent, but the ground worked overtime, because everything was sequenced by time…it's an astronomy thing. Whether we're on the dark side or the light side, all that had to be rewritten."

On the flip side, Bartoe felt that seven years of training worked in their favour. "One of the advantages of the fact that this mission took years from formation to launch was that there was a long time to fine-tune each instrument's observing plan," he said. "That was good, because a lot of things went wrong, but we really understood how the various observations fit together and it was easy to make quick changes."

CLOCKWATCHING

Spacelab-2's dual-shift system made the aft flight deck relatively roomy, with no more than four people on duty at any one time. But training proved problematic in the weeks before launch, since the fixed-base Shuttle mission simulator at JSC could not be reconfigured to mirror STS-51F's exact aft flight deck layout until quite late in the day. This was due primarily to the proximity of other Shuttle missions on the manifest, the most recent of which flew in June 1985. As a result, the simulator could not be 'de-configured' from its layout on that mission and reconfigured to its STS-51F layout for high-fidelity integrated simulations until shortly before launch. "This led to a more demanding schedule for an already fatigued crew," wrote Shayler and Burgess, "who recommended that a second simulator should be brought online to cope with the expected rise in launch rate and the overlapping crew training programs."

Working in orbit was noisy, remembered Musgrave, even with the benefit of middeck sleep stations. "They were banging around all night," he recalled of the on-duty shift. "I took a pill to help me to sleep and I forgot I took the pill, so I went off to sleep, floating around. You don't get the head nods in space. Your head doesn't fall—there's no gravity to make that happen—so I went off to sleep." As he slumbered, insensible to his surroundings, Musgrave floated through the access hatch from the middeck to the flight deck, much to the amusement of Bridges' on-duty red shift.

"Oh, a monster," one of them shrieked, then grabbed Musgrave and shoved him back into his bunk with their fingertips.

The work schedule was punishing. "During your 12 hours 'on', you ran all these instruments," said Fullerton. "During the 12 hours off, you had dinner, slept, had breakfast and then went to work. Two weeks before launch, we set that up. I anchored my schedule to overlap transitions, so if something came up on one shift, I could learn about it and carry it over to the next shift. I also had to stagger things, so I got on the right shift for re-entry [and] was in some kind of reasonable shape at the end of the mission. We had the red team sleeping up till launch time, so that once we got on orbit, they were the first one up and they'd go for it for 12 hours. The last week [before launch], we didn't see the other team or I only saw part of one and part of the other myself."

Breakfast, recalled Bridges, who rustled up Mexican scrambled eggs, coffee and orange juice for Henize and Acton before shifts, was a problematic affair, with sausage tending to crumble and float around the cabin. "I chased it," he said, "while they ate." Acton also took along some Louisiana hot sauce, a condiment highly appreciated by his crewmates.

For the scientists, Spacelab-2 was exceptionally professionally rewarding, not least as they participated as co-investigators or flew with hardware built at their former workplaces. Before his astronaut career, Henize worked for the Smithsonian Astrophysical Observatory and derived great satisfaction from operating their IRT instrument, which met more of its technical objectives than scientific ones. Its superfluid helium/porous plug cooling system exceeded expectations and demonstrated that extremely low operating temperatures could be established and maintained. But saturation of its mid-wavelength detectors by an intense infrared background compromised some results (Fig. 4.7).

Nonetheless, half of the galactic plane was mapped at short wavelengths. The IRT also unveiled the infrared 'background' of Challenger herself and determined the extent to which the mission's space plasma physics studies and the 'Shuttle glow' phenomenon affected its sensitivity. "We see the IRT back there doing its hickory dickory dock," Acton recalled of the telescope's

Fig. 4.7: His feet anchored in place by foot-loops, John-David Bartoe works at one of the Data Display Units (DDUs) on Challenger's aft flight deck during Spacelab-2.

sweeping motion in the payload bay. "It makes you think of one of those birds that you dip in a glass of water and it dips, dips, dips."

Alongside the IRT on Pallet Three was the PDP, deployed into free flight on STS-51F to examine the electromagnetic and particle environment and ionospheric plasmas surrounding the Shuttle. Data from OSS-1 yielded sufficient understanding of the payload bay's atomic 'cleanliness' to commit sensitive instruments to Spacelab-2. For four days, beginning on 31 July, the PDP was extended by Challenger's RMS and released into space to acquire wide-band spectrograms of plasma waves at frequencies up to 30 kilohertz at distances up to 1,200 feet (400 metres). Due to the lessened amount of propellant—depleted following the ATO—plans for Fullerton and Bridges to fly Challenger 'around' the PDP for a photographic survey could not be completed in full.

Still, valuable data was gathered. Two types of interference patterns were identified in the wide-band data: one associated with its passage through

VCAP electron-beam pulses, the other from lower 'hybrid' waves generated between the neutral gas cloud surrounding the Shuttle and ambient ionospheric plasmas. During six hours of PDP tests on 1 August, a momentum wheel spun-up the small satellite to effect stability. Among its findings was a region of intense broadband turbulence around the Shuttle at frequencies from a few hertz to ten kilohertz. The highest intensities occurred in the region 'downstream' of Challenger and along magnetic field lines passing close to the orbiter, which tended to increase during periods of high thruster activity.

In general, the PDP-VCAP observations showed that thermal ion distributions around the Shuttle were more complex than predicted with an unexpectedly intense background level of ion current due to incoming hot ions. Surprisingly, those ions often changed energies, indicative of high temperatures and turbulent plasma activity and demonstrative of the huge impact of a large, gas-emitting spacecraft on the ionosphere. Water vapour was detected in the orbiter's immediate vicinity and proved dominant in its wake. Whilst the PDP was held by the RMS, Challenger rolled to sweep the sensors through this wake. Ions from the 'ambient' ionosphere were accelerated into the wake from 'above' and 'below' the spacecraft; triggered, perhaps, by a strong electric field created by density differences between the two. Ground-based and PDP observations of Shuttle thruster firings yielded faint red airglow emissions that produced a cloud 190 miles (300 kilometres) wide. Even minute thruster firings affected ambient plasmas.

One particularly interesting investigation was the Plasma Depletion Experiment for Ionospheric and Radio Astronomical Studies, sponsored by Los Alamos National Laboratory and Boston University, which examined Shuttle thruster bursts on ionospheric dynamics. These firings and their brief alteration of plasma states and transmissional characteristics of the ionosphere (even producing short-lived 'holes', observable from the ground) were monitored by five Earth stations. Original plans called for seven 'burns'—two over Millstone Hill in Massachusetts, two over Arecibo in Puerto Rico, two over Roberval in Quebec and one over the University of Tasmania's low-frequency radio observatory at Hobart—lasting between 12 and 47 seconds. But Challenger's limited propellant after the ATO meant only four burns could be performed. Two were executed over Millstone Hill and one apiece over Arecibo and Hobart. Observers on the ground tried to conduct astronomical observations through artificial 'windows' temporarily opened in the ionospheric plasmas by the burns. Some cosmic signals were received and the Hobart team identified low-frequency radio emissions. Eugene Urban remarked that one burn over Millstone Hill produced a temporary hole that was "very large and bright" for radio observations. But radio waves in the band lower than three megahertz remained blocked by ionospheric plasmas.

"It was a great mission," said Fullerton of Spacelab-2. "Some of the missions were just going up and punching out a satellite and then they had three days with [very little] to do and came back. We had a payload bay absolutely stuffed with telescopes and instruments." At 11:45 a.m. PST on 6 August 1985, he and Bridges landed Challenger safely on Runway 23 at Edwards Air Force Base, after eight days in orbit.

At Spacelab-2's preliminary science review, held at MSFC in November 1985, Urban branded STS-51F "a complete success", despite the limitations of the low orbit and its IPS woes. He paid glowing tribute to SOUP, which came online late in the mission. "We were able to obtain long sequences of high-resolution photos of the solar surface," Urban was quoted in the *Marshall Star* on 20 November. "We now have the means to study the growth and fading of various solar features, like sunspots, over long periods of time." And for the first time with the deployable PDP-VCAP, interactive ionospheric plasma physics was explored. "Together, the diagnostics package and VCAP found some highly interesting correlations between the man-made electron-beam simulations of the ionosphere and naturally occurring auroras," Urban said. "The information will help us understand better how such auroras are formed from beams of charged particles from the Sun" (Fig. 4.8).

STS-51F's success prompted the solar physics community to opine that the mission would "stand until the era of the space station, because no payload now under consideration matches the complexity of Spacelab-2, which tested the limits of hardware, software and people, everywhere in the system". One of those people was Fullerton, who returned to Earth exhausted. "The pressure is higher when you're the commander," he said. "You're really tired after spaceflight, mostly because you elevate yourself to this high level of mental awareness that you're maintaining. Even when you're trying to sleep, you're worried about this and that. It's not like you're just lollygagging around and having a good time. You're always thinking about what's next and mostly clock-watching.

"Flying in orbit is watching a clock," Fullerton continued. "Everything's keyed to time and so you're worried about missing something, being late. We had 270 manoeuvres. Every sunrise and sunset, we had to go to a different attitude to put the right telescopes at the right stars or Sun. Those are all 'typing exercises'—typing long strings of numbers into the computer and the time to start the manoeuvre, so it goes to the right attitude. You mess one number and you're going to go to the wrong attitude, then you're going to miss that data. Every 40 minutes, you've got a new one."

Fig. 4.8: Commander Gordon Fullerton shakes hands with George Abbey after leaving Challenger on 6 August 1985. Following the commander down the steps are Tony England, Loren Acton, Karl Henize, Roy Bridges, John-David Bartoe and shaven-headed Story Musgrave.

5

Of monkeys, mice and men

GRANDPA'S PAINTING

For years, an unusual piece of artwork hung in Don Lind's house. Painted by the astronaut himself, he gifted copies to his children, hopeful that they (and their children) would comprehend its message. Through Lind's mind's eye, Challenger launched from Earth on 29 April 1985 on STS-51B, the first 'operational' Spacelab mission, with 15 experiments across five disciplines—materials science, life sciences, fluid mechanics, atmospheric physics and astronomy—from the United States, France and India. In Lind's painting, the Shuttle is cradled by two great celestial hands, as if God himself was guiding the mission to success. Lind titled his piece as 'Three-Tenths of a Second'. When Challenger exploded shortly after liftoff in January 1986, killing all seven crewmembers, the tragedy hit Lind hard. For when he flew that same vehicle, nine months earlier, his own mission was almost lost for the same reason. Rubberised O-ring seals in the SRBs, meant to stop a catastrophic leak of hot gases, had failed, cracked and come within a hair's breadth of burning through.

One day in 1986, Lind was summoned into the office of STS-51B Commander Bob Overmyer, a stern Marine Corps colonel. At the time, Overmyer was serving as NASA's representative on the presidential inquiry into the disaster. "Don," he said. "Shut the door."

Lind shut the door.

In measured tones, Overmyer explained that STS-51B had itself almost exploded, seconds after liftoff. "The first [SRB] seal on our flight had been totally destroyed and the [secondary] seal had 24 percent of its diameter burned away," Lind remembered. "All that destruction happened in 600 milliseconds and what was left of that last O-ring…if it had not sealed the crack and stopped the outflow of gas…if it had not done that in the next 200-300 milliseconds, we'd have exploded. *That* was thought-provoking!"

It made Lind realise that, for three-tenths of a second that day, his life hung from the shredded fragments of a single O-ring booster seal. And so too did the first operational mission of Spacelab, which would also have been lost in a conflagration. "Each of my children have a copy of that painting," he said, "because we wanted the grandchildren to know that we think the Lord really protected Grandpa."

GRAVITY GRADIENT

STS-51B would usher in an era of 'routine' Spacelab flights. Planned as a week-long mission, Challenger was set to fly at noon EDT on 29 April 1985, at the opening of a one-hour 'launch window', selected to provide maximum viewing opportunities for one of its two Earth-watching instruments. The California Institute of Technology's Atmospheric Trace Molecule Spectroscopy (ATMOS), affixed to a truss variously named 'Mission-Peculiar Equipment Support Structure' or 'Multi-Purpose Experiment Support Structure' (MPESS) at the rear of the Shuttle's payload bay, would examine the compositional variability of the upper atmosphere. Equipped with a Sun-tracker, telescope, interferometer and data-handling system, ATMOS applied infrared absorption spectroscopy to provide high-resolution spectral data and map the three-dimensional temperature structure of the upper atmosphere. Seventy-two ATMOS observations were planned—half during orbital sunrises, the rest during orbital sunsets—with the instrument afforded three minutes during each 'occultation' for data acquisition.

With STS-51B's launch window thus established, planning for another of the mission's primary objectives—flying albino mice and squirrel monkeys on a crewed spacecraft for the first time—could get underway, as handlers began to train the animals' circadian patterns to fit the timings of the flight plan. Assuming an on-time liftoff, Challenger and her seven-man crew would complete 108 Earth orbits, before landing on KSC's Runway 15 at 8:58 a.m. EDT on 6 May, after 164 hours and 58 minutes. STS-51B targeted an orbit 218 miles (352 kilometres) high, inclined 57 degrees to the equator. In a fashion not unlike Spacelab-1, activation of the module would occur two hours after launch, with experiment operations and around-the-clock shifts commencing five hours into the flight. Finally, the module would be powered down six days and 15 hours after launch, some 4.5 hours before re-entry. It was hoped that economical use of consumables might keep open an option for up to two extra days: the first yielding a landing on the 125th orbit after 189 hours (on 7 May), the second on the 140th orbit after 212 hours (on 8 May).

This flight was designated 'Spacelab-3': the second outing of the pressurised module having leapfrogged Spacelab-2, whose pallet-only 'train' and IPS had suffered multiple development difficulties (see Chapter Four). For the first 17 hours of STS-51B, France's Very Wide Field Camera (VWFC) would conduct high-quality ultraviolet observations from the Scientific Airlock (SAL) in the module's roof. The camera was a repeat flier from Spacelab-1 and would conduct a general survey of the celestial sphere. It was stored for launch on the Spacelab module's floor and assembled in the SAL for a half-day of observations from six to 17 hours into the flight, during the seventh through 12th orbits. The VWFC was then removed from the SAL and stowed.

The reason for the ATMOS and VWFC timeline being so tightly front-loaded into the first day of the mission was dictated by a need for multiple Shuttle manoeuvres to achieve proper orbital positioning. Spacelab-3's primary focus was microgravity research, which demanded an ultra-stable platform. Eighteen hours after launch, the pilots would orient Challenger into a 'gravity-gradient' attitude, with her tail pointing Earthward and her starboard wing facing into the direction of travel, affording high stability for several sensitive experiments by eliminating the need for thruster firings which might distort their results. For the remaining 136 hours of the mission, the Shuttle would occupy this gravity-gradient attitude, supporting crystal growth, fluid physics and animal observations. This intriguing flight originated in 1976 as an MSFC endeavour, sponsored by the Office of Space and Terrestrial Applications at NASA Headquarters and led by Mission Scientist George Fichtl. Two years later, the 15-member Spacelab-3 IWG outlined a mission whose research remit ran from materials science and life sciences to atmospheric physics and astronomy.

STS-51B's gravity-gradient attitude aided three materials science investigations. The Solution Growth of Crystals in Zero Gravity experiment, provided by Alabama A&M University in Huntsville, produced triglycine sulphate crystals—used in infrared detectors—using a low-temperature growth technique. It was expected that in microgravity, convective flows known to disturb Earth-grown crystals would be reduced and a slower, more uniform and more precise growth process would be attainable. Three crystal-growing 'runs' were planned: the first starting late on the first day of the flight, lasting 29 hours, a second beginning on Flight Day Three and running for 62 hours and a final sample, 19 hours long, to commence early on Flight Day Six (Fig. 5.1).

Fig. 5.1: Artist's concept of Challenger in her 'gravity gradient' attitude during the majority of STS-51B, tail directed Earthward. Note the pressurised Spacelab long module and the truss-like MPESS at the rear of the payload bay.

The Mercuric Iodide Growth/Vapour Crystal Growth System (VCGS), from defence contractor EG&G of Goleta, California, sought to grow a mercuric iodide crystal by diffusion-controlled vapour transport over 137 hours. Both the triglycine sulphate and mercuric iodide experiments occupied adjacent Spacelab payload racks, sharing a video system which allowed scientists to monitor the process. The last experiment, the Mercuric Iodide Crystal Growth (MICG), would run for over a hundred hours, using a two-zone furnace. It was provided by France's Laboratoire de Cristallographie et de Physique. 'Soft' mercuric iodide crystals, used in gamma-ray detectors and spectrometers for the nuclear industry, rarely approached anticipated performance levels when grown on Earth, due to gravity-induced defects. During Spacelab-3, they would be grown by vapourisation and re-condensation at 120 degrees Celsius (250 degrees Fahrenheit) in a furnace in the VCGS. "Delicate crystal growth and fluid mechanics experiments, which are dependent on a lack of gravity," NASA explained, "have been clustered near the spacecraft's centre-of-gravity, the most stable part of the vehicle." With Challenger's centre-of-gravity close to the rear of the module, all three materials science experiments sat aft of Spacelab-3's midpoint, to ensure the most quiescent environment for growth.

Also close to this centre-of-gravity 'sweet-spot' were two fluid mechanics experiments. The Dynamics of Rotating and Oscillating Free Drops (DROP) used a specialised Drop Dynamics Module (DDM) in a double rack to manipulate liquid drops acoustically, as part of efforts to develop 'containerless' materials processing methods. This was expected to produce new materials without touching the walls of their container, which might otherwise introduce structural defects. Provided by JPL, it investigated rotating drops and the large-amplitude, non-linear oscillation of liquid drops.

And the Geophysical Fluid Flow Experiment (GFFE), from the University of Colorado at Boulder, occupied the lower half of a single rack, at the very end of the module. It simulated fluid flows in planetary atmospheres to understand large-scale mechanics of global geophysical flows. GFFE used two concentric, rotating, electrically conductive spheres. One was the size of a baseball, made from nickel-coated stainless steel, mounted within a larger, transparent sphere of sapphire, both attached to a turntable. A thin layer of silicone oil filled the gap between the two hemispheres. Voltage was applied to identify conditions pertaining to fluid viscosities and rotational/gravitational dynamics. It was hoped that the experiment would provide correct vector relationships between the rotational vector and gravitational body-force vector of planetary bodies and GFFE's camera acquired views at 90 degrees longitude and from the equator to the pole.

Six life sciences investigations were also aboard Spacelab-3, four supporting engineering verification tests of hardware supplied by NASA's ARC in Mountain View, California. These included a pair of Research Animal Holding Facilities (RAHFs): one to house four male squirrel monkeys in four cages in an Ames Single Rack (ASR), the other containing 24 male albino rats in 12 cages (two per cage) in an Ames Double Rack (ADR). The RAHFs—which originated in a pair of study contracts, worth a combined $420,000, awarded by MSFC to Lockheed and McDonnell Douglas in June 1975—could support mammals sized from rodents to small primates by 'cage-module interchange'. Food and water were automatically dispensed and monitored and infrared observations conducted of behavioural patterns and measurements of system performance, environmental control, food/water delivery, lighting and animal activity. A Biotelemetry System (BTS) monitored output from transmitters surgically implanted in four rats before launch, gathering data on basic physiological functions (including heart rate, muscle activity and deep body temperature), telemetered to antennas in the RAHF cages. And a Dynamic Environment Measurement System (DEMS) monitored acceleration, vibration and acoustics from the RAHFs during launch and re-entry.

In addition to these four engineering verification experiments, a final pair of life sciences studies investigated the astronauts' adaptation to microgravity. ARC's Autogenic Feedback Training (AFT) demonstrated techniques to control bodily processes voluntarily, using training and mental exercises (rather than medication) to counteract space sickness. And JSC's Urine Monitoring Investigation (UMI) assessed the toilet's urine monitoring system and its ability to collect and sample urine, perform in-flight calibrations and monitor astronauts' water intake. This would aid the construction of advanced monitoring systems for future Spacelab life sciences missions.

Rounding out Spacelab-3's 15 experiments was Studies of the Ionisation States of Solar and Galactic Cosmic Ray Heavy Nuclei (shortened to 'IONS'), which sat on the external MPESS truss, alongside ATMOS. It was provided by India's Tata Institute of Fundamental Research in Bombay (today's Mumbai) and observed energetic ions from the Sun and galactic sources. Specific focuses included low-energy cosmic ray ions of carbon, nitrogen, oxygen, neon, calcium and iron and heavy elements from oxygen to iron in highly energetic particles emitted in solar flares. And the Auroral Imaging Experiment, built by the Geophysical Institute at the University of Alaska, acquired photography, videography and naked-eye observations of high-latitude aurorae to assess magnetic activity in the upper atmosphere and the three-dimensional auroral structures.

Preparatory work for Spacelab-3 got underway in December 1983, when the pressurised module returned to KSC after STS-9. Its racks were removed and modifications began for its second mission. The roof-mounted Scientific Window Adapter Assembly (SWAA) was removed and covered with an aluminium panel, whilst the SAL was retained to accommodate France's VWFC. In March 1984, a mission sequence test verified the compatibility of all 15 experiments and the first Spacelab-3 payload racks were rolled into the module's shell the following May. An integrated mission simulation was conducted at MSFC in August 1984 and on 27 March of the following year the full payload—module, tunnel and MPESS—was installed aboard Challenger (Fig. 5.2).

Another twist in Spacelab-3 pre-launch preparations was the unique procedure of loading the nameless male squirrel monkeys (*Saimiri sciureus*), which Lind described as "cute", and the "not so cute" male albino rats (*Rattus norvegicus*) into the module, as the Shuttle sat vertically on the pad. This required some interesting gymnastics from technicians. "They had to develop this boatswain's chair," recalled Bonnie Dalton, former acting chief of ARC's Life Sciences Division. ARC's Institutional Animal Caring Use Committee

5 Of monkeys, mice and men 115

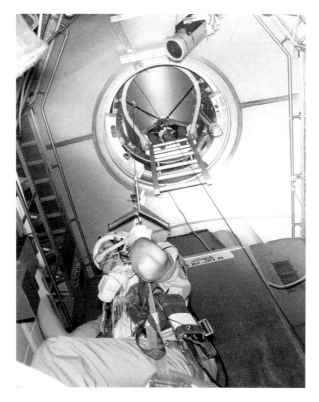

Fig. 5.2: Technicians lower themselves into the Spacelab-3 module to install the rodent and primate cages before launch. With Challenger in a vertical orientation on the pad, one technician waited in the tunnel's joggle section (top of frame), whilst a second descended into the module by means of a sling-like seat.

(IACUC) dictated that the primates and rodents be moved during their 'awake' time, to avoid undue stress, which meant they were transported in a specially air-conditioned van and loaded aboard Spacelab-3 only 20 hours before launch.

Working from the Shuttle's middeck, two technicians were lowered, one at a time, in these sling-like seats down the tunnel into the Spacelab module. One stayed in the joggle, whilst the other entered the module to activate the RAHFs and await the cages. These were lowered on separate slings, called module vertical access kits. The delicate, two-hour procedure ran smoothly, although with the Shuttle in a vertical configuration on the pad the entire complement of animals—the monkeys in a single rack in Rack Five, the rats in a double rack in Rack Seven, both on the module's port-side wall—were left resting against their cage 'walls' during Challenger's ascent.

Unfortunately, in addition to animal welfare concerns, human welfare concerns took centre-stage (and the international stage) in the weeks preceding launch.

SCREWING THE MONKEYS?

Spacelab-3 sought to evaluate how well the RAHFs could support animals in an environment comparable to a ground-based vivarium. Effective rodent and primate observations were impossible if their health and well-being were improperly managed. In addition to providing staples of water and food (rice-based bars for the rats, banana pellets for the monkeys), the crew's 12-hour shifts saw Mission Specialists Norm Thagard and Bill Thornton—both physicians—assigned to separate teams to watch the animals day and night. The monkeys had developed a reputation for being somewhat feisty, one having bitten a trainer's finger, and would go on to scuffle with the astronauts in orbit.

"The squirrel monkeys adapted very quickly," said Lind. "They had been on centrifuges and vibration tables, so they knew what the feeling of space was going to be like. Squirrel monkeys have a very long tail and if they get excited, they wrap the tail around themselves and hang onto the tip. If they get really excited, they chew on the end of their own tail. By the time we got into the laboratory, about three hours after liftoff, they were adjusted. They had…apparently chewed off a quarter of an inch of the end of their tails!"

Early plans called for four squirrel monkeys, but only two eventually flew. All were required to be free of various specified pathogens. "This didn't mean they were completely pathogen-free, because if you had a completely pathogen-free animal, that animal would not have a good immune system," said Dalton. "But there are certain organisms—salmonellas, some pneumococci—that people could get. When we have animals or plants in the Spacelab…they share the air with the people in the vehicle."

Six months before launch, NASA announced its requirement for the squirrel monkeys to be free of *Herpes saimiri* antibodies. Although this virus was not known to cause disease in squirrel monkeys or humans, problems had been documented in other species and a global search got underway. But a month before launch, NASA reduced the number of monkeys from four to two, having found a strain of herpes potentially transmissible to the astronauts. "The monkeys' herpes…was not the virus transmitted sexually by humans but was unique to New World primates whose natural habitat was the rainforests of South America," reported the *Washington Post* on 27 March. "Researchers suspected the virus of causing cancer in lower mammals, such as rats, and therefore classified it as potentially cancerous in humans." NASA replaced the four monkeys from colonies bred to be virus-free. Four monkeys recruited from the National Institutes of Health and one from Harvard University had been training since January 1985, with scientists searching for a sixth candidate, although it was later determined that only three might fly, with at least one being too small. "We finally got a pool of three squirrel monkeys and they had to be shipped directly to KSC and go into quarantine," said Dalton. But due to time limitations, NASA could prepare only two of them for flight, training them to reach food pellets and activate water-taps in their cages.

The possibility, however remote, of the Spacelab-3 crew getting infected by herpes was pounced upon by their peers, wrote astronaut Mike Mullane. Several Navy astronauts joked that as long as the Marine Corps and Air Force members of the crew—a prod at the respective military services of Commander Bob Overmyer and Pilot Fred Gregory—did not "screw the monkeys", all would be fine.

UNTOUCHABLES

Overmyer, Gregory, astrophysicist Lind and physicians Thagard and Thornton were assigned to Spacelab-3 on 18 February 1983, for a flight initially designated 'STS-18' and targeted to launch on 22 November 1984. Among the military astronauts, Spacelab was perceived generally negatively, as it involved few of the Shuttle's high-profile mission tasks, like spacewalking, satellite deployment or rendezvous. "Operating a robot arm had a lot more sex appeal and generated a lot more personal fulfilment than watching a volt-meter on some university professor's experiment," wrote Mullane, labelling the astronauts 'unlucky' enough to draw Spacelab crew assignments as 'Untouchables' (Fig. 5.3).

"They collected blood and urine and butchered mice and changed shit filters for primates," Mullane added. (Indeed, poor Overmyer would find an unwanted faecal 'gift' floating right under his nose as he sat in the commander's seat on Challenger's flight deck.) There also existed a pervasive sense that Navy pilots were getting the more challenging missions, whilst Air Force pilots received the 'untouchable' science flights: Spacelab-1's Brewster Shaw, Spacelab-2's Roy Bridges and Spacelab-3's Gregory were all Air Force officers. "This is bullshit, man," Gregory is said to have growled in protest upon learning of his Spacelab-3 assignment, according to Mullane. Even Shaw admitted

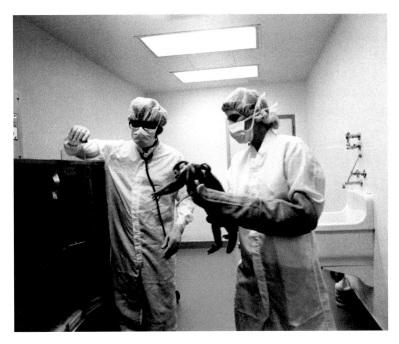

Fig. 5.3: One of STS-51B's squirrel monkeys is prepared for loading into his enclosure cage.

that despite its historic 'firstness', Spacelab-1 imposed relatively few piloting demands upon his time.

As much as some astronauts considered Spacelab an undesirable assignment, so the researchers were equally careful when considering who they wanted to run their experiments. According to John Charles, principal investigator for a lower body negative pressure experiment flown on a joint U.S./Japanese mission (see Chapter Ten), selecting astronauts for Spacelab was carefully orchestrated. "I think the astronauts were motivated to be successful, because they did have enough time to bond with the investigators and we certainly had enough face-time with the investigators and the projects and the astronauts, so they got to know what the purpose was, they got to know what the motivation was, what the end-goal was for the set of investigations," said Charles. "They may not have preferred to have been on this research mission. In fact, one of the things that we did…is that we essentially down-selected astronauts. I didn't get a chance to pick which astronauts would be on the mission, but I got a chance to pre-brief large groups of astronauts who might end up on the mission and to tell them what we were about." This was particularly vital for missions featuring invasive medical investigations. "I petitioned the astronaut management repeatedly," he added. "Please don't give me astronauts who don't like life sciences work, because that's a major part of what we're doing."

Alongside Overmyer and Gregory, Spacelab-3's untouchables included Lind, Thagard and Thornton, plus two payload specialists. In June 1983, the IWG selected four candidates: metallurgical engineer Mary Helen Johnston of MSFC, Dutch-born chemical engineer and materials scientist Lodewijk van den Berg of EG&G and fluid physicists Taylor Wang and Gene Trinh of JPL. On 5 June 1984, van den Berg and Wang were selected, with launch scheduled for January 1985. The selection surprised van den Berg, who only cast his name into the ring on a whim. He was an authority on vapour-driven crystal growth—once likening the process to "gardening"—and an expert in mercuric iodide crystal growth. He was a co-investigator on EG&G's VCGS experiment and as the IWG drew up its candidate list, van den Berg and his boss, Harold Lamonds, could only find seven names, rather than the required eight. Lamonds told van den Berg to add his name, joking that the quinquagenarian's age, huge spectacles and limited physical strength would likely eliminate him early in the selection process. But it did not. Four candidates were dropped in the initial screening for scientific competence. Van den Berg thus made it to the final four and two others fell by the wayside due to possible heart issues, meaning he and Johnston made the cut. As for the second Spacelab-3 payload specialist, Wang was principal investigator for the DROP experiment and builder of the DDM. During the mission, his expertise would literally 'save' both the experiment and the facility.

Flying principal investigators as payload specialists and affording builders of hardware the chance to physically operate their equipment in space was part of Spacelab's original pledge, articulated by NASA as early as February 1972. "Having a PI on board with his own experiments means that the PI is focused on that and everybody else is taking care of the rest of things," said Spacelab-1's Bob Parker. "On the seven-to-ten-day Shuttle flights, you can kind of understand that. It's a major experiment. That's a huge investment of resources." Before the Shuttle, everyone was a *bona fide* astronaut, but payload specialists enabled missions to benefit from what Parker called "real, more current scientists". This garnered great concern from the outset, particularly from JSC, which had monopolised astronaut selection and training and was unwilling to cede its vice-like control to MSFC and the IWGs. In JSC's view, for the first time in U.S. human spaceflight, an establishment outside Houston was infringing on 'its' sovereign soil by providing mission-specific training.

Pilot-astronauts had other worries. "There was a little bit of confusion in those days about who *are* these people and what's their real role," remembered

Bryan O'Connor. "Are they astronauts or not? Are we going to see them again? Will they fly repetitively?"

On O'Connor's first flight, a non-Spacelab mission, one of the payload specialists showed interest in the Shuttle's hatch; an interest of harmless curiosity, but one which raised eyebrows. The commander was so concerned that he discreetly requested a padlock—"due diligence," said O'Connor—to be added, its combination held by a NASA crewmember. "Why did he keep asking about that?" Shuttle commander Hank Hartsfield mused. "It turned out it was innocent, but at the time, you don't know. Some of us worried about these 'short-termers' coming along and joyriding, we call it." On his second mission, a Spacelab flight, O'Connor was upfront with his payload specialists about the padlock. "They probably thought, 'Well, this is a fine how-do-you-do. We train for two years together and they don't trust us'," he said. "Maybe it was bad judgement. There's a potential downside in that it creates concern among the crew about trust and all that. But I [erred] on the side of due diligence and kept it on there. And I was honest with them." Other astronauts struggled to hide their angst, having trained for years for mission assignments, only to see non-professional payload specialists flying ahead of them.

When he joined Spacelab-3 in February 1983, Lind anticipated launching in September 1984. By the fall of 1983, their launch had moved to November 1984 and by mid-1984 slipped again to January 1985, assigned to Shuttle Discovery. But when Challenger was grounded later that year, her high-priority Department of Defense mission, planned for December, was shifted onto Discovery and flew in January 1985. As a result, Spacelab-3 was remanifested onto Challenger and launch moved to April.

As a dual-shift flight, the crew started 'sleep-shifting' in the final days before launch. On the 'gold' shift were Overmyer, Lind, Thornton and Wang, with Gregory, Thagard and van den Berg on the 'silver' team. "I was responsible for all the support systems that keep the orbiter functioning," said Gregory of his role as silver leader. "Norm and I had respective jobs on board, but we…were the folks who supported the work of the payload specialists." As flight engineer, Thagard was part of Challenger's flight crew, but his workload crossed with the scientists in the Spacelab module. And as a physician he would care for the rodents and primates on his shift.

UNDER THE COMMANDER'S NOSE

In Gregory's recollection, it took a half-day to acclimate to weightlessness, following Challenger's rousing launch at 12:02 p.m. EDT on 29 April. "The body very quickly adapted to this new environment and it began to change," he said. "You could sense it when you were on orbit. You learned that your physical attitude in relation to things that looked familiar to you—like walls and floors—didn't count anymore and you translated floors and ceilings and walls to your head is always 'up' and your feet are always 'down'. That was a subconscious change in your response: it was an adjustment that occurred up there." On one occasion, Gregory tossed a pen to van den Berg and cackled as it twirled its way along the length of the Spacelab module into the payload specialist's waiting hand (Fig. 5.4).

Very soon, however, one of the squirrel monkeys exhibited space sickness symptoms—lethargic and dispirited, lacking in appetite, but no observed vomiting—for the first half of the mission, being hand-fed by Thagard and Thornton at one stage, then recovering completely for the last three days. (This was quite contrary to the flight plan, which did not foresee any direct contact between the astronauts and the animals.) The second monkey displayed no ill effects, even doing somersaults in his cage. The primates proved much less active in space than on Earth, although both they and the rodents

Fig. 5.4: Bill Thornton checks on one of the squirrel monkeys during his shift.

grew and behaved normally, were free of chronic stress and differed from their 'controls' on Earth only through gravity-driven variables.

The monkeys were spoiled, too. "I think the environment they had come from was a place where they received a lot of attention," remembered Gregory. "Norm and I would look into the Spacelab and see Bill Thornton attempting to get these monkeys to do things, like touch the little trigger that would release the food pellets. I could tell they expected Bill to do that for them, even though he was outside, looking in. We looked back one time and could see that the roles were kind of reversed and Bill was doing antics on the outside of the cage and the monkeys were watching!" Thornton and Thagard viewed the primates through a window in their cages, a perforated opening affording limited access to the interior.

The rodents' enclosures were not unlike those of the squirrel monkeys, with the exception that they housed two occupants per cage, divided by a partition. Half of the 24 rats were rapidly growing, eight-week-old juveniles and the remainder were mature 12-week-old adults. Four were implanted with transmitters three weeks before launch to monitor heart rates, deep body temperatures, muscle activity and other parameters. Those readings were transmitted through the BTS. The data was so good that it was even possible to monitor one rat for indicators of stress. Neither of the monkeys was outfitted with BTS sensors, although their cages included provisions for future flights. Typically, implant data was transmitted via a dedicated computer to scientists at the POCC.

Although the animals were maintained in healthy conditions, the rats proved not as 'savvy' as the monkeys in terms of adaptability. "They hadn't learned that this was going to last a while and when we got into the laboratory, they were hanging onto the edge of the cage and looking very apprehensive," explained Lind. "After the second day, they finally found out if they'd let go of the screen, they wouldn't fall and they probably enjoyed the rest of the mission." Despite their adaptive slowness, they showed no signs of sickness, although post-flight dissection identified a marked loss of muscle mass and increased fragility of their long bones. Investigators speculated that this was probably caused by microgravity, rather than the stress of the RAHF cages. Nonetheless, the animals returned in good physical condition, healthy and free of microbiological contaminants.

However, the crew returned home with several concerns, for the animal enclosures leaked food crumbs, faeces and unpleasant odours. The cause was traced to gaps in the cages' construction, coupled with positive pressure, which allowed small sources of debris to quickly accumulate. "The first time I cracked the food tray an eighth of an inch," said Thornton, "it was as if you had fired a gun with material that blew out." George Fichtl remarked on 2 May that vigorous motions by the 'healthy' squirrel monkey was a likely cause, the increased turbulence overwhelming the cage's capabilities. "The later analysis was that primarily it was food, though there may have been some contaminants in it," admitted Gregory. "It was a passing issue; not something that would have caused any disruption in the current activities."

But it hardly made for a conducive living environment. "Be advised we now have faeces in the crew compartment and it isn't much fun, guys," reported Overmyer, after a piece of animal dung floated right under his nose on the flight deck. "How many years did we tell them these cages would never work?" The crew donned surgical gowns, gloves and masks to remove the debris, although Thagard found even this to be inadequate. Thornton remarked that the animals' food bars tended to crumble, scattering clouds of particulates from the RAHF and into Spacelab's air. Only the presence of a seal over the cage, Thornton mused, might solve the problem.

Initial reactions on the ground were that the crew had 'confused' the debris and Overmyer had witnessed food pellets, not faeces. But Thornton disagreed. "I can absolutely tell you," he said, "that as a boy that grew up on a farm with chickens, pigs and so forth, I knew at the age of three years old [the difference between] faeces and food pellets."

On the ground, it was a big news story. One newspaper cartoon depicted a Shuttle astronaut telling his crewmate: "I'm not upset, I'm just glad we didn't have elephants on board!" But behind the humour, such issues required resolution before the RAHF could fly the Spacelab-4 life sciences mission. After landing, the animals were in good spirits and strikingly calm when handled, although the rats had a lot of dried urine and food powder on their coats. This was caused by variable airflow-rates in their cages, which prevented urine, faeces and food residue from being deposited in waste trays.

Spacelab-3 demonstrated that the RAHF was suitable for animals. But time was of the essence to fix issues pertaining to food, faeces and odours. NASA hoped to fly at least one RAHF on Spacelab-4, housing 24 (or even as many as 48) rodents and transferring them, in Spacelab's main work area, to a General Purpose Workstation (GPWS). This made containment of particulate debris essential. In the aftermath of the Challenger disaster, ARC modified the RAHF and a 12-day 'biocompatibility' test in August 1988 verified multiple improvements. Its ability to contain debris and deal with odours and micro-organisms were identified as key issues. A single-pass auxiliary fan was added to assist its environmental control system. Tests in March 1989 verified that Spacelab-3's problems were overcome. And when the dedicated life sciences mission eventually flew in June 1991 (commanded by Bryan O'Connor), tests confirmed that the hardware could indeed capture crumbs, flecks of rodent hair and faeces—simulated by black-eyed peas—with no noticeable contaminants. Moreover, when the crew moved rats from the RAHF to the GPWS, scientists could observe their behaviour and performance outside their cages.

SHIFTWORK AND *SEPPUKU*

It should be remembered, of course, that Spacelab-3 was a test flight of the RAHF. The mission's main 'operational' focus—the "big hitters", wrote Croft and Youskauskas—was fluid physics and crystal growth and the two payload specialists were internationally recognised experts in their fields. Wang ran his drop dynamics experiment whilst van den Berg focused on the growing of crystals. "One shift worked, the other slept," said Gregory of their shiftwork. "We had enclosed bunks on the middeck of the orbiter and that's where the 'off' shift would sleep, so we never saw them, really." Those bunks were ideally suited for their quietness. "Weren't nearly the complaints of backache and insomnia that you get on many flights," Thagard added (Fig. 5.5).

"There was a handover period, but once we began working, they were sleeping and we just wouldn't see them," continued Gregory. "There was a

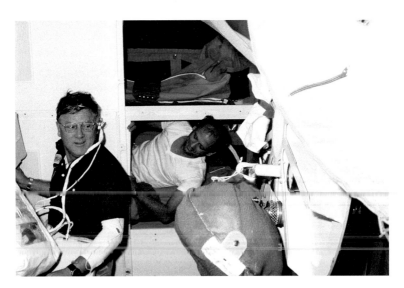

Fig. 5.5: Once the gold and silver shifts began operating, the two teams rarely saw each other. Here, silver-shift member Norm Thagard bales out of his coffin-like sleep station to greet gold crewmate Don Lind.

common portion of the training, and that was the ascent and re-entry, so Norm Thagard, myself and Bob Overmyer were always involved in the ascent and landing portion of the training. I'd say 75-80 percent of the training was on ascent and re-entry. The intent was to try to get us three in a kind of mindset like a ballet without music—individual, but co-ordinated activities that resulted in the successful accomplishment of these phases, regardless of the type failures or series of failures that the training team would impose on you. The intent was for us to learn this so well—understand the system so well—that we could brush through a failure scenario and 'safe' the orbiter in the ascent, such that we could get on orbit and then have time to discuss what the real problem was and correct it."

STS-51B was meant to be the first Spacelab flight to land on the concrete runway at KSC in Florida, but tyre and brake problems experienced on the previous Shuttle mission obliged NASA to opt for Edwards. "The decision will provide more safety margin for the Challenger's tyres and brake system," read a 24 April 1985 news release, "because of the availability of the unrestricted lakebed and the smoother surface. The Spacelab-3 payload will be a heavy return weight for an orbiter. The decision to land at [Edwards] for the next flight will enable engineers to determine what corrective actions are appropriate before returning to KSC for normal end-of-mission landings."

But landing was a long way off as STS-51B got underway. As Overmyer and Gregory tended Challenger, Lind and Thagard opened the hatch into the Spacelab module and set to work on almost a full day of co-ordinated ATMOS/VWFC activities. Only 19 of ATMOS' planned 72 observations were achieved when a power supply failure disabled its laser, although each data-collection period provided 150 independent spectra of more than 100,000 measurements of atmospheric constituents between the altitudes of ten miles (16 kilometres) and 175 miles (280 kilometres). The results highlighted the presence of five molecules—dinitrogen pentoxide, chlorine nitrate, carbonyl fluoride, methyl chloride and nitric acid—whose existence in the stratosphere had hitherto only been suspected. Analysis of the lower mesosphere showed it to be more 'active' than previously supposed, with many 'minor' gases typically split by sunlight to trigger other chemical reactions. The instrument's spectrometer measured changes in the infrared portion of sunlight as it passed through the 'limb' of the atmosphere. Since each of the trace gases under scrutiny was known to absorb sunlight at very specific infrared wavelengths, it was possible to determine their presence or absence, concentrations and altitudes, by identifying which wavelengths had been absorbed from the data. Furthermore, the instrument's ability to detect trace gases at concentrations of less than one part per billion meant its data could be exploited reliably to test theoretical models of atmospheric physics and chemistry.

Human influence on the atmosphere was one of the primary reasons for the decision to build and employ ATMOS. It would fly on three Earth-watching Spacelab flights in the early 1990s. On the first, in March 1992, it examined the effects of the previous year's Mount Pinatubo volcanic eruption in the Philippines and detected large amounts of crustal material and sulphur-based aerosols in the stratosphere. Additionally, many of the Spacelab-3 science team's predictions of atmospheric change between the first and second ATMOS missions were vindicated when chlorofluorocarbon quantities were shown to have increased dramatically and their role in atmospheric photochemistry became more pronounced. When the two sets of results were compared, they highlighted an increase in inorganic chlorine levels from 2.77 to 3.44 parts per billion, together with a fluorine rise from 0.76 to 1.23 parts per billion; the latter confirmed that the primary source of the increased chlorine level was indeed from industrial chlorofluorocarbons.

Other studies of Earth focused on aurorae. By examining changes in form and motion, great insights were derived into the dynamics of the magnetosphere. During Spacelab-3, observations were conducted from closer range—in low Earth orbit—than had been possible with previous, higher-orbiting missions. Five hours of video recordings and 270 still photographs were acquired to be 'overlapped' and viewed stereoscopically. The results included features never previously seen, including views from beyond the 'sensible' atmosphere of thin, horizontal layers of enhanced aurorae. Previously considered rare, these layers were recorded on two of Challenger's three orbital passes over the aurora, eliminating earlier suspicions that ground-based observations may have been optical illusions caused by atmospheric refraction. Of 21 scheduled opportunities, 18 were achieved. The experiment proved particularly satisfying for Lind, who proposed it and was its primary operator. "Before our mission, the aurora had only been photographed by some slow scan photometers, which gives you a blurred picture, like trying to take a picture of a waterfall," he explained. "We found out that there is a different component to the mechanism that creates the aurora, involving microwaves, that was not understood before."

The second time-critical experiment for STS-51B's first flight day was the VWFC, which Lind assembled in the SAL for ultraviolet observations of very young, massive stars at one end of the celestial scale and their ageing counterparts at the other. Such observations could be more rapidly achieved than by scanning individual points and offered the additional advantages of constant comparison with the background sky and 'reference' stars and easier

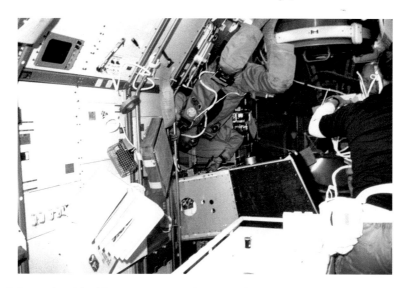

Fig. 5.6: Assisted by his crewmates, Taylor Wang labours to save the Drop Dynamics Module (DDM), which short-circuited and failed shortly after Challenger reached orbit. The drum-like Scientific Airlock (SAL) in the Spacelab-3 module's ceiling is visible near Wang's feet at the top of the frame.

interpretation. The camera yielded promising results on Spacelab-1 and should have duplicated or exceeded this achievement on Spacelab-3, but for a bent handle on the SAL. Mission Control examined photographs of the SAL and decided that a maintenance procedure by the crew would be inappropriate. This was a pity, because on its initial extension into space the VWFC flawlessly acquired its target and took a brief exposure (Fig. 5.6).

Eighteen hours into the mission, Overmyer and Gregory manoeuvred the Shuttle into her gravity-gradient attitude for six days of fluid physics and crystal growth research. Housed in a double rack on the module's starboard wall, Wang's DDM afforded an opportunity to levitate and manipulate drops in microgravity. It had already been theoretically demonstrated that space research could lead to advances in materials technology, including glasses, crystals, ceramics and alloys. But chemical mixtures proved to be highly reactive to the walls of their processing chambers and contamination levels as small as a few parts per billion could seriously degrade the final product. The DDM, said Wang, had applications in future 'containerless' materials processing methods to reduce such flaws. Certain fluoride glasses—particularly attractive for their infrared transmission properties—could be manufactured in ground-based laboratories, but imperfections induced by their containers prevented them from attaining theoretical performance levels. 'Effective' containerless processing, where acoustic and electromagnetic forces were applied to suspend and manipulate fluid drops, could only be practically achieved in space: terrestrial gravity rendered it impossible to levitate liquids without introducing forces that masked the very phenomena that fluid physicists desired to examine.

For the DDM's first flight, fluids used were water and glycerin, but when Shanghai-born Wang attempted to activate the facility during his shift on 30 April, it shorted out and failed, as circuit breakers continuously popped open. "I was the first person of Chinese descent to fly on the Shuttle and the Chinese community had taken a great deal of interest," he wrote later, "You don't just represent yourself—you represent your family—and the first thing you learn as a kid is to bring no shame to the family. When I realised my experiment had failed, I could imagine my father telling me, 'What's the matter with you? Can't you even do an experiment right?' I was really in a desperate situation."

Unlike Columbia, Challenger did not carry additional cryogenic tanks and her consumables were unlikely to stretch comfortably beyond seven days. Any time lost on failed experiments could not be made up. In his memoir, Mullane recalled the incident. "Its failure severely depressed [Wang] and he surrendered to episodes of crying, but this was just the beginning of his torture," Mullane wrote. "He turned out to be a cleanliness freak. Living aboard the Shuttle *doesn't* leave its occupants feeling springtime fresh!"

Despite his discomfort and upset, Wang gained Mission Control's blessing to repair the DDM and got to work, opening the Spacelab rack, isolating the fault and rewiring it. Several dramatic photographs showed Wang's legs sticking out into the module as the DDM rack appeared to swallow his upper body. Watching admiringly from the POCC, Mary Helen Johnston praised Wang's "good-looking legs".

He had already threatened not to come home if NASA refused to let him fix the DDM, so it proved fortuitous that his bluff was not called. "I hadn't really figured out how not to come back," Wang admitted later. "The Asian tradition of honourable suicide—*seppuku*—would have failed since everything on the Shuttle is designed for safety. The knife on board can't even cut the bread. You could put your head in the oven, but it's really just a food warmer. If you tried to hang yourself with no gravity, you'd just dangle there like an idiot!" So concerned were mission controllers for the payload specialist's mental wellbeing that on Flight Day Four a log entry titled 'Prayers for Wang' was added.

But with the DDM repaired—to whoops of delight from Wang, recorded in Mission Control's log at 77 hours and 26 minutes into the mission—there was no time for suicidal thoughts and he worked virtually non-stop to complete his experiments in the last three days of the flight, aided by his crewmates.

"You've never seen so much joy in Taylor's face," reported the crew after Wang finally got the DDM up and running. "Taylor is glowing!"

"Copy, sounds like a party going on," came the reply. "Taylor, are you going to be able to sleep tonight?"

"Oh, I don't know," replied Wang. "Champagne and dancing-girls?"

The results confirmed several assumptions about liquids in microgravity, although other findings were unexpected: the 'bifurcation point', when a rotating droplet assumes a dog-bone-like shape to hold itself together, occurred earlier than predicted under certain conditions. Another dog-bone returned to a spherical shape and stopped spinning more rapidly than anticipated, apparently from internal differential rotation. Wang positioned freely suspended liquid drops under the influence of their own surface tension and gently manipulated them with acoustic speakers inside the DDM. After a drop was observed as 'stable' and spherical, it was set into rotation or oscillation by acoustic torque or modulated radiation pressure force. In December 1986, Wang received NASA's Exceptional Scientific Achievement Medal in recognition of his "contributions to microgravity science and materials processing in space and for his exceptional contributions as Payload Specialist on Spacelab-3". His father would have been proud.

Elsewhere in the module, located in its own rack on the port-side wall, close to the module's aft end-cone, was GFFE, which simulated fluid flows and convective processes in terrestrial oceans and the atmospheres of the Sun and giant gaseous planets, particularly Jupiter. Simulations of atmospheric dynamics were first undertaken in the early 20th century, using oil and water in rotating pan experiments, but their effectiveness was limited. Supercomputers of the 1960s and 1970s offered greater advances by numerical modelling, but the only practical way to eliminate terrestrial gravity was to run the experiment in space. Before Spacelab-3, "there was a question of whether you could get convection patterns and wind distributions that resembled those on a gas giant planet," said GFFE's principal investigator, John Hart. This question was partially answered on Spacelab-3, by creating and observing 'banana cells'—rapidly rotating columns formed as differential heating was increased—which were thought to be a key feature of Jupiter's atmospheric structure. Not all phenomena could be fully investigated, due to time and film constraints, plus an inability to interact on a 'real-time' basis with the experiments. Yet GFFE operated well, completing its computer-run scenarios over 84 hours; another 18 hours of operations were also undertaken, yielding 46,000 images for post-flight analysis.

One of the principal reasons for the success of the DDM and GFFE was the high-quality microgravity environment afforded by the gravity-gradient attitude, which NASA described as "quite stable and conducive to the performance of delicate experiments in materials science and fluid mechanics". Experiments requiring this environment were clustered around the Shuttle's centre-of-gravity, from the midpoint to the aft end of the module. The French-supplied MICG shared the same Spacelab rack as GFFE and processed six cartridges of mercury iodide crystal seeds at differing pressures for 70 hours at a time, using a two-zone furnace. As with GFFE, it operated under computer control and crystals from Spacelab-3's two mercuric iodide experiments had fewer defects than their ground-grown counterparts. Typically, crystals the size of sugar cubes were grown from 'seeds' 20 times smaller. The VCGS controlled the growth process at less than 0.1 inches (3.1 millimetres) per day over 104 hours. Van den Berg concluded that vapour-driven growth could be effectively deployed in space, producing higher-quality specimens with better electronic properties (Fig. 5.7).

Two crystals of triglycine sulphate were produced, yielding the first three-dimensional laser holograms and video recordings of their growth. Visual observations by the crew provided invaluable descriptions of the slow-forming crystals, which were grown by slowly extracting heat at a controlled rate through a seed of triglycine sulphate in a saturated solution of the same substance. Variations in liquid density, solution concentration and temperature

Fig. 5.7: Engineers, technicians and scientists worked around-the-clock to support Spacelab-3 from the Marshall Space Flight Center (MSFC) in Huntsville, Alabama.

were carefully monitored. By extracting heat from the crystal, it was possible to maintain saturation at its 'growth interface', permitting slow but very uniform processing and a higher degree of perfection than could be achieved on Earth.

The astronauts viewed the crystals through a microscope and images were relayed directly to the POCC. This allowed them to be monitored across every growth phase and scientists could adjust temperature parameters to reduce defects. "Lodewijk van den Berg and I ran the crystal growing experiments, so we would brief each other on what was going on," said Lind, his counterpart on the gold team. "He'd brief me and then he'd go to sleep and when he woke up, I'd brief him on what I'd done during the last shift."

AROUND-THE-CLOCK SCIENCE

"I don't think there was competition, because the two shifts did two different kinds of science," said Gregory of relations between the silver and gold teams. "Each shift had its own area of interest and would pick up any unclosed item from the shift preceding them but would very quickly transition to the activities on orbit. There were really about four hours a day when there was an interaction between the two. During that time, it would just be a kind of status brief on orbiter problems or issues, any review of notes that had come up from Mission Control or some deviation to the anticipated checklist that we had."

Watching Spacelab-3 with great enthusiasm was India, one of whose experiments had been provided by astrophysicists at the Tata Institute of Fundamental Research. It initially refused to respond to commands and rotate its detector stack, but a procedure conducted by the crew enabled it to operate normally and it completed two-thirds of its planned observations. New data on the ionisation states of solar heavy nuclei helped to develop an understanding of the acceleration and confinement of energetic nuclei in the Sun. The experiment's detector comprised thin sheets of specialised plastics—including cellulose nitrate and lexan polycarbonate—which were efficient low-noise receptors for heavy nuclei. It was possible to ascertain the identities and energies of particles from measurements of the geometry of the tracks and through the sheets.

Oftentimes on dual-shift Spacelab flights, the only occasions that the entire crew got together were shortly after launch and just before re-entry. "It may have been anticipated that we would prepare a meal and everyone would eat at the same time," said Gregory. "In reality, that's not what actually happened."

I called it 'almost grazing'. You would go down and perhaps get a package of beefsteak and heat it and cut it open and eat it. You may stay on the middeck or you may go back up to the flight deck or you would go back into the laboratory and eat as you were doing your other routine duties."

Spacelab-3's results required years to analyse after Challenger came home and the remarkable success of the mission produced multiple reflight opportunities. But several scientists argued that one of its most significant achievements was biomedical research, most notably bone and muscle changes in the rats. And since muscle protein 'turnover' in rats is much more rapid than in humans, a week or two of microgravity exposure in their bodies was roughly equivalent to two months in ours.

Spacelab-3's biomedical research did not solely focus on the rodents and primates, but also on the astronauts—and van den Berg and Wang in particular, who served as 'subjects' for the AFT experiment, monitored by Lind and Thornton. Several techniques were used to counteract space motion sickness, including the wearing of electronic monitors to record physiological data such as sweat, pulse, heart and respiration rates. One astronaut exhibited a low heart rate and little sweating, indicating a lack of stress, although the other revealed less ability to control physiological responses and fell foul to space sickness. Nevertheless, AFT offered insights into crew workloads and behavioural responses to environmental stressors; 'baseline' information of importance when planning future flights. Spacelab-3 had proved to be a tremendous success, its scientific harvest more than ample to declare the system operational. Some 250 million bits of data were obtained from STS-51B's experiments, together with three million frames of video footage.

For the non-scientist pilots, this success proved ironically disappointing. "The only flying would be attitude adjustments," remembered Gregory of his limiting piloting role, "and those are generally keypunched in and then executed. In our training, we would simulate failures where you had to do that manoeuvre by hand, and it was quite possible to do it, but not as efficient as the automatic systems. I don't recall manually flying any of the manoeuvres in orbit and I don't recall Bob Overmyer doing it either. The only time we really put our hand on the stick was in the less-than-the-speed-of-sound descent for landing."

Overmyer guided the orbiter to a precision landing at Edwards at 9:11 a.m. PDT on 6 May. As soon as residual propellants had been drained from the orbiter, the first time-critical items—data tapes and film—were removed from Spacelab-3. Three hours later, the rats and monkeys were removed, the former heading directly for euthanasia. But Overmyer was scathing in his criticism of the RAHF. It was imperative, he argued, that improvements were made before

Spacelab-4. "NASA has a problem that NASA needs to solve if we're going to fly those cages again," he said. "I never dreamed that all that stuff would come out of those cages and escape into our atmosphere."

Watching the facility's woes on that inaugural RAHF flight was an aghast Bonnie Dalton. For her team, there were only two options. "We were told," she remembered, grimly, "you either change this, or you'll never fly again."

6

Deutschland-Eins

"NOT ENOUGH BALANCE"

Ulf Merbold was not a happy man.

West Germany's first astronaut returned from orbit on 8 December 1983, clearly impressed by Spacelab-1. He was not alone in praising Europe's achievement in building this world-class facility, which many observers regarded as nothing less than a steppingstone to a space station. Yet Merbold was dismayed that ESA—and his own homeland—received just half of that first mission. "The return to Europe should be better than it was for this particular flight," he said. "The politicians in Europe will not be able to sell European participation in the future, because there's not enough balance. The Memorandum of Understanding, signed by the Europeans and Americans, called for joint spaceflights by Europeans and Americans, not *one* European and *many* Americans. I think the Americans have to rethink this agreement to make it fairer."

After Spacelab-1, ESA figured highly in NASA's future plans. And unlike the early 1970s, when European industry lagged behind the United States and its ability to build Spacelab aroused scepticism and self-doubt (see Chapter Two), its role in a next-generation space station would come not as a mere contractor or 'lip service' associate, but as a fully-fledged partner, a partner which with Spacelab had cut its teeth and proven its mettle. When the Berlin Wall tumbled in November 1989, the reunified 'Federal Republic' of Germany continued to press its advantage as Spacelab's principal sponsor. It was no accident that when the Columbus laboratory arrived at the ISS in February 2008 as Europe's scientific showpiece, one of the astronauts who installed it was German.

But although Spacelab's 1973 Intergovernmental Agreement "contemplated" that a European crewmember would be aboard the first mission, it stressed that "subsequent flight crew opportunities" would occur in collaboration with ESA or individual member-states. Such missions would fly "on either a co-operative or cost-reimbursable basis". In April 1975, West Germany contemplated purchasing a Spacelab module for its own research under this 'cost-reimbursable' banner, which *Aviation Week* described as "a field of considerable interest" for the country. In February 1978, West Germany's research ministry made a financial deposit to reserve experiment volume on the first two Spacelab missions, but as Shuttle pricing policies soared the reusable spacecraft seemed no nearer to its maiden voyage. Hopes for a 'barter' agreement, whereby NASA would provide four 'free' Shuttle launches—including two for West Germany—in exchange for a second Spacelab flight unit, were turned down, notably by the West Germans themselves, who feared that if they could not use the spacecraft's full capabilities on 'their' missions they would be obligated to 'fill-up' the vehicle or absorb the losses.

However, in April 1981, after the Shuttle finally entered service, NASA Acting Administrator Alan Lovelace signed a Memorandum of Understanding with Hans-Hilger Haunschild of West Germany's ministry for research and technology for the first cost-reimbursable mission. Under its terms, West Germany would develop a pair of missions—one devoted to materials processing and life sciences, the other to astrophysics—and in October 1982, launch contracts worth $62 million were signed for the first flight. Designated 'Spacelab-D1' (or 'Deutschland-Eins'), it was slated for June 1985 (Fig. 6.1).

Fig. 6.1: Ulf Merbold, the first non-American astronaut to fly aboard a U.S. spacecraft, works inside the Spacelab-1 module.

But attitudes continued to swirl around perceived European 'ownership' of Spacelab. At Spacelab-D1's pre-launch press conference, questions and answers flowed, until the turn came for a particularly dogmatic member of the West German media. Ernst Messerschmid, one of the two West German payload specialists, replied in his native tongue. Sitting next to him, Commander Hank Hartsfield (who had studied German at college) caught the gist of the question but turned to Messerschmid to enquire what the journalist asked.

"He wants to know," said Messerschmid, awkwardly, "how much Germany has to pay the United States to use *their* Spacelab."

The incident underscored how sensitive the Europeans, and particularly West Germany, still felt about 'their' product. The journalist, admittedly, had taken an aggressively nationalistic stance—"We built it, now we have to *pay* for it"—but in reality, the Spacelab-D1 hardware was wholly purchased and 'owned' by the United States. "This one is *not* ours," the Spacelab-D1 program manager was forced to admit. "The United States bought this one from us. It's *theirs*."

NEW TERRITORY

Spectators at KSC's Operations and Checkout Building beheld an unusual sight on 30 October 1985, as eight astronauts from three nations headed to the launch pad. Commanded by Hank Hartsfield, the crew included Steve Nagel as pilot and Guy Bluford, Jim Buchli and Spacelab's first female astronaut, Bonnie Dunbar, as mission specialists. Rounding out the octet were West German physicists Reinhard Furrer and Ernst Messerschmid and Dutch physicist Wubbo Ockels, the latter having backed-up Ulf Merbold on Spacelab-1. Although Spacelab-D1 was funded principally by West Germany, ESA had contributed 38 percent of the mission's costs in return for flying one of its astronauts as a unique third payload specialist and, according to Croft and Youskauskas, in light of the mission's "heavy workload…to ensure that the payload objectives could be accomplished". Original plans called for a seven-member crew with two payload specialists, likely Ockels and one of the West Germans. (Furrer feared that, as the only bachelor among the trio, his candidacy might be viewed less favorably than Messerschmid and Ockels.) In the aftermath of the Challenger tragedy, safety demands would restrict future crew sizes to seven and STS-61A remains the only Shuttle flight to have launched and landed with eight people.

And like the three prior Spacelab missions, its crew operated around-the-clock, which required them to begin sleep-shifting a week or two before launch. "The red team of Jim, Ernst and I had to do a circadian rhythm shift, so for us the launch was coming near the end of our workday," said Bluford, an Air Force aerospace engineer. "While in quarantine, one team was up while the other was in bed. A new lighting system had been installed in the crew quarters to facilitate the shift in circadian rhythm. Once we got on-orbit, the blue team activated Spacelab, while the red team went to bed."

Also like the earlier missions, the starboard wall of Challenger's middeck had bunk-like sleep stations and the two shifts—the reds consisting of Bluford and Furrer, led by Buchli, the blues comprising Dunbar and Messerschmid, led by Nagel, with Hartsfield anchoring his schedule across both and Ockels assigned to neither—met during handovers at breakfast and dinner. "Each of the crew shared a sleep bunk with a member from the opposite team," said Bluford. "Only Hank had a bunk to himself, which gave him the flexibility to work on either shift." Conditions were cramped. "Middeck is our living area, sleeping area, the whole nine yards," Nagel recalled. "It would get pretty congested at handover times."

But whereas three bunks flew on earlier missions, Spacelab-D1's super-sized crew needed four. And those stations, though dark and quiet, could prove disorientating. On one shift, Hartsfield was startled by a loud commotion, as if someone was tearing up the middeck. It came from Bluford's bunk. As his crewmates slid open the privacy door, Bluford suddenly started in shock—"Oh…"—then, half-asleep, hastily tugged the door shut and resumed snoring. "Apparently, he had awakened and didn't know where he was," said Hartsfield. "He had a little claustrophobia…and he was completely disorientated." Dual-shifts on the Shuttle, the commander concluded, were "a little confusing, if it's new territory for you".

Shift handovers typically ran for a few minutes. "We had a flight plan to follow…that Guy and I would hand off that would have notes in it, plus we had our notebooks," Dunbar remembered. "It could be ten, twenty minutes. They were actually very efficient. I'd tell him where we were on the timeline." Breakfast and dinner were rehydrated, heated and gobbled quietly, to avoid disturbing their sleeping shift-mates. But there was no competition between the teams. "We were working different things," she added. "Competition is against the clock. Time is money and it's not in our interest that we look at the clock, it's in the interest of all the investigators on the ground. It's very important that you're well trained and understand their objectives, but we're just part of the research team."

Spacelab-D1's 76 experiments almost entirely emphasised life and materials science for West Germany and even Challenger's launch at midday EST ("banker's hours," Dunbar joked) was carefully timed, in Bluford's words, for "maximum TV coverage to Germany". In February 1984, Bluford, biomedical engineer Dunbar and Steve Nagel were assigned to the mission, initially designated STS-51K. And the following August, the other five crewmembers were announced, with launch anticipated on Columbia on 14 October 1985. As Spacelab entered maturity, it was becoming commonplace to assign 'science' crewmembers ahead of 'flight' crewmembers to iron out payload-related issues with principal investigators. Two other astronauts—Terry Hart and Bill Lenoir—were reportedly considered for seats on Spacelab-D1 but opted to leave NASA; both men disliked the notion of years preparing to fly the mission. ESA's own Claude Nicollier was also an early Spacelab-D1 mission specialist candidate, hinted in the STS-9 press kit.

Dunbar's assignment as the first woman on a Spacelab mission brought back unpleasant memories of the difficulties she endured as a young engineer. Many of the West German life sciences experiments were not intended to include female blood, there being concerns that it might ruin their statistics. "I was actually told, in front of my face, but any time you're at the point of the pathfinder, there's going to be things happening," Dunbar said. "I was told that maybe NASA had done this intentionally, to offend the Germans, by assigning a woman." Other Spacelab-D1 experiments included the Vestibular Sled, which did not fit her petite body size. Eventually, George Abbey, JSC's director of flight crew operations, insisted that the West Germans redesign their equipment to cover a percentile 'spread' that included Dunbar. She would go on to develop exceptional working relationships with Furrer and Messerschmid. The pair were selected for Spacelab-D1 in December 1982 and both were university professors, Furrer at Berlin and Messerschmid at Stuttgart. "That's a very well-esteemed position in Germany," Dunbar said. "It's better than being president of a company if you're *Herr Professor*." Serving as backup to all three payload specialists was Spacelab-1's Ulf Merbold, who acted as crew interface co-ordinator for the mission.

"Our primary training was conducted at Porz Wahnheide, a small, very picturesque town, south of Cologne," said Bluford. At the site were mockups of several Spacelab-D1 experiments. These included Werkstofflabor ('materials-laboratory'), a multi-purpose, multi-user facility for materials processing with three furnaces—a mirror heating facility, gradient heating facility and isothermal heating facility—and a fluid physics module, crystal growth unit, cryostat and high-temperature thermostat to investigate semiconductors, fluid-boundary surfaces and heat-transfer phenomena. The Prozesskamer ('process-chamber') examined flows, mass transportation, heat and temperature distribution during melting and solidification of materials. Biowissenschaften ('life-sciences') housed three life sciences experiments: a small 'botanical garden' with corn and lentils, a study of frog larvae and an investigation to measure internal pressures of the astronauts' eyes. Finally, Biorack was an ESA-provided life sciences facility with a pair of incubators, a freezer and a sealed glovebox for biological research.

Costing $180 million (inclusive of the $62 million Shuttle launch fee), it took five years to prepare Spacelab-D1 from concept to flight and the mission was managed by the Deutsche Forschungsanstalt für Luft- und Raumfahrt (DFVLR, the Federal German Aerospace Research Establishment) on behalf of the Bundesministerium für Forschung und Technologie (BMFT, the German Federal Ministry of Research and Technology). The Spacelab long module and MPESS truss structure housed 76 experiments, over half of which came from West Germany and the remainder from the United States, Spain, the United Kingdom, the Netherlands, Belgium, France, Switzerland and Italy. Notably, the module used for this mission represented the first flight of the second Spacelab flight unit, ordered by NASA at a cost of $183.96 million in January 1980. It arrived at KSC aboard a C-5A Galaxy airlifter on 1 May 1985 and when fully outfitted with experiments, utilities and the tunnel/MPESS, Spacelab-D1 tipped the scales at 30,541 pounds (13,850 kilograms). By the time the module reached Florida, the mission had been redesignated STS-61A and reassigned to Columbia for launch on 16 October 1985 (Fig. 6.2).

In addition to their Werkstofflabor, Prozesskamer and Biowissenschaften training at Porz Wahnheide, the science crew also trained at MIT in Cambridge, Massachusetts, to operate the Vestibular Sled, which was mounted in the module's centre aisle. Driven by an electromotor and traction rope, the sled included a seat for an astronaut subject and moved backwards and forwards with precisely adjusted acceleration on rails to support a multitude of experiments to understand the human vestibular and orientation systems and adaptation processes in space. Acceleration of the astronaut subjects was combined with thermal stimulations of their inner ears and 'optokinetic' stimulations of their eyes.

"Translating that training into procedures…is something I learned to do fairly well as an engineer," said Dunbar. "How do you capture the intent of a researcher and translate that into an operational procedure that several crewmembers with many different backgrounds have to be able to read…and come out of the other end of that with the right scientific result? Or…be able to use human judgement to determine if something is not following the way

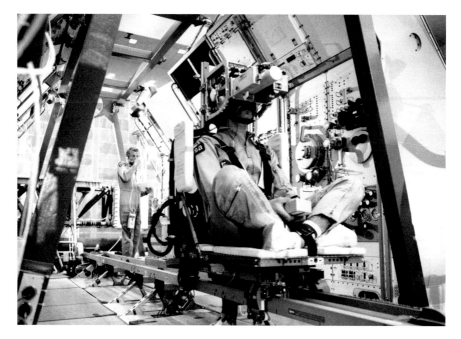

Fig. 6.2: Wubbo Ockels trains on the Vestibular Sled in a ground-based mockup of the Spacelab-D1 module, as crewmate Reinhard Furrer looks on.

you thought it was going to follow and know that you have to contact the investigator on the ground."

One important facility was the Materials Science Experiment Double Rack for Experiment Modules and Apparatus (MEDEA), which had three furnaces: one for long-duration crystallisation, another that used 'directional solidification' to process metallic crystals at extremely high temperatures and a third with a high-precision thermostat to examine the behaviour of metals under carefully controlled conditions.

Situated outside the module, atop the MPESS, was the Navigation Experiment (NAVEX), which included a pair of canisters and an antenna to test a precise clock synchronisation to evaluate highly precise, one-way distance measurements and position determination, and the Materials Experiment Assembly (MEA). The latter examined atomic diffusion and transport processes in various liquid metals.

With 76 experiments from nine nations, the crew trained at multiple locations, ranging from the European Space Technology Centre (ESTEC) at Noordwijk in the Netherlands and the University of Paris in France to the Universities of Tubingen and Bremen in Germany. "Bonnie and I spent about six months in Europe—three to four weeks at a time—training on the experiments," recalled Bluford. "Our astronaut office scheduler and training co-ordinator ensured that the Europeans had a training plan and syllabus for us to work from for this flight." In addition to performing experiments, they were also test subjects. And Bluford and Dunbar took time to learn German, too.

INSPIRATION AND HOPE

Although previous Spacelab missions were jointly overseen by teams in Houston's Mission Control and the POCC, this flight was notable in that its payload operations control center was not in Texas, but at the German Space Operations Centre (GSOC) at Oberpfaffenhofen, on the southwestern outskirts of Munich. There, an estimated 250 scientists monitored the mission. This made STS-61A the first Shuttle flight to be 'controlled'—at least from a payloads standpoint—from outside the United States. And the result made for some interesting radio transmissions, with the crew calling not "Houston, Challenger" when contacting the POCC, but rather "Munich, Challenger" or, properly, "München, Challenger".

The new setup worked exceptionally smoothly, although GSOC's limited data-transmission capabilities meant that several functions had to be managed from Houston. Moreover, due to the presence of only one operational TDRS satellite in geostationary orbit, Spacelab-D1 received communications coverage for around 30 percent of each 90-minute orbit and the Intelsat V satellite was used in tandem to relay data firstly to a station at Raisting in Bavaria and thence to Oberpfaffenhofen via microwave link. By the time the second German mission, Spacelab-D2, took place in April 1993, four TDRS were fully operational and the communications architecture was far smoother.

According to Nagel, training in Germany did not differ substantially from the United States. "You could say it's more complex and there are more issues to be resolved when you're working an international program," he said. Spacelab-D1 had no U.S. mission manager, instead being headed by a West German leadership team of Mission Manager Hans-Ulrich Steimle and Operations Manager Hans-Joachim Panitz of DFVLR and Mission Scientist Peter Sahm of the Institute of Technology in Aachen. "Not having a U.S. mission manager made it more complex, but I see that mission was an early lead-in to the space station," Nagel said. "It was hard for Hank to pull together and complicated when you're dealing overseas. We got along fine with the Germans, but we butted heads about things and the long-distance parts made it more complex."

One unforeseen incident surrounded a West German documentary being filmed about the mission during one of the crew's visits to Oberpfaffenhofen. "Things are different in Germany about drinking," remembered Hartsfield. "To have wine and beer at lunch is a common thing. In the basement of the control centre, they've got beer machines. The flight controllers had all gone down and got a beer and here is this crew—all of them with a beer—sitting on console, eating lunch."

After the training, Hartsfield caught up with Steimle.

"Hans-Ulrich," he began, "I know how things are here in Germany, but you're filming for posterity here. If this film goes outside of Germany, some people may not understand your flight controllers drinking on duty."

Steimle nodded and walked away. After that, none of the controllers drank beer again on console. And Hartsfield, by his own admission, "became *very* unpopular".

Inspiring West German youth and connecting the country was an unexpectedly important offshoot of Spacelab-D1. "The mission manager at that time told me that they hoped to use this mission not only to advance their science and human spaceflight," said Dunbar, "but to inspire a generation of young people…that really hadn't had inspiration since the war. So it had very much a…political flavour to it, not just a scientific flavour."

Despite the flight's common language currency being English, on a few occasions German was spoken over the airwaves, including an opportunity for Messerschmid and Bluford to talk to the minister-president of Bavaria, Franz Josef Strauss. "The conversation was conducted in German, with Ernst doing all the talking," said Bluford. "Although the mission's dialogue was conducted primarily in English, infrequently the payload specialists would revert to German during on-orbit discussions" (Fig. 6.3).

Years later, Hartsfield remembered this decision as controversial, with the West Germans adamant that their language ought to be spoken whilst over Bavarian ground stations. "I opposed that for safety reasons," said Hartsfield. "We can't have things going on in which my part of the payload crew can't understand what they're getting ready to do. It was clear upfront: the operational language *will* be English. We finally cut a deal…that in special cases, where there was real urgency…we could have another language used, but before any action is taken, it has to be translated into English so that the commander or my other shift operator lead and the payload crew can understand it."

Fig. 6.3: Challenger launches on her last fully successful mission on 30 October 1985.

SUSPENDED IN A GONDOLA

By Halloween, the second day of the STS-61A mission, Spacelab-D1's medical and biological experiments were progressing far more smoothly than their materials science counterparts. A problem with MEDEA's pressure sensor was corrected via an in-flight maintenance procedure and a faulty lamp on the furnace was replaced. Unfortunately, the result was several lost hours of 'run-time' on the facility. Another repair was conducted on Werkstofflabor's fluid physics module, one of whose valves had been positioned at an incorrect setting. The crew tried to reset it using a wrench, but initial efforts proved fruitless, leading to authorisation to cut away a plastic cap and a successful fix was effected.

With early plans to fly STS-61A on Columbia—with her additional cryogenic tanks—delays bringing the fleet's oldest orbiter back online after a protracted period of modification meant Spacelab-D1 was reassigned to Challenger, whose consumables for a longer flight were more limited. Discussions to extend STS-61A from seven days and 111 orbits to eight days and 127 orbits were turned down, primarily because the Spacelab module's power levels could not be reduced substantially enough to eke out another 24 hours.

Results from ESA's Biorack offered striking evidence of the influence of low gravity upon bacteria, unicellular organisms and white blood cells. Fourteen cellular and developmental biological investigations were aboard Spacelab-D1 and the mission marked the first time that specimens were 'fixed' for preservation and post-flight analysis. Two experiments confirmed observations that bacteria tend to reproduce more rapidly in space than on Earth, suggesting that astronauts might be exposed to higher risks of infection. An investigation featuring the common pathogenic organism *E. coli* demonstrated its increased resistance to antibiotics. Other bacteriological studies indicated that some cells exchanged genetic material through physical 'bridges', perhaps leading to novel methodologies to introduce human genes into bacteria cells to synthesise useful pharmaceuticals.

Vestibular experiments involving humans and tadpoles were also conducted, the latter exhibiting a pronounced alteration in swimming behaviour. The tadpoles swam in small circles, around fixed centres, until their behaviour returned to normal a few days after returning to Earth. Later examinations of the morphology of their vestibular gravity receptors revealed no structural deformities, indicating they developed normally in space. This was consistent with earlier studies of amphibians and rodents. Seedlings developed roots which curved in odd directions. And the *Bacillus subtilis* bacteria exhibited a higher growth-rate than expected. ESA's Vestibular Sled accelerated its occupants at up to 0.2G along the length of the module. Bluford recalled that although the rapid back-and-forth shuttling motion of the sled "looked very provocative from the spectator's point of view…it was actually very benign from the rider's point of view".

This remarkable mission had gotten underway at high noon EST on 30 October 1985, executing a 135-degree roll program manoeuvre—"considerably larger" than earlier flights, noted Hartsfield—to achieve an orbit 200 miles (320 kilometres) high, inclined 57 degrees to the equator. From the commander's seat, Hartsfield had good reason for having anticipated favourable weather; the previous night he observed spiral bands of clouds and thunder over the Cape, usually an indicator of satisfactory conditions for the next morning. Three and a half hours into the flight, Dunbar and Nagel started activating the Spacelab module; Dunbar remarked on its similarity to the simulator and the ease of translating down the tunnel, which she compared to a silo.

Neither Hartsfield nor Ockels were assigned to either shift but tended to align their workdays with Nagel's blue team. According to Dunbar, Ockels 'bridged' the two shifts, middling between day and night; they nicknamed him 'the purple shift'. "Wubbo tended to freelance," said Hartsfield. "His shift would overlap the other two. It was kind of a weird arrangement. He chose to sleep in the airlock. He had a sleeping bag—a design of his own—and the only trouble was that people going back and forth would bump him as they went through there." As well as a transit-route to the Spacelab module, the tunnel was used for storage, but one of its stores drew particular attention from the rest of the crew. In honour of his Dutch homeland, Ockels brought along a large bag of gouda cheese in his personal food allowance, which he taped to one of the tunnel's walls. "It was so convenient that anybody that went there…reached in," recalled Hartsfield. "About the second or third day, Wubbo was upset because two-thirds of his cheese was gone!" (Fig. 6.4).

Fig. 6.4: Wubbo Ockels climbs into his self-designed sleeping bag.

Like Spacelab-3, Challenger was oriented in a gravity-gradient attitude to provide quiescent conditions for her materials processing and fluid physics experiments. "There's a little bit of atmospheric drag, even at those altitudes, and there's a gravity effect from one end of the Shuttle to the other, which will cause it to change attitudes, so you get it in a stable attitude before you turn the jets off," said Nagel. "This is interesting, because usually you want the long axis pointed at the Earth, either tail-to-Earth or nose-to-Earth, and the wing oriented in some way that it'll be fairly stable. And we would get it in this attitude, which was nose-to-Earth, and the right wing pretty well forward. You 'slide' along like that and get it all stable and turn off the jets, and it would just stay there. It would slowly wander around a little bit and roll over a long period, like half-hour or so, kind of oscillate."

The result was like being suspended in a gondola. "We flew that attitude for eight or ten hours a day and the other time we were called -ZLV, which is 'top-of-the-Earth', with tail forward," said Nagel. Years later, he remained unconvinced as to how useful the gravity-gradient attitude really was for highly sensitive experiments. Nagel was the only crewmember to fly both Spacelab-D1 and Spacelab-D2 and on the second mission a gravity-gradient attitude was not selected. "I don't think you could tell the difference that the microgravity was significantly better when we had the jets turned off," he said, "because on the next mission, we didn't do that."

Activation of Spacelab-D1 was complete by five hours into the mission and towards the end of the first workday of Buchli's red shift, 73 of the 76 experiments were up and running. Perhaps demonstrative of the monotony of Spacelab flights (at least, for the pilots), Nagel's job involved purging fuel cells, dumping wastewater, photography and preparing meals. "But the good thing about the mission was the high inclination," he said. "We flew 57 degrees, which means you cover most of the inhabited part of the world. It was just a bonanza of Earth observations. We shot all of our film." For Hartsfield, the comparatively relaxed pace for the 'orbiter crew' allowed the pilots to capture 3,300 photographs.

Dunbar also remembered her occasional forays out of the Spacelab and back down the tunnel to the middeck, which always seemed darkened to ensure fitful rest for the off-duty shift. Floating through the access hatch to reach the flight deck, she invariably caught sight of the (thankfully clothed) backsides of Hartsfield or Nagel, positioned awkwardly in one of Challenger's overhead windows, shooting photographs. "Usually had this view as I came up," she told the post-flight press conference, to which her immediate remark to the embarrassed pilots was "Hi, I'm here!"

One of STS-61A's few problems was a persistent cabin leak, which triggered alarms on several occasions. "We discovered, later on, the leak was due to one of the experiments inadvertently venting into space," said Bluford. "We also had a false fire alarm go off on us during flight." Nevertheless, some time was granted to each spacefarer simply to gaze upon the home planet; particularly the payload specialists, for whom the opportunity to fly in space would come only once.

"We were flying into darkness, passing over Tasmania and heading down toward Antarctica," Buchli recalled. "The southern aurora was just unbelievable! It looked like an octopus sitting over the South Pole, with tentacles of light coming out. The orbiter was flying upside down, with the nose pointing toward the pole, and the tentacles shimmered a fluorescent blue-pink. It was like the whole nose was bathed in aurora. Even though we were much higher, you could still see the glow off the front of the nose. I knew what was coming, because I had seen the same geometry when we passed over the pole the day before."

Buchli floated downstairs to Challenger's middeck, grabbed Furrer, pulled him up to the flight deck and shoved him, face-first, into the windows. For ten minutes, with switches and circuit breakers bristling against their chests, the two men watched silently as shimmering auroral bands cavorted around the South Pole.

Then, Furrer spoke.

"Jim, that was fantastic," he breathed. "That was the most beautiful thing I've ever seen." But the clock was forever their enemy. The timeline took precedence. No sooner had those words left his mouth, Furrer checked his watch and hurried back to work (Fig. 6.5).

Watching the aurora and these fleeting periods of momentary reflection afforded some peace for the crew as their busy mission entered its home stretch. And as the Shuttle neared the end of the fourth dedicated Spacelab flight, Bluford's description of their return to Earth on 6 November 1985 seemed almost businesslike. "We closed up Spacelab and readied the vehicle for re-entry as the blue team was getting up. I rode upstairs in the cockpit, next to [Buchli] as we came home. Hank and Steve flew us home to a safe landing at Edwards Air Force Base in California."

Little could they have known, but Challenger had just concluded her last successful flight. Two months later, she was lost, seconds after liftoff, along with her entire crew and the Shuttle fleet was grounded for almost three years. And although the Shuttle and Spacelab would fly more missions after the tragedy, neither would ever truly be the same again. An era of innocence had ended, but a Golden Age lay ahead.

Fig. 6.5: Reinhard Furrer and Bonnie Dunbar are pictured at work inside the Spacelab-D1 module during blue-shift operations.

"YOU WOULDN'T BELIEVE ME"

One late spring day in 1993, a shopper in a Houston supermarket parking lot received an unusual call from a friend via hand-held amateur radio. At first glance, such a call might not appear out of the ordinary, but for one thing: the shopper's friend was Steve Nagel, commander of Shuttle Columbia. Before the STS-55 flight, Nagel arranged to call his friend, if time permitted. "I gave him the number in Mission Control to call a contact back there to update that time," he said. Mission Control enabled the brief communique and the pair spoke for a few minutes, before the Shuttle passed out of range. By now, the unusual nature of the call had piqued the interest of fellow shoppers and a small group of listeners gathered.

"Who were you talking to?" one of them asked.

The friend grinned. "If I told you, you wouldn't believe me!"

In fact, they had been talking about a mission which, for the reunified Federal Republic of Germany, was an exceptionally expensive scientific venture, costing more than a half-billion dollars to bring to fruition. Columbia carried a Spacelab long module and Unique Support Structure (USS) in her payload bay and was loaded with 88 experiments in life sciences, radiation physics, materials science and technology. "I'm not fluent in German," said Payload Commander Jerry Ross, the lead mission specialist in charge of payload activities. "Fortunately, most of the international science people work in

English anyhow, because you've got all the other languages in Europe that have to find some common language and, fortunately, *that's it*. I tried to learn some German, but I'm not good in foreign languages to start with and trying to do all the other things I was doing there wasn't time to learn much German."

Designated 'Spacelab-D2'—'Deutschland-Zwei'—it was the second mission sponsored by the former West Germany. Even as Spacelab-D1 flew in late 1985, plans were already afoot for several follow-on flights. Spacelab-D2, identified as a "dedicated application and technology science mission", was formally requested in June 1984 as a payload opportunity for emplacement into the flight manifest for launch in July 1988. But by the time the first post-Challenger manifest appeared in March 1988, Spacelab-D2 had moved to STS-52, then planned as a nine-day mission aboard Columbia with a seven-person crew in December 1991, its 186-mile (300-kilometre) orbit inclined 44 degrees to the equator. Two years later, in the January 1990 manifest, it had slipped again, this time onto Shuttle Endeavour and STS-55, scheduled for May 1992 with a revised orbital inclination of 28.5 degrees.

In a sense, these flights were intended as precursors for long-duration space station expeditions, although Spacelab could only remain in orbit for a week or so; Nagel described them as miniature marathons. "You want to load them up as much as you reasonably can, because you want to get as much for your money, but they're like sprints," he said. As the only astronaut to fly both German Spacelabs, he experienced firsthand the feverish, round-the-clock nature of such missions.

Ross, who was named payload commander for Spacelab-D2 in April 1991, agreed that training was excessively hectic and "not a very viable way to do business". And Ross' assignment was unusual as he was not a physician or a materials scientist or a physicist, but an Air Force mechanical engineer. When Chief Astronaut Dan Brandenstein asked him to be Spacelab-D2's payload commander, the request gave Ross pause.

"Dan, you want a scientist for that," he said. "You don't want me, an old engineer."

"No, we want you," retorted Brandenstein. "We want somebody that will work well with the Germans and get the flight pulled together."

Payload commanders originated in February 1988, when six NASA astronauts (including Ross) created a science support group for direct interaction with prospective experimenters on future Shuttle missions. "The group believes that increasing crew involvement in the design, development and operation of experiments will improve the return of data and simplify repair of equipment in space," NASA explained. "This is particularly important in maximising the scientific return from each experiment." The first payload commanders were assigned to Spacelab missions from January 1990, their remit identified as "long-range leadership in the development and planning of payload crew science activities" and "overall crew responsibility for the planning, integration and on-orbit co-ordination of payload/Space Shuttle activities". In essence, the enormously complex Spacelab missions demanded a single point of contact among the mission specialists between NASA leaders and the scientific community. Payload commanders also helped to "identify and resolve training issues and constraints prior to crew training" and their role was "expected to serve as a foundation for the development of a space station mission commander concept".

Ross described his duties on Spacelab-D2 as the most demanding of his astronaut career. "The payload commander is basically the guy that's responsible for interfacing with the payload sponsors and the crew to make sure that what the payload sponsors want to happen on-orbit are things that the crew can physically do, both from the interfaces to the payloads, the checklists and the timeline. I had to do all the co-ordination, all the dealings—*everything*—from the safety community, the medical community, the science community. I had to work all that in addition to trying to get three rookies ready to go fly on 90 very different, very complex experiments" (Fig. 6.6).

Fig. 6.6: The Spacelab-D2 pressurised module is pictured in Columbia's payload bay, awaiting the installation of the tunnel adapter.

Pulling together such an intricate flight was easier said than done, for Spacelab-D2's multitude of experiments imposed severe timing constraints and Ross found that just keeping up with the mission timeline proved challenging. Like Spacelab-D1, the flight had no U.S. mission manager, but rather a German one: Hauke Dodeck. "One of the things I learned fairly late in the training flow was the fact that D1 and D2 Spacelab missions were the only flights flown that did not have a NASA mission manager," Ross recalled. "I wondered why I was struggling so hard and having to do so many things myself. *That* was why."

An added complication was that although the mission was run by the Deutsches Zentrum für Luft- und Raumfahrt (DLR, the German Aerospace Centre, previously DFVLR) and the Deutsche Agentur für Raumfahrtangelegenheiten (DARA, the German Agency for Space Flight Affairs), it included substantial external collaboration. Sixteen experiments were financed by DLR, alongside 21 others from ESA and NASA, France's CNES and Japan. "The way the flight was set out, the Germans bought the flight, but to offset their expenses, they had sold back a lot of the research time and space to the U.S. and they'd also sold quite a bit of it to ESA, but yet they wanted to do all the things that they decided that they wanted to do in the first place," said Ross. "The flight was well overbooked in terms of the number of experiments and the number of hours required and it was, at times, more of an adversary-type of arrangement than I wanted it to be. But sometimes I had to take on the attitude that I'm not going to let the crew fail." Often, this required difficult judgement calls, refusing experiments if procedures were not ready, if the crew had not trained on them or if they were improperly safety-reviewed. "And that was very tough for me, because I like to work as a team," he added. "I'm always a positive kind of guy and want to make things happen, but at the same time there was a couple of points in there where we just had to lay down the law."

Already training for two payload specialist seats were meteorologist Renate Brümmer and physicists Hans Schlegel, Gerhard Thiele and Ulrich Walter. They were selected in August 1987 by DLR and in July 1989 the United States and West Germany agreed "to send two German astronauts and German scientific payloads" on Spacelab-D2. A year later, in October 1990 Brümmer, Schlegel, Thiele and Walter were named as candidates for Spacelab-D2. In August 1991, NASA physician astronaut Bernard Harris was selected as a mission specialist and the following February the remainder of the crew was in place with the announcement of Nagel as commander, Tom Henricks as pilot, Charlie Precourt as flight engineer and Schlegel and Walter as payload specialists, backed-up by Brümmer and Thiele. By this stage, launch had moved from May 1992 to February 1993.

The payload was heavily biased in favour of life sciences, as well as being longer than Spacelab-D1: nine days over seven. Targeting an altitude of 186 miles (300 kilometres), inclined 28.45 degrees to the equator, STS-55 would fly for 214 hours and 143 orbits. Its life sciences focus was aided by Harris' presence. "The Germans had requested a medical doctor mission specialist and since I wasn't one, they were certainly hoping that the next one would be," recalled Ross. "While I didn't personally think that a medical doctor was mandatory, I did think that it was not a bad idea."

Another person the Germans hoped would be aboard was Spacelab-D1's Nagel. "I was hoping he may come back and we could fly together, so I could harass him some more," Ross joked. "The system thought that would be a good thing to do to have that kind of continuity and he came back as commander." Nagel had already heard similar noises. One day, in the months before February 1992, he bumped into Hauke Dodeck. "Hey," the German said. "We'd really like to have you on this mission." Years later, Nagel wondered if NASA brass intended him for Spacelab-D2 all along, but in any case, he "went and volunteered for it, even though it's not as glamorous a mission as some", adding that he always enjoyed working with the Germans.

DEUTSCHLAND-ZWEI

Spacelab-D2 might have stayed aloft beyond nine days, ideas having been floated for a two-week mission. But as circumstances transpired, the Germans turned down the option of a lengthened mission in favour of the USS and four external payloads.

The Materials Science Autonomous Payload (MAUS) comprised a pair of experiments to investigate complex boiling processes and the diffusion phenomenon of gas bubbles in salt melts. It processed 16 individual specimens in four separate furnaces, attaining temperatures as high as 1,250 degrees Celsius (2,300 degrees Fahrenheit). The Atomic Oxygen Exposure Tray (AOET) exposed 124 samples—cut into circular or rectangular pieces—to assess the harsh reactive and erosive effects of atomic oxygen and cosmic radiation in low-Earth orbit on candidate materials intended for a future space station. These investigations spilled into the pressurised module, with 'biostacks' of plant seeds, insect eggs and bacterial spores, radiation detectors worn by the astronauts and containers of biological specimens emplaced in areas of Spacelab where the substantiality of radiation shielding differed.

The Galactic Ultra-Wide-Angle Schmidt System Camera (GAUSS) acquired wide-angle ultraviolet images of the Milky Way galaxy, young stars and gaseous nebulae. With a field-of-view of 145 degrees, it was hoped that GAUSS would capture a hundred or more images during STS-55. The final USS payload was the Modular Optoelectronic Multispectral Scanner (MOMS), which gathered high-resolution imagery of Earth's surface in furtherance of remote-sensing applications. MOMS' capacity to discriminate between differing classes of vegetation, rock and soil cover. During Spacelab-D2, it imaged irrigation ditches, rice crops and roads in Vietnam, underwater reefs off the Egyptian coast and reddish-pink 'outgassing' from Mexico's Colima volcano. The astronauts complemented MOMS with their in-cabin photography, using hand-held Hasselblad and Linhof large-format cameras. This revealed the redistribution of ash from the Lascar volcano in the Chilean Altiplano, which erupted only days before Columbia's launch.

Inside the pressurised long module was a raft of research spanning six disciplines: materials science in the areas of fluid physics, nucleation and solidification, as well as biological sciences, technology, Earth observations, atmospheric physics and astronomy. Of the 88 experiments, almost a quarter were devoted to materials science. In many ways, Spacelab-D2 was a carryover from Spacelab-D1, again flying MEDEA and Werkstofflabor, plus the new Holographic Optical Laboratory (HOLOP). MEDEA's three furnaces—an elliptical mirror furnace, a gradient furnace with quenching and a high-precision thermostat—supported ten experiments in long-duration crystallisation and high-temperature directional solidification, carefully controlling the temperatures of metallic samples. During the mission, the largest crystal of gallium arsenide ever grown in space, measuring 0.8 inches (20 millimetres), was produced using MEDEA; such materials have Earth-based applications in the electronics industry to build light-emitting diodes, semiconducting lasers, photo-detectors and high-speed switching circuits. Other materials under study included gallium-doped germanium and alloys of copper-aluminium, copper-gold, aluminium-lithium, copper-manganese and aluminium-silicon (Fig. 6.7).

Werkstofflabor deployed five furnaces—an isothermal heating facility, a turbine blade facility, a gradient heating facility, an advanced fluid physics module and a high-temperature thermostat—in support of 20 experiments in metals processing, crystal growth for electronics applications, fluid-boundary surfaces and transport phenomena. Notably, its turbine blade facility was used to process specialised metallic alloys and cast them into the shape of turbine blades, part of efforts to create future aircraft structural parts, highly resistant to heat and stress. Materials studied in Werkstofflabor included alloys of

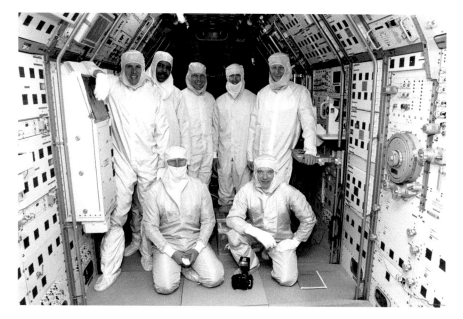

Fig. 6.7: The STS-55 crew, clad in clean-room garb, inspects the Spacelab-D2 facility, prior to final closeout for launch. Standing, left to right, are Ulrich Walter, Bernard Harris, Hans Schlegel, Steve Nagel and Tom Henricks. Kneeling are Jerry Ross (left) and Charlie Precourt.

nickel, aluminium-lithium and aluminium-silicon and compounds of indium-antimony, with scientists observing convection flows, oscillations in fluids and the heating/cooling/melting process.

Results from Werkstofflabor included a 'monotectic' alloy of bismuth-aluminium. The newest facility, HOLOP employed 'holography', using laser light to create three-dimensional imagery and render more easily visible the physical processes of heat and mass transfer and cooling in transparent materials for use in metallurgy and casting. One of its four experiments demonstrated 'telescience', offering ground controllers an ability to command instruments in space for future space station tasks. Commanding was overseen from DLR's microgravity life support centre at Cologne-Porz. All told, over 600 telescience commands were transmitted to HOLOP during the mission.

Also benefitting from telescience was the Robotic Technology Experiment (ROTEX), a six-jointed robotic arm, equipped with tactile and torque sensors, laser range-finders and stereo and fixed cameras. During Spacelab-D2, it assembled a small 'tower' of cubes and retrieved, connected and disconnected an electrical plug to demonstrate its autonomous functionalities. Operating within its own sealed workspace, ROTEX was used both by the crew and ground personnel.

If materials science encapsulated nearly a quarter of Spacelab-D2's payloads, then about 40 percent was devoted to life sciences. Ross sought to transfer as much of the mission's 'basic' medical training as possible from Europe to the United States, to lessen a workload which saw himself and Harris routinely travel overseas for three to four weeks at a time. "After looking at it for a while, and understanding exactly what was required in the training, I was able to get the Germans to offload some of the training back here," Ross remembered. "Things like being able to put in a catheter, draw blood, give shots…take pulses and respiratory rates. They allowed me to offload most of that training and then just do a little bit over there to demonstrate that I'd got what they needed. I also tried to consolidate the amount of training we were doing over there, so that we didn't have to travel to Europe any more than was required."

Central to this life sciences thrust was a trio of research facilities: Baroreflex, Anthrorack and Biolabor. Televised images at various points of the mission offered a distinct impression that these facilities were a modern-day torture chamber, with Ross, Harris, Schlegel and Walter subjected to frequent saline injections and blood draws. Since the orbiter crew had to remain in peak physical condition during the flight, they were immune to such punishing tests. As such, Nagel, Henricks and Precourt tended the non-invasive experiments: monitoring GAUSS and MOMS on the USS pallet, participating in Urine Monitoring System (UMS) investigations and testing the Crew Telesupport Experiment (CTE). The latter resembled a child's Etch-a-Sketch toy and sought to establish a two-way communications feed with text and graphics to Mission Control.

Another (reasonably) non-invasive experiment with which the pilots engaged was Baroreflex. This explored relationships between post-flight cardiovascular 'deconditioning' and the baroreflex, which is responsible for maintaining blood-pressure levels in the human body. Such deconditioning was typically accompanied by light-headedness, a decrease in blood pressure and even fainting when astronauts attempted to stand immediately after landing and was thought to be associated with the heart's decreased workload in microgravity. Crewmembers wore a silicone rubber neck-cuff and electrodes, through which pulses of pressure and suction were transmitted to baroreceptors, with heart-rate measurements taken before, during and after the mission. In one televised view, Precourt put on a brave face and a half-smile, the neck-cuff wedged uncomfortably between his shoulders and chin. Nagel also wore the collar during his infrequent visits to the Spacelab module. "The orbiter crew cannot get involved in any invasive experiments, and not one that would tie you down…where it would be hard to get up and get back to the front in a real hurry, if you were needed," he said. "This thing, you could just take the collar off and leave."

Anthrorack's 20 experiments enabled the first comprehensive, integrated screening of the entire human body in orbit, with simultaneous measurements of the respiratory, cardiovascular and endocrine systems, as well as trials of an ultrasound device. Its research explored fluid shifts, the hormonal system, lung circulation and ventilation and the body's reaction to various physical states. It included a blood collection kit and integrated UMS to collect urine for studies of protein metabolism, fluid electrolyte regulation and pathophysiology of mineral loss, together with a high-speed centrifuge, an ergometer for exercise, a blood pressure tracking system, a tonometer to measure intraocular pressures in the eyes and a device to measure limb volume. The ergometer was unstowed from Anthrorack in space and assembled in the module's centre aisle.

Biolabor focused upon 'electrofusion' of cells, cell cultivation, botany and zoology. Equipped with a workbench, microscope, a pair of cell cultivation incubators, a cooler and two cooling boxes in the Shuttle's middeck, its 12 experiments supported plant and mammalian cells. Under study were the vestibular and gravity-perceiving systems of aquatic vertebrates, the gravity-sensing mechanisms of cress roots and various types of fungus and metabolite production, growth and regeneration of yeast cell cultures. Larvae of coloured perch fish and tadpoles of the South American clawed frog were also aboard (Fig. 6.8).

Five other experiments, worn by astronauts or placed close to biological experiments, examined the effects of cosmic radiation and its potential health risks. And a Microgravity Measurement Assembly (MMA) comprised six accelerometers—four situated in research racks, the other pair placed elsewhere in the Spacelab module—to assess transfer functions or responses in the spacecraft's structure and how they induced disturbances in another area, to better understand dynamic behaviours. All told, the Spacelab-D2 payload—module, tunnel and USS—weighed 25,025 pounds (11,340 kilograms). As with Spacelab-D1, the POCC again resided at the GSOC in Oberpfaffenhofen, which by the spring of 1993 benefitted from substantial data-handling enhancements and could accommodate all communications traffic from the Shuttle.

Fig. 6.8: After a month-long delay, Columbia roars into orbit on 26 April 1993.

RED LIGHTS AND FAILED FREEZERS

But if Spacelab-D2 had already waited for years to fly, that wait was made a little longer by technical maladies in early 1993. Obsolete tip-seal retainers on the high-pressure oxygen turbopumps in Columbia's main engines required replacement, postponing the launch from 25 February until mid-March. During this down time, GAUSS had its ultraviolet camera film changed and other payloads required servicing of batteries. Then, a flex hose in the Shuttle's aft compartment burst, spilling hydraulic fluid, and STS-55 slipped towards the month's end. This provoked great unease from the Germans, who were reportedly paying around a million dollars daily to keep Spacelab-D2 and its hardware in a state of launch readiness.

The mission's estimated $560 million price tag had raised eyebrows in the years following reunification, the costly reconstruction of the country after the fall of the Berlin Wall imposing a brake on Germany's space ambitions. A Spacelab-D3 mission, consisting—like Spacelab-D2—of a pressurised long module and USS, came close to fruition, before being cancelled. "There was intended to be a third, and maybe beyond that," Nagel said, sadly, "but their funding became a problem with that…so lots of things changed there." In NASA's August 1988 Shuttle manifest, Spacelab-D3 was 'requested' for October 1992. And by the January 1990 manifest, it had taken form as a nine-day mission aboard Columbia on STS-82, scheduled for August 1994 and aimed at an orbital altitude of 186 miles (300 kilometres), inclined 28.5 degrees. Three years later, Spacelab-D3 vanished, in favour of limited participation in a pan-European Spacelab-E1 mission. Also baselined as a module/USS format, Spacelab-E1 was requested for launch by September 1994 and a Spacelab-E2 in the first quarter of Fiscal Year 1998. Both missions fell to the budget-cutter's axe.

Another potential German mission, Spacelab-D4, was envisaged much earlier as a pallet-only flight. According to a paper in the journal *Advances in Space Research* in 1982, it was alternately known as the German Infrared Radiation Laboratory (GIRL) with a 'train' of two Spacelab pallets, plus an IPS/igloo. GIRL's optics comprised a 19.6-inch (50-centimetre) telescope, cooled by 66 gallons (300 litres) of liquid helium and equipped with four focal-plane instruments—a detector array, photopolarimeter, spectrometer and interferometer—for astronomical and aeronomical observations, including studies of star-forming regions and spurious gases in Earth's atmosphere. Requested in April 1984 for a Shuttle flight in October 1988, after Challenger's loss and German reunification Spacelab-D4 also disappeared from the manifest like a blip from a radar screen.

Though these two follow-on German missions met unfortunate ends, cancellation was out of the question for Spacelab-D2. But delayed the flight certainly was. By mid-March 1993, conflicts on Florida's eastern range with a military Atlas rocket waiting on the pad at Cape Canaveral Air Force Station pushed STS-55 back to 9:51 a.m. EDT on the 22nd. That morning, the crew headed to the pad and took their seats aboard Columbia. With 31 seconds remaining on the countdown clock, the Shuttle's on-board computers assumed control of all critical functions.

Six and a half seconds before liftoff, the three main engines roared alive with their familiar, low-pitched rumble…then, within three seconds, and accompanied by gasps from assembled spectators, all three shut down. The right-hand engine had incompletely started, having sustained a leak in a liquid oxygen preburner check valve, which caused its purge system to overpressurise beyond its maximum redline. (The culprit turned out to be a tiny shard of rubber, trapped in a propellant valve.)

With unburnt hydrogen lingering underneath the still-hot engines, the risk of a fire or explosion remained acutely real. Seconds after the shutdown, the 'white room' was automatically moved into position alongside Columbia's hatch, to allow the crew to evacuate. Although the astronauts had trained to escape from a pad abort and slide to a fortified bunker, the danger of having them run through invisible hydrogen flames led controllers to instruct them to stay inside the orbiter. For 40 tense minutes, Nagel and his men remained strapped into their seats as ground personnel deactivated electronic components. Paramedics headed to the pad as a precautionary measure, but for the crew all was well—albeit eerily *still*—as the Shuttle's cabin gently swayed in the breeze, the only sound apart from their racing hearts being the screech of seabirds outside.

"I'd convinced myself," Nagel said, "and all my crew that we were going to fly." As the noise of the engines died, he called over the intercom to announce the abort. "I wouldn't call it fear," he said of those adrenaline-charged seconds. "There's a couple of moments wondering what's happened, because all you see on-board are red lights, indicating an engine shutdown. You know the computers shut down the engines, but you don't know why or exactly what went wrong with them." The situation was equally tense in the Launch Control Center, where ashen-faced Launch Director Bob Sieck watched the proceedings. "Your initial reaction is to make sure there are no fuel leaks or that there's nothing that's broken that's causing a hazardous situation," he said. "Really, it was one of those nice, boring countdowns...until the last few seconds. What did work, and worked very well, were the safety systems on-board. As a result, the crew is safe."

A month-long delay until the end of April ensued, due to the need to return the STS-55 stack to the VAB and replace the engines. Several of Spacelab-D2's time-critical experiments were removed and replaced, but another launch attempt on 24 April was also scrubbed, due to a problem with an inertial measurement unit. Two days later, at 10:50 a.m. EDT Columbia roared smoothly into space and within hours of reaching orbit the crew divided themselves into their two shifts—Nagel, Henricks, Ross and Walter on the blue team, Precourt, Harris and Schlegel on the red—for nine days of research.

But in those opening hours, Ross realised that there was not one Jerry Ross on this mission, but two. Before launch, Nagel had craftily gotten hold of a picture he took of Ross grinning through the Shuttle's aft flight deck windows on a spacewalk during their previous mission together. He had the photograph enlarged to fit one of the windows, which was covered by a protective panel for ascent. Nagel knew that Ross would be removing that panel and would be first to see his own face looking back at him. He hoped his friend would get a kick out of it. But by the time Columbia reached orbit, Nagel had forgotten all about the prank. And Henricks and Precourt, who were busy setting up the Shuttle for orbital operations, knew nothing.

Suddenly, Ross burst out laughing. Henricks and Precourt raised their eyes from their checklists in surprise. "It's just *me*, looking at myself," Ross said later. "They thought Ross had lost it back there. I was just laughing hysterically."

After the laughter subdued, Nagel's blue team set to work. Although the seven men worked two teams, Ross later admitted that "there *was* a Crew C… the Shuttle crew as a whole" and as payload commander he saw to it that time existed at shift handovers for them to gather and discuss experiments, problems and summaries of notes. "I think a couple of times during the flight I gave little pep talks," he said. "I also made sure they'd go to sleep on time, so they'd get the right amount of rest, even though I wasn't. I probably didn't get more than five hours sleep a night for the whole time that we were up there and when I got back on the ground I was just flat wiped out." In Ross' recollection, he only left the Spacelab module to eat, sleep and use the toilet. The bunks, which Guy Bluford found so claustrophobic, were like "coffins", Ross reflected, and any chance of fitful sleep was lost in an incessant ambient din from the middeck: people eating and working, tapping and banging. Ross had a "kinda hyper" personality and his mind raced at a million miles per hour. "They're really small," he said of the bunks. "You can't even roll—my shoulders—I couldn't even roll over, turn over in them" (Fig. 6.9).

All seven crewmen participated in 'saline loading', drinking heavily salted water to rapidly expand their intravascular volumes and enable physicians to comprehend the mechanisms responsible for post-mission deconditioning and 'orthostatic intolerance' experienced by many astronauts after returning from space. "I think we will find that space adaptation syndrome comes in phases," said Harris. He concluded that acclimatising to the unusual environment occurred in at least three stages, spanning undetermined lengths of time: typically, it began with dizziness and nausea, then progressed through loss of blood volume and decreased bone and muscle mass.

Spacelab-D2 ran exceptionally smoothly, with the most minor of glitches. A small hole in a wastewater tank yielded a tiny nitrogen leak which required the astronauts to use burlap-type rubber-lined bags to collect urine. "Every once in a while, we'd have to sit there and squeeze this bag to shoot the urine back out the side of the orbiter into space," said Ross, wryly adding "*that* was a delightful thing to do." Years later, Nagel was philosophical. "It got us through the mission," he said of the bags, "so who argues with success?"

Fig. 6.9: All four Spacelab-D2 payload crew members are pictured at work during a shift changeover inside the pressurised module. Bernard Harris and Hans Schlegel work on a life science experiment, whilst Ulrich Walter winces with Baroreflex fastened about his neck and Jerry Ross (far left) tends to other duties.

Before launch, the experiment freezers responsible for housing Spacelab-D2's biological samples gave Ross cause for concern. "The freezers we were carrying…were failing at a fairly rapid clip," he said. "It became apparent to me that probably 40-50 percent of the science was counting on that freezer working and bringing back those samples that we collected over the ten days in orbit." On his recommendation, a backup freezer was flown. Although it came at the cost of forfeiting two middeck lockers, Ross' recommendation proved a wise one when Spacelab-D2's primary freezer overheated and failed only two days after launch. This required the astronauts to transfer their biological samples with gusto to the backup unit. "Well, wouldn't you know it, the freezer that was powered up to launch failed by the time we got to orbit and never did come back to life," Ross said, "so the other one that was our spare was the one we used throughout the entire flight. We did all kinds of things to try to nurse it along." That 'nursing' included cutting and taping plastic bags over the freezer's intake fans to keep it cool.

Nagel was convinced that the freezer problems prompted an anxious NASA to keep Columbia in orbit no longer than ten days, lest the backup also fail. Originally baselined for a couple of hours shy of nine full days, U.S. and

German personnel desired to conserve as much electrical power as possible to allow for a tenth. And the crew did it, easily accumulating 25 hours of added margin in their cryogenics. On 3 May, the go-ahead was given to extend STS-55 to ten days. But that extension would go no further. Early on the 6th, the red team closed Spacelab-D2's hatch, bade the Oberpfaffenhofen controllers "auf wiedersehen" for the very last time on any Shuttle mission and Harris shut the hatches between the middeck and tunnel.

Their return to Earth later that morning, to land at KSC, was called off due to unacceptable weather. An orbit later, conditions still failed to improve and the Shuttle was diverted to touch down at Edwards at 6:29 a.m. PDT. Nagel was certain that Columbia could have remained aloft at least another day, but the importance of Spacelab-D2 and the presence of only one freezer for all the biological samples troubled him. "The Germans were hanging it all on one freezer to save these samples," he said, "and I'm sure that's why, when it came time to land, we were weathered out of Florida and went to Edwards."

7

Stargazers

"THE SUN KEPT RISING"

Florida's weather in the pre-dawn darkness of 6 March 1986 convinced Bob Parker that—had the hands of fate played a different card—death likely awaited himself and his STS-61E crewmates. Thirty-seven days before Parker's crew were due to launch, Challenger had exploded, as freezing conditions conspired with faulty O-ring seals in one of her SRBs to produce one of the most public disasters ever witnessed. Observed with first-hand horror by the astronauts' families, and with millions more watching on television, the Shuttle disintegrated in a fireball. It was the first time that U.S. astronauts had been lost during a space mission. It would not be the last.

Had Challenger survived that day, the next flight on NASA's manifest was STS-61E, targeting liftoff at 5:45 a.m. EST on 6 March for the nine-day ASTRO-1 flight. It carried three ultraviolet telescopes, a pair of wide-field-of-view cameras and an IPS to observe the Universe and one celestial wanderer in particular: Halley's Comet, which in the late winter of 1985/1986 was making its 75-yearly visitation to the inner Solar System.

ASTRO originated in July 1975, with an *Interim Report of the Astronomy Spacelab Payload Study*, which articulated the Shuttle's ability to examine celestial sources across the electromagnetic spectrum. Specific emphasis was placed upon a 3.3-foot-diameter (1.1-metre) Spacelab Ultraviolet Optical Telescope (SUOT), envisaged to fly in 1981. "A one-metre diffraction-limited telescope from the ultraviolet…will provide high angular resolution imaging over relatively wide fields of view," the report noted. "Such a capability is required…for photometric studies of stellar evolution in globular and open clusters and to supply observations of nearby galaxies." In February 1978, NASA issued an announcement of opportunity an astronomical observatory and a trio of ultraviolet instruments evolved as payload originally managed by the agency's Office of Space Sciences. Since the IPS and Spacelab was needed for the mission, by March 1983 its leadership responsibility had passed to MSFC and the program was renamed 'ASTRO'. A Wide Field Camera (WFC) joined the payload in 1984, specifically for Halley's Comet observations.

One NASA astronaut assigned to follow its development in 1982 was astronomer Jeff Hoffman. Of particular concern to George Abbey, JSC's director of flight crew operations, was whether the mission's complexity warranted payload specialists. "We knew George didn't like the idea of payload specialists," Hoffman recalled. Abbey felt qualified scientist-astronauts from JSC's mission specialist corps were more capable than MSFC-picked 'outsiders'. JSC Director Chris Kraft was similarly convinced that Spacelab crewmembers ought to be "selected from the present corps of mission specialists", arguing that they were better suited through their training and experience. As early as July 1976, Kraft told MSFC that JSC already had "outstanding individuals in the scientist-astronaut group and are in the process of selecting 15 more mission specialist astronaut candidates who are expected to be equally capable and from whom payload specialists could be selected". Kraft felt strongly that future payload developers should "be made aware of the advantages of using trained NASA [mission specialists]", rather than payload specialists, and indeed in late 1977 scientist-astronaut Bill Thornton was proposed as a 'professional' payload specialist candidate for Spacelab-1. "It was one of the bitterest fights to select their own people," Thornton said of Kraft's eagerness to have his own astronauts in payload specialist seats. But MSFC would have none of it. "They wanted their own people," Thornton continued, "and there was no way in the world the boards…were going to say I was qualified for it."

"Payload specialists were [MSFC's] astronauts and they were always pushing to get payload specialists on flights," Hoffman said. "There was a lot of potential resentment at JSC, because if a payload specialist flies, then a career astronaut doesn't fly." Abbey wanted an astronaut's perspective to assess the need for payload specialists on ASTRO, "to see if they were for real," said Hoffman, "or if they were blowing smoke". But in compiling his report to Abbey, Hoffman became convinced that payload specialists on ASTRO were not only useful, but essential. He could spend two years training and still not know a payload in as much depth as a payload specialist did. Still, Hoffman wondered if championing the payload specialists' cause might hinder his own career prospects.

Fortunately, that proved not to be the case and in June 1984 Hoffman and Spacelab-1 veteran Parker were named to STS-61E as mission specialists, scheduled to fly Columbia in March 1986. Joining them was a third mission specialist, Dave Leestma. Interestingly, STS-61E was identified as "a crew of six", implying—with a yet-to-be-named commander and pilot, plus Hoffman, Parker and Leestma—only a single payload specialist. That situation changed a few months later and STS-61E expanded into a seven-man crew.

Later in June 1984, three candidates were selected for two ASTRO-1 payload specialist seats: astrogeophysicist Sam Durrance of Johns Hopkins University in Baltimore, Maryland, astronomer Ken Nordsieck of the University of Wisconsin at Madison and astronomer Ron Parise of GSFC. Years later, Shuttle commander Steve Oswald—who flew with Parise on ASTRO-2—expressed high regard for this Italian-American scientist. "I said [he] had been assigned to ASTRO since the Earth cooled," joked Oswald. "That's not really completely accurate. However, he has been working ASTRO since he graduated from college, which was shortly *after* the Earth cooled!"

In January 1985, NASA chose Jon McBride to command STS-61E, with Dick Richards as pilot. Initially planned for seven days, by NASA's March 1985 manifest the launch date had narrowed to 6 March 1986 and the flight extended to nine days, with insertion into a 230-mile (370-kilometre) orbit, inclined 28.5 degrees to the equator. ASTRO-1 was baselined as a 'train' of two Spacelab pallets, an igloo and the IPS to guide its instruments. It became the first Spacelab mission to focus upon just one scientific discipline. In October 1985, Durrance and Parise were assigned as prime payload specialists for ASTRO-1, the first of three dedicated missions of the ultraviolet observatory: after the March 1986 flight, the second would occur in January 1987 and a third the following July. With Durrance and Parise aboard ASTRO-1, it was expected that Durrance and Nordsieck would fly ASTRO-2, with Parise and Nordsieck on ASTRO-3. Hoffman and Parker were attached to all three missions, with McBride commanding the first two. "Good group," remembered McBride. "Gelled well together, worked well together, liked each other, no problems whatsoever."

Had Columbia launched early on 6 March 1986, her crew faced an ambitious timeline, working around the clock with a red team of Richards, Parker and Durrance, a blue team of Leestma, Hoffman and Parise, as McBride anchored his schedule across both shifts. When the mission's crew activity plan was published in November 1985, STS-61E looked to be the second-longest Shuttle flight to date (after Spacelab-1), with landing planned early on 15 March after 214 hours and 144 orbits.

By 28 January 1986, ASTRO-1 had completed pre-launch processing and its ultraviolet telescopes had been mounted atop the IPS. The crew entered JSC's Shuttle mission simulator early that day and spent a couple of hours practicing ascents and re-entries. Around mid-morning, they took a break to watch Challenger's launch on television. Having seen copious amounts of ice on the launch pad, and hanging from Challenger herself, and having judged the coldness of the weather, the STS-61E crew was surprised when NASA issued a 'go' for launch. Seventy-three seconds after liftoff, the men's hearts sank when the Shuttle exploded. Then, one of the crew piped up, quietly: "Well, there goes *our* flight." Decades later, memories of that dreadful day still caused tears to well in McBride's eyes. "You could feel everybody's heart just kind of…hit the concrete," he said (Fig. 7.1).

Fig. 7.1: Playful portrait of Jon McBride's STS-61E crew, illustrating the astronomical focus of their mission, which should have been the next Shuttle flight after the Challenger tragedy. Standing, from left to right, are Ron Parise, Jon McBride, Sam Durrance, Jeff Hoffman and Dick Richards. Seated are Bob Parker (left) and Dave Leestma.

Years later, Hoffman recalled that the STS-61E crew kept training for a while. "As horrible as it was, there was always a possibility that they would figure out very quickly what the problem was and it would be easy to fix," he said. "They told us we had to continue to train. It was very strange the next morning, going into the simulator to do an ascent simulation." So traumatised were the astronauts, having just lost their friends, that McBride asked the simulation supervisor: "How about just giving us *no* malfunctions on the first run? Let's just get safely up to orbit."

But there would be no quick fix and the Shuttle fleet was indefinitely stood down. Parker mused over why unbearable schedule pressures and cavalier attitudes had been permitted to overrule operational flight safety. "Mission 61E…must be launched by 10 March to achieve maximum science return," *Flight International* reported in December 1985. "A slip to March 20 would result in the flight's cancellation." Such immovable, inflexible targets astounded Parker. "It's amazing, when you look back at that…and the rate at which we thought we had to keep pumping this stuff out," he said. "You'd have thought the world was going to end. My favourite expression is: Guess what? The Sun kept on rising and setting. The Sun didn't even notice if we missed our launch windows!"

FOUR HIGH-ENERGY 'EYES'

The selection of Durrance, Nordsieck and Parise as payload specialists was by no means accidental. Not only were all three men highly accomplished, but all were intimately involved with ASTRO-1's instruments: Durrance was co-investigator for the Hopkins Ultraviolet Telescope (HUT), Nordsieck a co-principal investigator for the Wisconsin Ultraviolet Photopolarimeter Experiment (WUPPE) and Parise developed the flight hardware and software and designed the electronics system of the Ultraviolet Imaging Telescope (UIT). Together, the suite would observe the invisible, high-energy Universe, examining faint astronomical objects such as quasars, active galactic nuclei and galaxies at far ultraviolet wavelengths, tracing the 'polarisation' of hot stars and acquiring wide-field-of-view images of cosmic sources in broad ultraviolet wavelength bands. All three instruments stood 12.4 feet (3.8 metres) tall when deployed atop the IPS and the payload totalled 17,276 pounds (7,835 kilograms). An MSFC-built Image Motion Compensation System (IMCS) afforded fine pointing stability for UIT and WUPPE, negating the distorting effect of Shuttle thruster firings. Gyroscopic stabilisers transmitted data to the IMCS electronics, where pointing commands were computed and sent to the instruments' secondary mirrors to make automatic adjustments and attain stabilities of less than an arc-second. A JPL-provided star tracker 'fixed' onto bright stars to correct gyroscopic 'drift'.

Developed by Johns Hopkins University, HUT comprised a 36-inch (91-centimetre) primary mirror—coated with iridium to enhance its capacity to reflect far and extreme ultraviolet light—together with a spectrograph and electronic detector. It was expected that the 1,735-pound (790-kilogram) instrument would gather 300,000 seconds of data on 200 celestial targets, from aurorae and the magnetospheric dynamics of giant planets in the outer Solar System to quasars, billions of light-years away. Covering a wavelength range from 850-1,850 ångströms (first order) and 425-925 ångströms (second order), HUT's gaze would take in faint sources in a little-explored patch of the ultraviolet spectrum.

WUPPE, built by the University of Wisconsin at Madison and easily identifiable on the ASTRO-1 payload thanks to its large square baffle, surveyed hot stars, galactic nuclei and quasars. Its 20-inch (51-centimetre) dual-mirror telescope and spectropolarimeter measured the brightness and polarisation of ultraviolet emissions to provide chemical composition and physical data for objects across 1,400-3,200 ångströms. Lightest of the telescopes, it weighed 980 pounds (445 kilograms).

UIT, provided by GSFC, included a telescope and a pair of image intensifiers and cameras for observations in broad ultraviolet bands across 1,200-3,200 ångströms. It recorded data directly onto sensitive astronomical film, with capacity for 2,000 exposures, some lasting up to 30 minutes. The 1,043-pound (473-kilogram) instrument would focus upon nearby galaxies, large stellar clusters and more distant objects and UIT's advertised capabilities allowed it to resolve stars a hundred million times fainter than were perceptible with the naked eye on a clear, dark night.

During the post-Challenger downtime, ASTRO-1 was kept in storage and the three instruments were removed from their Spacelab pallets. In addition to periodic health checks, NASA recertified all three before committing them to flight. A total of 298 bolts on the ASTRO 1 support structure and electronics attachments were replaced in 1987 and the WFC was deleted that spring, Halley's Comet by now long gone. In its place appeared GSFC's Broad Band X-Ray Telescope (BBXRT), added in March 1988 following the dramatic appearance of Supernova 1987A in the Large Magellanic Cloud, 170,000 light-years away. One of its objectives was to examine the supernova's expanding cloud of gas and dust and identify its chemical constituents. It was not part of the ASTRO-1 payload and was controlled from the ground. BBXRT consisted of dual telescopes, each with an eight-inch (20-centimetre) aperture and 17-arc-minute field of view, which focused X-rays onto solid-state spectrometers to measure photon energies in the 'soft' X-ray regime between 380-12,000 electron volts. Weighing 1,500 pounds (680 kilograms), it was affixed to GSFC's Two-Axis Pointing System (TAPS) at the rear of the payload bay which pitched BBXRT forwards and backwards and 'rolled' it from side to side. A built-in star tracker used bright reference stars to achieve TAPS stability.

BBXRT was the progeny of another payload, the Shuttle High Energy Astrophysics Laboratory (SHEAL). It moved onto ASTRO-1 as its hardware was ready for flight sooner than anticipated. Although SHEAL was not a Spacelab mission *per se*, it was a research flight and its final form included the BBXRT/TAPS stack and a pair of Diffuse X-ray Spectrometers (DXS) attached to mounting 'plates' on opposite sides of the payload bay. Managed by GSFC, the 10,327-pound (4,685-kilogram) SHEAL would perform X-ray observations of active galactic nuclei, supernova remnants, galactic clusters and stars, with BBXRT expected to acquire over 188,000 seconds of data and DXS around 40,000 seconds of data during a seven-day mission.

An earlier, expanded version of SHEAL, described in a September 1983 NASA report, also included a reflight of the University of Chicago's Cosmic Ray Nuclei Experiment (CRNE) from Spacelab-2 (see Chapter Four) and a single Spacelab pallet, holding the Large Area Modular Array of Reflectors (LAMAR). The latter was intended to conduct highly sensitive X-ray observations of extragalactic sources, as well as imaging galactic clusters, mapping diffuse, 'soft' X-rays and measuring dust and granular distributions in the interstellar medium. "Use of a dedicated pallet" for LAMAR, noted Paul Gorenstein, "greatly simplifies the integration procedure and reduces its cost, because the experimenter group can actually assemble the instrument within their own laboratories and integrate it onto the pallet." The instrument virtually filled the pallet, its roughly cube-shaped bulk measuring about 7.8 feet (2.4 metres) by 8.6 feet (2.65 metres) and weighing 5,600 pounds (2,550 kilograms). LAMAR was led by the Smithsonian Astrophysical Observatory, NASA's MSFC, the University of California at Berkeley and the University of Chicago. Its array of detectors—configured into 32 or 64 'modules'—created an effective imaging area of between 5.4 square feet (0.5 square metres) and 10.7 square feet (1.1 square metres). The presence of payload specialists on SHEAL missions was considered "optional".

In the March 1985 Shuttle manifest, NASA envisaged three SHEAL missions—in October 1986, October 1988 and October 1990—with each flight baselined for a five-member crew (implying no payload specialists) at an orbital altitude of 186 miles (300 kilometres), inclined 28.5 degrees to the

Fig. 7.2: Technicians prepare the ASTRO-1 payload for flight in early January 1986, only three weeks before the Challenger tragedy. The circular base and yoke of the Instrument Pointing System (IPS) is visible at centre, with the igloo in the foreground.

equator. But following Challenger, in the March 1988 Shuttle manifest SHEAL's first mission was shifted to April 1992. And with the decision to divide the payload into two halves (with BBXRT flying alongside ASTRO-1 and DXS reassigned to another flight), SHEAL as a standalone mission disappeared entirely from NASA's planning charts by January 1990. LAMAR was also rescoped initially as a Space Station Freedom payload, before its funding was withdrawn that same year (Fig. 7.2).

As BBXRT was readied for flight, the ASTRO-1 instruments were removed from storage for their own mission. HUT remained at KSC throughout the post-Challenger downtime, although its spectrograph was returned to Johns Hopkins University in October 1988 for attention. Testing revealed that, although gaseous nitrogen ordinarily protected it from air and moisture contamination, the telescope's sensitivity had degraded and its spectrograph was replaced. But the second spectrograph also failed tests and was itself replaced in January 1989, together with an aging television camera the following May. The other instruments also underwent recalibration. WUPPE was not shipped back to the University of Wisconsin, but a portable calibration facility was set up at KSC, allowing it to sail through testing with flying colours in April 1989. Its spectrometer's power supply also received modification to reduce output noise. Similarly, UIT remained in Florida and received a new power unit for its image intensifier. By December 1989, all three instruments were declared flight-ready and were mated onto the IPS. In total, ASTRO-1 weighed 25,900 pounds (11,750 kilograms), one of the heaviest scientific cargoes ever carried aloft by the Shuttle.

In addition to the instruments' discrete research focuses, the complete ASTRO-1 payload had its own targets. Processes whereby ancient red giant stars steadily shed their outer atmospheres to leave dense 'embers', no larger than Earth (known as white dwarfs) were imperfectly understood. As these white dwarfs emitted much of their radiation at ultraviolet wavelengths, they were ideal candidates for ASTRO-1. The instruments would also survey suspected 'black holes' by resolving hot gas swirling into their clutches. Other targets included 'binary' stars and colossal stellar clusters. In visible light, it was difficult to discern individual stars in these clusters, but under ASTRO-1's stare they would blaze brightly. And the observatory would analyse the 'interstellar medium' between stars.

ASTRO-1 was the first mission run from MSFC's new Spacelab Mission Operations Control Facility, which superseded the POCC at JSC. The new facility, staffed by three teams and led by a payload operations director, could transmit commands directly to the orbiter, monitor the instruments, oversee observations, receive and analyse data and adjust schedules to take advantage of unanticipated events. Payload operations for BBXRT were conducted from a supporting payload operations control facility at GSFC.

LONG ROAD TO LAUNCH

But with the lengthy grounding of the Shuttle fleet following Challenger, ASTRO-1 had to await its turn to fly. Even after the reusable spacecraft returned to flight operations in September 1988, a pair of TDRS communications satellites, two planetary probes and a growing backlog of classified Department of Defense missions took priority over Spacelab on the manifest. In March 1988, ASTRO-1 was officially assigned to fly on Columbia on STS-35, then planned as a nine-day mission in March 1990, targeting insertion into an orbit 218 miles (352 kilometres) high, inclined 28.5 degrees to the equator.

The crew complement remained uncertain until the following November, when McBride reprised his command, accompanied by Hoffman, Parker, Durrance and Parise. As for the others, Richards and Leestma had joined other crews and for STS-35 the pilot's seat was taken by Guy Gardner, with Mike Lounge as flight engineer. Circumstances shifted again when McBride resigned from NASA in April 1989, one of few astronauts to do so whilst in dedicated mission training. Veteran Shuttle commander Vance Brand took his place. Years later, McBride rationalised his thinking: ASTRO-1 had come within a hair's breadth of cancellation and even as it reappeared on the Shuttle manifest in 1988, rumours of cancellation were never far away. The decision was tough—"I might be the only person in history who was assigned to a mission that pulled out of it"—and, indeed, ASTRO-1 did fly. But McBride knew he had made the right choice.

By early 1990, STS-35 had slipped from March until May, delaying in lockstep with other delayed missions before it, but that summer the Shuttle fleet succumbed to an intractable series of technical woes. Deep inside Columbia's belly were a pair of 17-inch (43-centimetre) valves, through which liquid oxygen and hydrogen flowed from the ET into the main engines' combustion chambers. Both sides possessed mechanical 'disconnect' fittings and shortly before the giant tank's separation, at the edge of space, a pair of flapper valves were commanded shut to preclude further propellant discharge. The criticality of the valves was paramount, for any inadvertent closure whilst the main engines were running would cut propellant flow and trigger an early shutdown, perhaps even an explosion. Columbia's launch had already moved from early to mid-May, thanks to a glitch with her freon coolant loop, before slipping again to the 30th. This impacted two more Spacelab missions (also by Columbia), scheduled for 29 August and 12 December 1990.

But worse was to come. As engineers loaded liquid oxygen and hydrogen into the ET, a tiny leak was detected, close to the tail service mast on the mobile launch platform. Further investigation revealed a more extensive—and worrisome—leak from the disconnect. The launch attempt was scrubbed and the ET emptied of propellants and inerted. Since the astronauts would work in shifts, they had already begun 'sleep-shifting' and the blue team was awake when news of the delay materialised. Blue team crewman Hoffman woke up his 'red' counterparts with the bad news.

A miniature 'tanking test' on 6 June to identify the leak's source proved inadequate and the Shuttle was rolled back to the VAB on the 12th for inspections, forcing a delay until mid-August. Disconnect hardware for the Shuttle and ET 'sides' of the fitting, as well as faulty seals, were replaced and in early August the STS-35 stack returned to the pad. Two more launch attempts in early September were scrubbed, firstly when an avionics box failed, then following a second hydrogen leak. Recirculation pumps in the Shuttle's aft fuselage were removed and replaced, as was a seal on one of the main engines, but to no avail. The only option was to assign a 'tiger team' to totally retorque Columbia's liquid hydrogen system. Over the next three months, MSFC engineer Bob Schwinghamer received a one-way ticket to Florida and was told by his bosses, only half-jokingly, not to return to Huntsville until the leak was fixed. By the time they concluded their work in late October NASA declared that Columbia was "the soundest leak-free orbiter at that time in the fleet".

Finally, at the end of November 1990, the STS-35 crew arrived in Florida. The delays to ASTRO-1 yielded fortune and misfortune; for astronomy as a science often produces phenomena that are neither predictable nor plannable. Losing the chance to observe Halley's Comet was acutely disappointing, but the delays would enable a study of Supernova 1987A…and another celestial event—a 'blazar'—which occurred late in November 1990. Blazars are distant, highly luminous objects, now known to be the cores of active galaxies, which suddenly flare several dozen times brighter than normal (Fig. 7.3).

On 29 November, Launch Director Mike Leinbach reported that teams were "right on the timeline", with no problems being tracked as the countdown began for a 237-hour, 158-orbit mission. Weather conditions on 2 December were favourable, although forecasters kept watch on a tropical storm south of Cuba. When the crew arrived in Florida on the evening of the 29th, they were jubilant. "We're back!" yelled Brand. "We're ready!"

Hoffman was equally optimistic. He knew that historically comets were perceived to be bringers of ill-fortune and many comets had been linked with ASTRO-1; Halley being the obvious one, but a string of other, lesser-known celestial wanderers had also made the rounds as STS-35 struggled to fly. By December 1990, there were no comets. "We all know comets are harbingers of bad news," Hoffman rationalised. "This time, we have no comet, so we're going to go!"

A TROUBLED MISSION

And go they did. At 1:49 a.m. EST on the 2nd, Columbia lit up Florida's night sky, after a brief delay caused by marginal cloud cover. Sitting on the flight deck, behind Brand and Gardner and jostling shoulders with Lounge, Hoffman never forgot the awe-inspiring sensation of his first nocturnal launch. As the vehicle rose, cleared the tower and began its roll program manoeuvre, he glanced over his shoulder through the Shuttle's overhead

Fig. 7.3: The Hopkins Ultraviolet Telescope (HUT) is lowered gently into position aboard the ASTRO-1 payload.

windows to see the ground illuminate and the coast drop away at astonishing speed. Later, when the SRBs burned out and were jettisoned, Hoffman was astonished as the entire cockpit was bathed in a brilliant orange glow.

"You had the feeling you were lighting up all that part of Florida," Brand said. "It was like night flying in an airplane. You had to really have your lights adjusted and pay attention to your gauges. You couldn't really tell much about what was going on outside. You couldn't see the horizon very well." Launching at night ensured that Columbia's passage through the South Atlantic Anomaly—a region in which the home planet's Van Allen radiation belt 'dips' toward the ionosphere—occurred mainly during orbital daytime. High-energy particles were known to trigger increased 'background' radiation levels in particularly sensitive scientific instruments. Since this natural background, which consisted of scattered light and ultraviolet airglow emissions, was higher on the daylit portion of each orbit, it preserved the nighttime passes for ASTRO-1 to focus on its faintest celestial targets.

Payload activation began almost as soon as the Shuttle was established in orbit. The 'red' team, consisting of Parker and Parise, led by Gardner, took charge of activating ASTRO-1 and BBXRT. Meanwhile, the 'blue' team of Hoffman and Durrance and shift leader Lounge bedded down for an abbreviated sleep period. They would awaken for their first 12-hour shift at 1:00 p.m. EST on 2 December. Brand anchored his schedule across both teams.

By this stage, the red team had switched on BBXRT and Parker received a go-ahead to unlatch and raise the IPS and deploy ASTRO-1's instruments. The process took less than seven minutes and by the time STS-35 was 16 hours old and Lounge's blue team came on duty the telescopes had sailed through initial checkouts and were ready for calibration. One observation of the star Beta Doradus using HUT was lauded by Durrance. Located in the constellation of Dorado (the Swordfish), the star had been chosen for its suitability when aligning and focusing the telescopes. The sighting was part of the Joint Focus and Alignment process, whereby the ASTRO-1 instruments were trained on a common target as a prelude for upcoming observations. Unfortunately, a computer failure in WUPPE prevented it from participating in the alignment. The problem was caused by an unactivated heater and when this was powered up, WUPPE's checkout got underway.

Typically, the mission and payload specialists, situated on the Shuttle's aft flight deck, overlooking the payload bay, used a pair of Spacelab keyboards and two DDUs to command the IPS and telescopes. Closed-circuit monitors provided images of starfields under observation and enabled the astronauts to check the data. Observations typically required between ten minutes and a full hour to complete.

Yet the mission quickly stumbled. Early on 3 December, the IPS began struggling to 'lock' onto guide stars. An alternate plan was devised to allow the astronauts to manually point the telescopes and track targets on HUT's television camera with a hand paddle. This worked reasonably well and allowed them to achieve an aiming accuracy of about three arc-seconds. "The mood is one of concern," said Flight Director Bob Castle. "We'd certainly like the system to work perfectly, but there is no panic. People are working to solve the problem and we have confidence we will solve [it] in a fairly short time" (Fig. 7.4).

Then, Brand noticed a scent of warm electrical insulation, which was traced to an overheating DDU. Since its displays enabled the crew to control the IPS and instruments, it was bad news and as the mission's second day wore on Castle admitted that the timeline was at least six hours behind schedule, with two dozen high-priority astronomical observations lost. Slowly, the pace quickened and moods brightened as the telescopes scrutinised the bright

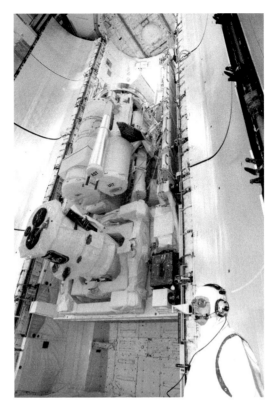

Fig. 7.4: Technicians prepare to close Columbia's payload bay doors, ahead of STS-35. ASTRO-1 is visible on its two Spacelab pallets at the top of the frame, with the Broad Band X-Ray Telescope (BBXRT) at the bottom.

galaxy NGC 4151, thought to contain a black hole, whilst BBXRT—whose pointing system was independent of ASTRO-1 and largely unaffected—successfully acquired detailed X-ray spectra of the Crab Nebula.

With its computer issues finally rectified, WUPPE came fully to life late on 4 December, examining a variable binary system and imaging the rapid-rotating 21 Velpecula, a white-hued star about 1.6 times larger than the Sun. As the telescopes found their feet, efforts continued to achieve full operational capability from the IPS' optical sensor package, whose star trackers offered one means of locking onto celestial targets. Support from MSFC and JSC enabled refinements to its pointing geometry and Durrance was able to accomplish the first operational identification of a desired target: a distant white dwarf.

On the whole, recovery seemed to reside on the horizon, with ASTRO-1 Mission Scientist Ted Gull of GSFC daring to admit that the mission was "really coming alive" and feeling able to smile at long last. After extensive troubleshooting, a string of refined calibrations of the star-tracker optics were uplinked and the observatory performed its first automatic target acquisitions. Hoffman and Durrance were able to utilise this capability by acquiring one target, followed by another, with no need to recalibrate the instruments between each sighting. Pointing stability, however, was not where it should have been, since control of the Spacelab pallets, IPS and instruments were being conducted through only a singular DDU. But it was hoped that both situations would improve, leading to an increase in the overall quality of the ultraviolet data.

All those hopes seemed dashed early on 6 December, when the second DDU overheated and failed. Initial efforts to restart it proved fruitless and Mission Control asked the astronauts to remove its lower panel and check the air-intake filters for the presence of lint. Hoffman used a vacuum cleaner to remove a small quantity of debris, but his efforts did not pay dividends and the DDU remained out of action. By now, ASTRO-1 had completed a paltry 70 of its planned 250 observations. Years later, Parker recalled the mission's grim prognosis. "Suddenly, we couldn't point," he said. "The ground came up with a scheme where they could control the telescopes…but what they couldn't do in near-real-time was guide the telescopes. Everybody put a brave face on it, but it was a far cry of what we had intended it to be." In fact, for a few hours that day the astronauts had little to do but enjoy watching the home planet drift serenely beneath them. Views of home, though, were an infrequent treat on STS-35, the focus of which tended to be in the other direction. "We weren't pointed at the Earth very much," admitted Lounge, "which was a little unusual, because the other flights [had their] payload bay down at the Earth all the time. We were pointed at the stars, so we could be, depending on the target, any attitude."

Eventually, ground specialists bounced back from the DDU failures and by 7 December were able to command all but the final motions of the telescopes. It was then left to the astronauts to effect fine-tuning ahead of each observation. Revised procedures involved operating the telescopes sequentially—first UIT, which had the widest field of view, then HUT and finally WUPPE—but even BBXRT was indirectly affected, since it had to cease its own activities whenever Columbia entered a 'safe' attitude. In one observation, alternate payload specialist Ken Nordsieck guided Durrance to acquire Supernova 1987A with UIT. As Durrance controlled the telescope with a joystick, Nordsieck provided pointing instructions and six high-quality minutes of ultraviolet spectra were collected. More was to follow with the observation of the radio-quiet, extremely luminous quasar, Q1821+643. The crew were by now established into a routine, working efficiently with ground personnel to maximise scientific output. Typical clipped exchanges with Nordsieck were crisp and terse: "Sam, you're within an arc-minute…Okay, give me a 'mark' when you're happy…The data's looking real good, Sam…We're seeing lots of photons down here."

Little by little, and observation by observation, ASTRO-1 slowly recovered to make 231 surveys of 130 celestial objects. And despite being adversely impaired by the observatory's woes, BBXRT also returned a colossal volume of data. One of its most important targets was Markarian 335, a Seyfert galaxy 325 million light years from us, which harbours a supermassive black hole at its heart. The X-ray imagery enabled astronomers to observe emissions from material being drawn into the black hole. Yet even BBXRT suffered issues with its TAPS supporting truss: an improperly compensated gyroscopic drift-rate meant that no stable pointings could be conducted for the first 60 hours of the mission and its original timeline was dropped in favour of a less efficient, day-by-day method. It also proved sluggish in acquiring celestial targets.

Still, the mission was an outstanding example of triumph over adversity, achieving far more than expected. In their final ultraviolet science briefing, ASTRO-1 Mission Scientist Ted Gull of GSFC asked Art Davidsen what the final results were in the 'match' between 'The Huntsville ASTROs' and 'The Universal Secrets'.

Davidsen grinned, a twinkle in his eye. The 'ASTROs', he retorted, had won by a mile (Fig. 7.5).

As the mission entered its final days, the gremlins were not yet done with STS-35. Originally scheduled to run for almost ten days, managers were forced to confront the reality that Columbia might return to Earth some 24 hours early, with weather at Edwards—the primary landing site—expected to be unfavourable between 12-14 December. Mission Control advised Brand that his crew would be coming home overnight on 10/11 December, forcing the ASTRO-1 scientists to hurriedly reprioritise their plans for the last observations. The three ultraviolet telescopes were finally deactivated and stowed about 12 hours prior to touchdown, followed by BBXRT six hours later. A cold front on the California coast was being tracked, Mission Control told the crew, and rain showers and a heavy pall of cloud were predicted to impact the area around Edwards.

So it was that Brand and Gardner guided Columbia through the darkness to alight on concrete Runway 22 at 9:54 p.m. EST on 10 December, after almost nine days. "We're home," radioed Brand as the Shuttle rolled to a halt, "and glad to be back."

Fig. 7.5: The ASTRO-1 instruments stand proud after deployment.

ASTRO-2

At the time of Challenger's loss, three ASTRO missions were anticipated, their payload specialist seats alternating between Durrance, Nordsieck and Parise, with each man flying twice. In its March 1985 Shuttle manifest, NASA listed ASTRO-1 in March 1986, ASTRO-2 the following October and ASTRO-3 in July 1987, all outfitted with two-pallet trains, an igloo and the IPS and all flying on Columbia. By January 1986, that schedule had slipped to anticipate ASTRO-2 (with Durrance and Nordsieck) in January 1987 and ASTRO-3 (with Parise and Nordsieck) retaining its original placeholder the following July.

The two follow-on missions were seven-person flights, lasting seven days apiece, with Hoffman and Parker assigned to both. In hearings before the Subcommittee on Space Science and Applications of the Committee on Science and Technology at the House of Representatives, ASTRO-1's status was listed as ready to fly by January 1986, with ASTRO-2 at 75-percent-complete and ASTRO-3 at 50-percent. Cancelling these back-to-back-*to-back* uses of the observatory proved intensely disappointing for Hoffman. "ASTRO was a payload that was developed at the time when people were saying that the Shuttle would be able to fly these payloads once or twice a year," he recalled. "It was a very expensive payload to develop and in the end it only flew twice. It would have been much more cost-effective had the Shuttle been capable of the sort of flight rates that were originally anticipated."

But plans for at least one follow-on ASTRO mission endured well into the post-Challenger timeframe. In its March 1988 manifest, NASA identified ASTRO-2 aboard Shuttle Discovery on STS-53 in December 1991 as a seven-day flight with a crew of seven and a two-pallet/igloo/IPS layout. But by January 1990, ASTRO-2 had vanished from a formal mission number and was obtusely listed as being 'required' by July 1993, with no definitive launch date. All that changed on 19 May 1991, when NASA Associate Administrator for Space Science and Applications Lennard Fisk announced that ASTRO-1's ability "to acquire high-quality scientific data" had prompted a decision to fly it again. Even ASTRO-1's problems did not prevent NASA from claiming a success rate as high as 80 percent, with 135 of 250 planned celestial targets acquired (Fig. 7.6).

No target launch date was initially given for the second mission—"as soon as the NASA flight schedule can accommodate it," was the rather vanilla summary —but in the January 1992 Shuttle manifest ASTRO-2 reappeared as STS-67 in September 1994. But its complexity (and duration) would steadily increase to such an extent that when it did finally fly in March 1995 it seized the record for the longest Shuttle mission at that time. Originally assigned to Discovery for a seven-day flight, by April 1993 it had shifted onto Columbia, whose Extended Duration Orbiter (EDO) capabilities permitted a longer mission of 13 days. Liftoff was targeted for December 1994.

Crew positions started filling on 3 August 1993, when astronomer Tammy Jernigan was assigned as payload commander. She was followed by fellow mission specialist John Grunsfeld, a physicist, the following 28 October. By this point, three candidates—ASTRO-1 veterans Durrance and Parise, plus electrical engineer Scott Vangen—had been identified for a pair of payload specialist seats. Finally, on 10 January 1994 the STS-67 crew was completed with the selection of Durrance and Parise as payload specialists, Steve Oswald as commander, Bill Gregory as pilot and Wendy Lawrence as flight engineer. Even four years after ASTRO-1, more than a hundred scientific articles were still being published in academic journals, whetting many appetites for what ASTRO-2 might offer.

Fig. 7.6: These three men were originally slated to rotate through three ASTRO missions between March 1986 and July 1987. Sam Durrance (right) and Ron Parise (centre) would fly ASTRO-1, with Durrance joining Ken Nordsieck (left) for ASTRO-2 and Parise joining Nordsieck for ASTRO-3. In the aftermath of the Challenger tragedy, Nordsieck departed the program and Durrance and Parise wound up flying both ASTRO-1 and ASTRO-2. A third mission was later cancelled.

But as 1994 wore on, Columbia was withdrawn from service for refurbishment and ASTRO-2 shifted onto her sister ship, the youngest member of the Shuttle fleet, Endeavour. This delayed the launch into early 1995, but Endeavour had been built from the outset with EDO capabilities and could support a long mission. According to the STS-67 press kit, published in February 1995, the flight was scheduled for 15 days, 13 hours and 32 minutes and 248 orbits, the longest Shuttle mission so far.

"INSTANT DAYLIGHT"

As darkness fell that evening, Endeavour sat silently under the glow of xenon floodlights at KSC's Pad 39A. It seemed unlikely that she would fly at all, with weather forecasters predicting only a 40-percent chance of acceptable conditions. Nevertheless, Oswald led his crew out of the Operations and Checkout Building at 10:20 p.m. EST. They boarded the Astrovan and arrived at the base of the launch pad 25 minutes later, to begin strapping into their seats. By midnight, all seven astronauts were aboard the Shuttle and communications checks had concluded.

As clocks ticked into 2 March, the Shuttle's middeck hatch was sealed and remaining technicians and pad personnel evacuated the launch pad. With liftoff scheduled for 1:37 a.m., at the start of a 2.5-hour 'window', everything proceeded normally until 1:26 a.m., when a problem was detected with the secondary heater of the flash evaporator system, responsible for cooling the orbiter's electronics. The data seemed to indicate that the heater was approaching a 'redline' condition. The countdown proceeded and was briefly held at T-5 minutes to clear the problem before the clock resumed counting.

"Instant daylight" was Gregory's recollection of their rousing 1:38 a.m. liftoff. Using a hand-held mirror, Lawrence glimpsed the launch pad, the exhaust plume and the rapidly receding Florida coastline through the Shuttle's overhead windows. Like ASTRO-1 before it, the nocturnal launch enhanced ASTRO-2's observing timeline by minimising the time that Endeavour would reside in the South Atlantic Anomaly during nighttime orbital passes. According to Oswald, the ASTRO-2 science team and Jernigan as payload commander pushed very strongly for a night launch for this precise reason.

On-orbit, there was little time to acclimatise, for ASTRO-2 was an around-the-clock operation, involving red and blue shifts. Oswald, Gregory, Grunsfeld and Parise formed the red team, with Jernigan, Lawrence and Durrance on the blue team. And for Oswald, those first few hours were so busy that they afforded him precious little time to check on his first-time fliers. "As the commander, you try to figure out how much extra time you need to add to the schedule, based on how many rookies you've got," Oswald continued. "Sometimes guys are semi-Velcroed to the wall, throwing up, while the folks you least expected to be heroes are just chugging along, executing the plan." That plan was enormously complex in scope; so enormous, in fact, that all seven astronauts would receive two half-days of free time at various points during the long mission.

Outside, in Endeavour's cavernous payload bay, sat the 17,380 pounds (7,885 kilograms) of HUT, WUPPE and UIT, affixed to the IPS on a pair of Spacelab pallets. Within four hours of reaching orbit, the IPS was rotated into its upright orientation, whereupon Durrance applied power to the three telescopes for ASTRO-2's inaugural observations. Early activities proceeded normally, despite a thruster leak which twice forced the closure of the telescopes' aperture doors to safeguard their optics from contamination.

The mission suffered none of the pointing troubles that plagued its predecessor. According to the STS-67 press kit, an MSFC test team "extensively

Fig. 7.7: Unusual view of ASTRO-2, as seen through Endeavour's aft flight deck windows.

modified and tested the IPS software and made other improvements to ensure the IPS works properly for ASTRO-2". Specifically, the IMCS had received substantial upgrades to refine pointing and stability. This was particularly vital in the case of UIT, whose images were recorded on film with individual exposures lasting up to 30 minutes. WUPPE experienced a slow start, the activation and verification of its detector delayed by problems keeping it aligned onto a 'test' target. Twelve hours into the mission, JSC controllers declared the IPS fully operational and transferred control to the payload team at MSFC. This enabled Grunsfeld and Parise to start the Joint Focus and Alignment procedure to ensure that all three instruments were capable of pointing in precisely the same direction (Fig. 7.7).

Despite the day-long calibration, astronomical observations got underway with pace and gusto. Early on 3 March, the HUT and UIT science teams had locked their instruments onto the ancient supernova remnant of the Cygnus Loop, gathering temperature, density and chemical data and imaging 'filaments' of excited gas and energising shockwaves. WUPPE demonstrated that its optics were in perfect working order by observing a calibration star, Beta Cassiopeiae, then the hypergiant luminous blue variable star P Cygni. Meanwhile, HUT examined EG Andromedae, a 'symbiotic' system of a relatively cool orange giant star and a tiny, exceptionally hot blue star. White dwarfs, globular clusters and 'Wolf-Rayet' stars took precedence over the following days. The latter included EZ Canis Majoris and are thought to represent one of the final phases in the evolution of supermassive stars, whose luminosities extend up to a million times as bright as our Sun. Their powerful ionised-gas emissions were thought to accelerate the aging process. Bright-centred Seyfert galaxies, distant quasars and interactive binary systems also received attention. In the latter case, WUPPE was directed to observe an X-ray binary, Vela X-1, with astronomers speculating that a neutron star was gravitationally 'stripping' material from a companion star, causing a large oval disk to form. Polarisation measurements by WUPPE measured the size and shape of this disk, as well as calculating the quantities of mass transferred between the two companions.

Closer to home—and illustrated on the STS-67 crew's mission patch—the largest planet in the Solar System, Jupiter, also came under scrutiny. HUT investigators examined its immense magnetosphere, as well as the planet's curious, volcanically active moon, Io. A recent eruption on Io had deposited material onto the surface and into its tenuous atmosphere, prompting HUT co-investigator Paul Feldman to seek evidence of changes in the number of sulphur and oxygen ions in its environment.

The 'strangeness' of the Universe was illustrated by the peculiarities of many targets. Phi Persei, a hot, rapidly spinning star, exhibited an unusual ultraviolet spectrum, possibly due to a 'shell' of gas which may have been an outer layer shed by its fast rotation. Two active Seyfert galaxies, both strong emitters of ultraviolet radiation, were studied; one of them, NGC 4151, was five times brighter in March 1995 than it had been when ASTRO-1 observed it. Indeed, the galaxy exhibited a notable increase in luminosity over a matter of days during ASTRO-2. Ancient and young stars and stellar graveyards and nurseries came under the observatory's gaze, including an open cluster called N4, whose youthful occupants were believed to be less than ten million years old. Elsewhere, M104—a distinctive spiral galaxy, nicknamed 'The Sombrero'—was observed, revealing potential star-forming regions within its 'brim' and older stars (and maybe a black hole) at its 'crown'.

Periodically, Jernigan's professional background in astronomy was overtaken by pure childlike wonder. "There was a lot of time when the cockpit was darkened," she remembered. "You would monitor observation of an object for maybe 20 minutes before you had to regroup, repoint the Instrument Pointing System and set up the instruments again. There were whole blocks of time where you could just look out and reflect, talk to the other crewmembers who were awake on your shift and really have a sense of what a beautiful Universe we inhabit."

In sharp contrast with the technical difficulties of its predecessor, ASTRO-2 was also a beautiful payload from a standpoint of systems performance, with the IPS performing "in an outstanding manner", according to NASA's post-mission report. "The IPS and IMCS for the first time achieved operational capacity," exulted ASTRO-2 Mission Manager Robert Jayroe, then quipped: "In my estimation, the IMCS and IPS teams have done everything but make the hardware stand up and do a tap dance!" The WUPPE team gathered three times as much data as ASTRO-1 did, whilst UIT investigators reported that all planned celestial targets had been acquired and HUT scientists announced more than a hundred successful observations.

Already planned for 15 days, the ASTRO-2 crew quietly eclipsed the previous Shuttle endurance record on the evening of 16 March, with the expectation that Endeavour would land at KSC on the 17th. But it was not to be. Unacceptable weather in Florida forced mission managers to scrub the attempt and reschedule landing for the 18th. With the situation on the East Coast showing little sign of improvement, STS-67 was diverted to Edwards. Sweeping across the Pacific Ocean, the Shuttle entered U.S. airspace over the California coast and alighted on Runway 22 at 1:47 p.m. PST, concluding a remarkable mission of 16 days and 15 hours, with 262 revolutions of Earth.

But the mission's length and its months of training beforehand imposed an extreme toll. "The training is structured such that it trains to the lowest common denominator and it just takes forever," Oswald explained. "You're going through all the stuff again for those that haven't flown before. It got to be kind of a long, drawn-out deal. It was a great flight, great crew; I had a great time, but afterwards, I was just done." He was not the only astronaut left physically and mentally burned out by the intensity of a space mission.

THE LOST FLIGHTS

When Endeavour's six wheels kissed the Edwards runway three decades ago, the final curtain descended upon the IPS after only three missions. For a piece of hardware into which so much complexity and cost had been invested, its troubled first pair of flights and its flawless performance on ASTRO-2 were a sad footnote for a career which carried such great promise. Had circumstances panned out differently, the IPS might have supported a multitude of Spacelab missions, all of which breathed their last after Challenger.

Following its inaugural outing on Spacelab-2, plans were well established to refly it, not only aboard ASTRO, but also on a series of solar physics missions called Sunlab. In January 1985, the paper *Max '91: The Active Sun*, jointly commissioned by NASA and the solar physics division of the American Astronomical Society, outlined Sunlab as a reflight of the Naval Research Laboratory's HRTS instrument, Lockheed's SOUP and the Rutherford

Appleton Laboratory/Mullard Space Science Laboratory's CHASE, all mounted atop an IPS with a single Spacelab pallet and igloo. "We anticipate that the Sunlab complement will fly repeatedly as it evolves," noted the report, "perhaps with the addition of new instruments." Notable research emphases for Sunlab included measurements of accelerated protons in the solar wind and coronal mass ejections and flares from the Sun's visible 'surface'. It also included a guest observer program. According to NASA's March 1985 manifest, Sunlab-1 was baselined as a seven-day flight in September 1987, with a crew of six, implying only a single payload specialist. Candidates for that seat included John-David Bartoe and Dianne Prinz. Sunlab would have operated in a 295-mile (475-kilometre) orbit—one of the highest non-Hubble orbits ever attained by the Shuttle—inclined 57 degrees to the equator. In the same manifest, and again in the November 1985 manifest, follow-on Sunlab-2 and Sunlab-3 missions were requested as payloads of opportunity for March 1989 and January 1990 (Fig. 7.8).

Fig. 7.8: Pictured during final closeout activities ahead of Spacelab-2, several of the IPS-based payloads from that mission were scheduled to fly the three Sunlab flights in 1987, 1989 and 1990. All three fell foul to budget cuts in the aftermath of the Challenger tragedy.

But in hearings before the Subcommittee on Space Sciences and Applications of the Committee on Space and Technology in the House of Representatives in mid-1986, it was noted (without elaboration) that Sunlab-1 had slipped until October 1988 at the soonest, even before the loss of Challenger. Moreover, according to the hearings, the payload's state of readiness, as of January 1986, amounted to only 40-percent-complete. Sunlab's end was clearly nigh. In September 1986, the *Washington Post* reported that NASA had cancelled 18 Shuttle missions "over a five-year period for Spacelab", as a direct consequence of the Challenger tragedy.

Two weeks later, *Aviation Week* noted that the Spacelab integration budget was being reduced from $90 million per year to about $55 million per year, as NASA aimed to cut its workforce. A Shuttle/Spacelab integration team of 800-900 personnel had already been slashed by ten percent, with further cuts expected to lay off up to 350 employees. With these numbers in mind, it is not difficult to comprehend why so many Spacelab flights met their end via the budgetary axe. Specific 'at-risk' flights were unidentified, but Sunlab was plainly one of them. In a September 1986 letter to NASA Administrator James Fletcher, the National Solar Observatory's John Leibacher expressed dismay from solar physicists as Sunlab teetered on this precipice of cancellation. "The participants were concerned by the possible suspension of Sunlab," Leibacher wrote. "While we appreciate the difficult times facing U.S. space science, we respectfully urge you to maintain the momentum of this exemplary program."

Sunlab's woes extended back to Spacelab-2, with the performance of SOUP raising specific concern. It had lost power only 4.5 hours into the STS-51F mission and all power-up commands failed for the next five days, until it suddenly came to life only 36 hours before landing. Power was lost again as SOUP was being deactivated for re-entry and efforts to restore it proved unsuccessful. After the flight, tests pointed to a relay driver circuit as a possible culprit. A March 1986 letter to Spacelab-2 Program Manager Louis Demas of NASA Headquarters from SOUP Principal Investigator Ted Tarbell of Lockheed Solar Observatory laid bare the plans to refurbish SOUP for Sunlab. "The first serious problem was, of course, the power loss and reappearance," Tarbell wrote. "A redesigned and redundant power-switching and distribution system will be flown on Sunlab. The second problem was overheating of the focal-plane package, a problem shared by many experiments on Spacelab-2." Tarbell added that an updated charge-coupled device with a four-times-greater field of view, faster readout, greater sensitivity and uniformity of operation, reduced noise levels and lessened contamination risks would fly on Sunlab.

But it was not enough. In October 1986, Burt Edelson, associate administrator of NASA's Office of Space Science and Applications, told Tarbell that "difficult decisions" faced by the agency after Challenger had produced "33 fewer equivalent Shuttle flights through 1992". The inevitable result was that "the Sunlab mission has been put on indefinite hold until such time as the implications of the Shuttle manifest are understood and the possibilities for additional flight opportunities are exhausted". Edelson added that NASA would continue to support SOUP "at a nominal level" through 1987, as options to fly the instrument on a high-altitude balloon or aboard a deployable Shuttle Pallet Satellite (SPAS) or as part of a new-start program were explored.

However, in a March 1987 Congressional budget request paper for Fiscal Year 1988, Sunlab's launch date was listed as "unscheduled" and it would never reappear on Shuttle planning charts. Edelson's successor, Lennard Fisk, announced Sunlab's definitive cancellation in November 1987. He advised Tarbell that the "significant reduction in flight opportunities and increased cost associated with the stretch-out of our flight programs" had led him to "regretfully inform you that the Sunlab mission is cancelled". Fisk noted that "all reasonable possibilities" were exhausted in seeking a place for Sunlab, "including the potential improvements in the Shuttle manifest made possible with the increased orbiter down-weight capability, but without success". The post-Challenger flight rate, he added, "will be sufficient to accommodate only a modest number of the many payloads that exist or else are ready for launch". By the time that NASA's March 1988 manifest emerged, Sunlab was gone, both from the perspective of a mission number and target launch date and even as a requested payload of opportunity.

Another potential IPS mission in this timeframe was the Shuttle Infrared Telescope Facility (SIRTF), originally a pallet-based Spacelab payload, before being rescoped and redesignated the Space Infrared Telescope Facility—neatly retaining its original acronym—to fly instead as a free-flying satellite, later renamed the Spitzer Space Telescope and launched in August 2003. But three decades earlier, it began life as a contender for Spacelab. It was presented by GSFC in July 1973 as a small, cryogenically-cooled infrared telescope with a primary mirror up to five feet (1.5 metres) in diameter. Its sensitivity requirements dictated that its optics should be cooled to below -250 degrees Celsius (-420 degrees Fahrenheit) and SIRTF would conduct deep sky astronomical surveys and observations of quasars, galactic composition, gaseous dust clouds and conditions in the early Universe. At around the same time, the National Academy of Sciences heartily recommended a Shuttle-borne infrared telescope, expecting such a mission to fly in 1981.

But concerns about the molecular 'cleanliness' of the Shuttle's payload bay abounded, with risks of dust and moisture contamination threatening to ruin sensitive detectors. Added to that grim picture, thruster firings could induce hardware 'jitters', rendering SIRTF's potential scientific harvest negligible. There were other worries, too. "Even if the Shuttle were contaminant-free, a major scientific concern remained as to the amount of viewing time available on a Shuttle-borne telescope," wrote Renee Rottner. "With astronauts on board, the Shuttle could stay in orbit for only a few weeks. The scientists wanted a month or more of observing time."

However, the reusable spacecraft had an ace up its sleeve in that it was touted as capable of flying 50 times annually. This vaunted goal, of course, was never met, but still failed to tip the scientific community's preference for a longer mission over multiple short ones. As these Shuttle-based and free-flying SIRTF concepts did the rounds within NASA, the anticipated costs correspondingly rose. Some scientists agitated for a free-flyer, periodically visited by the Shuttle to replenish cryogenic coolant, but this caused costs to soar still higher. Similarly, a Spacelab-based SIRTF would sustain such extreme contamination that it would have to be disassembled, meticulously cleaned and essentially rebuilt after every Shuttle mission.

Nonetheless, an announcement of opportunity for a Shuttle-attached SIRTF was issued by NASA in May 1983, envisaging it as "an evolving scientific payload", with "several flights" planned, followed by a "probable" transition into a long-duration system aboard a future space station. A Ritchey-Chretien telescope, with sensitivity to a wavelength range from two to a thousand microns, its instrument chamber, optics, baffles and radiation shields were cooled by a dewar of super-critical helium in SIRTF's Multi-Instrument Chamber (MIC). Up to six scientific instruments were identified, with each SIRTF flight baselined for a "nominal 14-day mission life". Three instruments in particular were the Infrared Array Camera (IRAC), provided by the Harvard-Smithsonian Center for Astrophysics, Cornell University's Infrared Spectrograph (IRS) and a far-infrared camera, known as the Multi-band Imaging Photometer (MIPS), supplied by the University of Arizona (Fig. 7.9).

"SIRTF will be mounted on the [IPS], which will provide a basic pointing stability of a few arc-seconds," noted Michael Werner and Fred Witteborn. "Fine pointing to achieve the required stability is achieved by an image motion compensation system, in which high-frequency disturbances sensed by the gyros, but not totally corrected by the pointing mount, a 'fed forward' to the secondary mirror." Led by NASA's ARC in Mountain View, California, its project manager was Lou Young and the project scientist was Michael Werner.

Fig. 7.9: STS-51F roars to orbit on 29 July 1985, carrying the IPS on its inaugural test flight. Had launch schedules and funding profiles been kinder, the pointing system may have flown many more times than the three missions it actually did.

Sadly, SIRTF did not endure for long as a potential Spacelab payload. By September 1984, NASA concluded that a free-flyer was the best option for the mission.

Also under early consideration as a pallet/IPS payload was the Solar Optical Telescope (SOT), a high-resolution instrument with a 4.5-foot-diameter (1.25-metre) primary mirror to observe the Sun with sufficient angular resolution to investigate basic structural mechanisms and dynamics. SOT was proposed to NASA by the Association of Universities for Research in Astronomy in 1976, thanks principally to solar physicists Harold Zirin and Robert Howard. It was top-ranked out of four candidate solar observatories two years later. The concept was green-lighted as a GSFC-led program in October 1979 and after an announcement of opportunity in February 1980 proposals were received the following August. In the first quarter of 1981, proposals to construct SOT were solicited from U.S. industry, with Perkin-Elmer of Waltham, Massachusetts—which later earned notoriety as builder of Hubble's ill-fated optics—selected to assemble and integrate it. However, two years later, Spacelab budget restrictions caused a deferment of the program and the final deletion of funding in 1986, in favour of a rescoped design which envisaged a free-flying observatory.

SOT's downfall was overwhelmingly cost-driven. Initially proposed as a $25 million program—an estimate which even solar scientists in 1976 deemed far too low—it morphed into a financial monstrosity, swelling to $360 million by mid-1985. These rising costs were triggered by repeated delays to SOT's design and construction and hampered by development problems with Hubble but were harshly criticised by Congress. Finally, in February 1986 SOT's funding was deleted by the Office of Management and Budget. Efforts to redesign the telescope, removing one of its instruments, reducing its aperture size to 39 inches (100 centimetres), eliminating its ultraviolet-sensing capability and downsizing it to a targeted $100 million program reportedly did not get close to achieving their ends. (Even a revised estimate came in no lower than $189 million.) The Naval Research Laboratory and MSFC proposed to complete SOT at a cost of $86 million, but NASA Headquarters baulked at the notion of removing the mission from GSFC's auspices.

This was a pity, for SOT might have provided unique insights into the Sun's behaviour. Capable of providing solar surface resolutions as fine as 0.1 arcseconds, the telescope weighed 7,930 pounds (3,600 kilograms) and its observations were expected to produce insights into the source of the Sun's intrinsic magnetism, the nature of radiative transfer mechanisms between atmospheric layers and how its outer 'chromosphere' and glowing corona were heated. 'Coarse' pointing towards the Sun would have been accomplished by the IPS, with fine pointing furnished by SOT's pointing and control subsystems. Scientific instrumentation for SOT-1 consisted of a photometric filtergraph and a co-ordinated filtergraph/spectrograph from Lockheed Missiles and Space Company and California Institute of Technology, integrated into the combined payload. It was expected that SOT-1 would have been a high-inclination mission of 57 degrees, with the Shuttle inserted into an orbital altitude of about 280 miles (450 kilometres).

Had SOT flown, the mission's most optimum launch opportunities lay near the summer and winter solstices, wrote GSFC's Stuart Jordan, "since at these times a morning launch in winter or an evening launch in summer…provides the largest possible orbital angle for maximising solar observing time". In a 1981 paper, Jordan outlined SOT-1 as a ten-day flight, with launch-window opportunities of about two months per year, centred upon the

December and June solstices, producing "excellent potential flight opportunities…for SOT on Spacelab for four months out of every twelve". He added that, although the IWG process of picking payload specialists would follow the pattern of previous Spacelab missions, their presence on SOT-1 was "not currently planned", implying that the first flight would feature an all-NASA crew.

One other pallet/igloo/IPS-based mission which arose as a payload study in 1981 and was simulated in a Honeywell engineering assessment dubbed 'Treetops' in 1985, yet never secured a formalised place on the Shuttle manifest was the MSFC-led Pinhole/Occulter Facility (P/OF). This 9,250-pound (4,200-kilogram) payload was designed to conduct high-resolution X-ray imaging and simultaneous extreme ultraviolet and white-light spectroscopic observations of the Sun and numerous other celestial targets. Specific focuses included the structure of the corona and the mechanisms responsible for generating the solar wind, using instruments sensitive to photon energies above ten keV.

The P/OF comprised a coded-aperture device (in essence, a multiple pinhole camera), —whose encoder and detector were separated by a flexible boom, described in several sources as measuring between 105 feet (32 metres) and 164 feet (50 metres) in length, deployed from a Spacelab pallet in the Shuttle's payload bay and targeted to within a handful of arc-seconds by the IPS control electronics. "If the aperture is a large distance from the detector," wrote Hugh Hudson in a 1986 paper on the P/OF, "images can have excellent angular resolution, limited only by diffraction if problems of fabrication and deployment can be solved." At the tip of the P/OF's "self-deployed" mast was a 110-pound (50-kilogram) X-ray 'mask' and occulter disk, with the X-ray and extreme ultraviolet detectors and coronagraph optics situated on the detector plane at the base. This intriguing payload's extremely long baseline was touted as possessing far higher sensitivity and improved angular resolution—on the order of 0.2 arc-seconds—than earlier solar instruments and its capabilities were such that it could even observe cosmic X-ray sources as they transited the corona. Active galaxies, galactic clusters, supernovae, stars and the obscured centre of our own Milky Way were also targets of opportunity for P/OF.

According to J.R. Dabbs, Einar Tandberg-Hanssen and Hudson in an October 1982 paper, the P/OF had a total X-ray detecting area of 16 square feet (1.5 square metres) with optical 'grids' on both the occulter and detector planes. The white light coronagraph had an aperture of 19.7 inches (50 centimetres) and the extreme ultraviolet spectrograph—with a resolution of 300-1,700 ångströms)—had a 17.3-inch (44-centimetre) aperture. In

Honeywell's Treetops simulation of P/OF deployment dynamics, published in May 1985, it was noted that standard rate gyroscopes on the outboard portion of the IPS, coupled with accelerometers at the pointing system's base, would enable precise determinations of errors pertaining to mast-tip deflection, line-of-sight and mask-tilting.

Two other non-IPS, pallet-based missions which also vanished from the manifest during Spacelab's post-Challenger 'cull' were Dark Sky and two flights of the Space Plasma Laboratory (SPL-1 and SPL-2). The former had reached a state of 40-percent hardware/software completion by January 1986. Dark Sky was requested as a NASA payload of opportunity in March 1985, identified as requiring two Spacelab pallets and an igloo, and targeted for launch no earlier than September/October 1988. In the final pre-Challenger manifest, issued in November 1985, some additional detail was 'fleshed' onto the payload, noting that it would "conduct sky survey for extended infrared sources, X-ray imaging of galaxy clusters and…cosmic-ray measurements". Little more detail ever came to light, although informal discussions on Internet fora suggested that Dark Sky was likely a reflight of Pallets Two (the X-Ray Telescope, XRT) and Three (the Small Helium-Cooled Infrared Telescope, IRT)—and perhaps the University of Chicago's Cosmic Ray Nuclei Experiment (CRNE)—from the Spacelab-2 mission. The presence of only three principal investigators and seven co-investigators, according to data issued as part of the Rogers Commission report in June 1986, suggests that Dark Sky was plausibly a three-instrument payload. With the enforced stand-down of the Shuttle fleet following Challenger, it was reported that launch delays to Dark Sky ran in the order of $100,000 per month.

Yet even that figure was dwarfed by SPL-1, which reportedly cost $360,000 for every month that its hardware sat unflown on the ground. Although proposed in the early 1980s, SPL-1 made its first appearance in the June 1985 Shuttle manifest, baselined as a single pallet and an igloo, with launch targeted for December 1989. A second mission, SPL-2—"with augmented scientific instrumentation capability planned," according to Rogers testimony—would then fly in an expanded format of two pallets and an igloo in December 1991. By the time of Challenger's loss, SPL-1 had "completed the definition phase" and its hardware/software sat at a level of about 35-percent preparedness. A September 1983 NASA paper identified SPL-1 as a 6,600-pound (3,000-kilogram) payload, targeting a high-inclination orbit (presumably 57 degrees) at an altitude of 186 miles (300 kilometres) for a mission of seven to nine days. Its 11 experiments—three 'active' investigations and eight *in-situ* payloads, affixed to the Spacelab pallets—would utilise radio frequency transmitters and particle-beam accelerators to perturb the surrounding plasma environment and observe its response to those controlled perturbations. "NASA, Canada, France and Japan are supporting investigations for the Space Plasma Laboratory," it was noted of the mission, which originally appeared on Shuttle planning charts as Spacelab-6. "The transmitters, antennas, particle-beam accelerometers and several of the diagnostics instruments are carried on the pallets. Other sensors are carried on subsatellites to investigate the plasma effects produced by the transmitter, particle beams and the Shuttle wake."

The three active experiments included Japan's SEPAC and Utah State University's Vehicle Charging and Potential (VCAP), the first of which flew on Spacelab-1 and the second aboard OSS-1 and Spacelab-2. A third, Waves in Space Plasma (WISP), provided by TRW, would have employed a thousand-foot-long (300-metre) dipole antenna and instrumentation to generate plasma waves at radio frequencies from 300 hertz to 30 megahertz to investigate wave-particle interactions, propagation paths, map regions of near-Earth space accessible for communications and remotely sense variations in electron densities. Meanwhile, the *in-situ* experiments included a recoverable Plasma Diagnostics Package (PDP) from the University of Iowa, not unlike the payload deployed during Spacelab-2, but capable of attaining a greater distance—up to 60 miles (100 kilometres)—from the Shuttle. The University of Texas at Dallas provided its Energetic Neutral Atom Precipitation (ENAP), Lockheed's Palo Alto Research Laboratories in Palo Alto, California, supplied the Atmospheric Emissions Photometric Imager (AEPI) and France's Centre National de la Recherche Scientifique added its Atmospheric Lyman-Alpha Emissions (ALAE), all of which had either flown earlier Spacelab missions or would later fly aboard ATLAS (see Chapter Eight). The other *in-situ* experiments were the deployable Theoretical and Experimental Study of Beam Plasma Physics (TEBPP) to examine particle-beam interactions with ionospheric plasmas, Southwest Research Institute's Magnetospheric Multiprobes (MMP) to deploy a group of non-recoverable subsatellites to gather plasma data at up to six locations and Canada's Energetic Ion Mass Spectrometer (EIMS) to explore energetic ions in the magnetosphere and the GSFC-led Wide-Angle Michelson Doppler Imaging Interferometer (WAMDII) to measure temperatures, airglow, weather dynamics, wind speeds and upper atmospheric aurorae, specifically in the stratosphere, mesosphere, thermosphere and ionosphere.

Of added note, WAMDII found its way into the post-Challenger Shuttle schedule as a co-manifested payload, mounted atop TAPS, not unlike BBXRT. Standing 9.5 feet (2.9 metres) tall and weighing 995 pounds (433 kilograms), WAMDII was devised by Canadian researchers from the universities of Calgary, Saskatchewan, York and Western Ontario, together with the Herzberg Institute of Astrophysics at the National Research Council of Canada. Originally targeted to fly in May 1989 on a four-day Shuttle mission, the instrument would conduct 50 hours of observations from an altitude of 186 miles (300 kilometres), inclined 28.5 degrees to the equator. At the time of Challenger's loss, and three years out from launch, WAMDII had reached a state of 25-percent flight readiness, according to testimony laid down in mid-1986 during the presidential inquiry. By the August 1988 manifest, it was rescoped as a seven-day flight, with a five-member crew, due to fly Shuttle Atlantis on STS-48 in May 1991. But with the budgetary axe forever the enemy of so many would-be Shuttle payloads, funding eventually was withdrawn and by the January 1990 manifest WAMDII had vanished, never to return.

8

Earthgazers

A FLIGHT UNFLOWN

Years after it should have taken place, Owen Garriott did not recall doing much training for a Shuttle mission he never flew. On the day that Challenger vanished in a fireball in the cold Florida sky, Garriott was eight months away from launching on a flight like no other: the only Spacelab mission ever manifested with a 'short' pressurised module—the nine-feet-long (2.7-metre) 'core' segment, minus the 'experiment' segment (see Chapter Three)—and single pallet, a mission which began life as a pair of missions, later merged into one. The Earth Observation Mission (EOM), first planned in 1983, foresaw multiple flights of instruments previously demonstrated on Spacelab-1 for solar physics, astronomy, space plasma physics, atmospheric physics and Earth observations.

The first mission envisaged EOM as a primary payload, the short module equipped with data-processing utilities, a workbench and floor-mounted racks. It was connected to the Shuttle's crew cabin by an 8.7-foot-long (2.7-metre) tunnel and configured with a single pallet and MPESS truss at the aft end of the payload bay. According to NASA's March and June 1985 Shuttle manifests, this series' first mission was listed aboard Columbia with a seven-member crew and a seven-day flight, targeting a 186-mile-high (300-kilometre) orbit, inclined 57 degrees to the equator. Led by Mission Scientist Marsha Torr of MSFC, it was identified as a reflight of nine space plasma physics, solar physics, atmospheric science, astronomy and Earth observation instruments from Spacelab-1. Combined with a small life sciences payload, EOM-1 covered six scientific disciplines with 15 experiments involving the United States, Belgium, France, Japan and West Germany.

Columbia was reportedly chosen for EOM-1 on account of her desirability for a high-powered, long-duration mission, facilitated by a fifth cryogenic tank. But in testimony provided to the Rogers Commission in the aftermath of Challenger, as the EOM payload matured its evolving configuration produced an end-of-mission landing weight of 217,000 pounds (98,430 kilograms), substantially higher than the maximum allowable 214,000 pounds (97,000 kilograms). As a result, in the second half of 1985 the heavier Columbia was substituted for her younger sister Atlantis—27,000 pounds (12,000 kilograms) lighter—and by November of that year, in its final definitive pre-Challenger manifest, STS-61K was listed as a seven-crew, seven-day flight on Atlantis, creating "a reasonable solution to this problem", according to the commission. The orbital altitude was reduced to 155 miles (250 kilometres), but the 57-degree high inclination remained.

Follow-on EOM flights would adopt a pallet-only approach, with a single pallet, MPESS and command and control subsystems housed inside an igloo, rather than a short module. In the March and June 1985 manifests, up to 12 EOMs were either baselined or definitively requested as payloads of opportunity: a combined EOM-1/2 on 3 September 1986, followed by EOM-3 in November 1987, EOM-4 in November 1988, EOM-5 in November 1989, and EOM-6 through EOM-12 flying yearly each October from 1990 through 1996. Whereas EOM-1/2 was categorised as a 'primary' payload, EOM-3 onwards were co-manifested alongside other payloads. By the November 1985 manifest, EOM-1/2's launch on Atlantis had slipped to 27 October 1986, with the dates for subsequent EOMs unaltered, save for EOM-3 which moved to February 1988 as a seven-day flight with a crew of five or six. Interestingly, in the final pre-Challenger manifests EOM-3 also featured a lower orbital inclination of 28.5 degrees.

The mission's Spacelab-1 heritage made it not unsurprising that EOM should carry over some crew members from that first flight. In May 1984, Byron Lichtenberg and Mike Lampton (previously prime and backup payload specialists on Spacelab-1) were assigned to EOM-1, then scheduled for the summer of the following year. A few weeks later, Shuttle commander Vance Brand was assigned to lead the mission, designated STS-51H and rescheduled for November 1985. Joining him were pilot Mike Smith and mission specialists Garriott, Bob Springer and ESA's Claude Nicollier, with Lichtenberg and Lampton rounding out the seven-man crew. But Smith was named to another crew in the meantime and as the gap narrowed between his two flights to a point that he could not train for both, in September 1985 he was replaced as Brand's right-seater by Dave Griggs. And Springer, having drawn another Shuttle assignment, was himself replaced by veteran astronaut Bob Stewart.

In addition to crew changes, mission alterations also impacted the first EOM. In November 1984, STS-51H/EOM-1 was removed from the manifest, reportedly due to payload delays. In February 1985, *Flight International* reported that NASA had cancelled EOM-1, reassigned several of its crewmembers to other missions and merged its objectives into the second flight to create a peculiar nomenclature of 'EOM-1/2' for the first mission. Launch correspondingly slipped from November 1985 to September 1986 on STS-61K. Finally, in December 1985 MSFC physicist Rick Chappell—who served as Spacelab-1's mission scientist—and Belgian physicist Dirk Frimout of ESA were identified as alternate payload specialists, backing up Lichtenberg and Lampton.

By this point, largely due to the shift from Columbia to Atlantis, STS-61K's launch date moved to 27 October 1986. But more changes were afoot. Delays in the delivery of the Hubble Space Telescope, scheduled to fly Atlantis on STS-61J on 18 August 1986, prompted NASA to switch the mission order. On 6 January 1986, three weeks before the loss of Challenger, STS-61K was correspondingly moved into STS-61J's former launch slot. On 17 February 1986, STS-61K's preliminary crew activity plan was released—poorly timed, one might think, so soon after the tragedy—to reveal that EOM-1/2 would have launched at 10 a.m. EDT on 18 August and landed at 7:40 a.m. EDT on the 25th, after a mission of 165 hours and 109 Earth orbits.

Moving STS-61K earlier on the manifest "was made to provide additional contingency time for the delivery of the Space Telescope from the West Coast, through the Panama Canal to the Kennedy Space Center," NASA noted. "A concurrent benefit obtained as a result of the change will be a significant improvement in the scientific return from the EOM." Specifically, ESA's large-format Metric Camera from Spacelab-1, accommodated in the Scientific Window Adapter Assembly (SWAA) in the short module's roof, would "benefit significantly from a higher Sun-angle and an improved chance of better weather over the primary landmasses of interest in the Northern Hemisphere". NASA pledged to refly the camera after Spacelab-1 and for EOM-1/2 it was upgraded with high-resolution films and forward motion compensation system to facilitate considerable improvements in ground resolution. The mission's astronomy and solar physics experiments, explained Marsha Torr, also stood to make substantial gains from additional observing time provided by longer periods of orbital darkness in the Southern Hemisphere.

All these plans came to nought in January 1986, when the Shuttle fleet was grounded. But EOM-1/2 found its way into the Rogers Commission's final report, for precisely the wrong reasons: not only on account of its excessive landing weight when assigned to Columbia, but due to inadequate training provision afforded to its crew. By January 1986, the EOM-1/2 mission hardware was identified as 85-percent-complete and in his Rogers testimony astronaut Hank Hartsfield spoke about the impossibly short training times available to crews as Shuttle mission rates soared. "Had we not had the accident, we were going to be up against a wall," Hartsfield told the investigators. "STS-61K would have to average 33 hours" per week of crew time in the simulator, he added, to prepare for flight. "That is ridiculous," Hartsfield scoffed. "For the first time, somebody was going to have to stand up and say we have got to slip the launch, because we are not going to have the crew trained."

With EOM-1/2 not categorised as a critical payload when the Shuttle returned to flight, it became increasingly probable that—if the mission flew at all—it would certainly not do so before 1991. That reality made up Owen Garriott's mind (Fig. 8.1).

"It was clear that before my next chance to fly, it was going to be another five years," Garriott said, sadly. "In '86, I was fifty-six years old, and five years

Fig. 8.1: The dual Spacelab pallets of the ATLAS-1 payload are readied for installation into Atlantis' payload bay. Visible are two of the three spheres for the Space Experiments with Particle Accelerators (SEPAC) and the mission's igloo.

would have put me at sixty-one and I was getting to the point where I was a little doubtful that I would have enough time to really finish with that crew. It was going to be too long before I had a chance to fly again." Had EOM flown in 1986 or 1987, Garriott would have stuck around, but the lengthened delay made it "better to leave that to somebody else".

That 'somebody' turned out to be geologist Kathy Sullivan.

THE ATLAS METAMORPHOSIS

In the year after Challenger, former astronaut Sally Ride chaired a task force to formulate a new strategy for NASA. One of its recommendations was the implementation of a 'Mission to Planet Earth', intended to foster new technologies for studying the home planet, its atmosphere and climate. It was around this time that EOM-1/2 underwent a figurative and literal metamorphosis, changing both its payload configuration and its very name. Instead of the Spacelab short module, single pallet and MPESS, the new mission would utilise a pair of pallets and an igloo. A reduction in the maximum allowable landing weights of the Shuttle in abort scenarios, mandated following the presidential inquiry, drove a NASA decision to never fly the heavy module/pallet combination together again. This decision was made at some point early in 1986, for the Rogers Commission added a footnote to one of its hearings in June of that year that EOM-1/2 had already been "reconfigured from Spacelab module reflight to a pallet-only series of missions".

However, flying EOM's successor in this configuration was not too dissimilar to the pre-Challenger plans for EOM-3 through EOM-12, which were always manifested as pallet-only layouts. The successor program was renamed 'Atmospheric Laboratory for Applications and Science' (ATLAS), folded under the Mission to Planet Earth umbrella. "The nature of its scientific work," said Kathy Sullivan, payload commander on ATLAS-1, "really genuinely did align with the purpose of that new program, which was gaining momentum and clarity."

Early plans called for up to ten ATLAS missions, launched every 12-18 months for research in atmospheric chemistry, solar and space plasma physics and ultraviolet astronomy. Of added import was a comprehensive assessment of the extent of human-driven impacts upon the atmosphere through agriculture, forestry and heavy industrial practices. It had already become apparent by the late 1980s that chlorofluorocarbons, as well as halons and naturally occurring chemicals (including methane and nitrous oxide), together with increasing concentrations of carbon dioxide, were contributing to a significant depletion of stratospheric ozone and impacting global atmospheric temperatures. ATLAS was designed to explore the features, gaseous constituents and solar effects upon the troposphere, which extends from the surface to about 12 miles (20 kilometres), and above it the stratosphere, which rises to 30 miles (50 kilometres), the mesosphere, which ascends to 53 miles (85 kilometres) and the thermosphere, which tops-out around 430 miles (690 kilometres).

According to NASA's March 1988 Shuttle manifest, ATLAS-1 was baselined as a nine-day mission with a two-pallet 'train' and igloo, scheduled aboard Columbia on STS-44 in December 1990, targeting a 57-degree-inclined orbit at an altitude of 155 miles (250 kilometres). ATLAS-2 and ATLAS-3, scheduled respectively for STS-58 in June 1992 and STS-67 in January 1993, were listed as single-pallet/igloo combinations, but unlike ATLAS-1 they were not 'primary' payloads but co-manifested with others. According to this earliest incarnation on the Shuttle manifest, ATLAS-1 targeted a 57-degree high-inclination orbit, with ATLAS-2 and ATLAS-3 both at lower-inclination orbits of 28.5 degrees. By the time the January 1990 manifest was released, planning had changed only slightly. ATLAS-1 moved onto Columbia for STS-45, scheduled for April 1991, with ATLAS-2 following on STS-56 in June 1992, ATLAS-3 on STS-66 in April 1993, ATLAS-4 on STS-78 in March 1994, ATLAS-5 on STS-94 in September 1995 and ATLAS-6 on STS-106 in August 1996. Follow-on missions beyond ATLAS-6 were anticipated, but not formally requested. All six were slated for seven crewmembers, although it was left unclear if payload specialists would fly them all. Their orbital altitudes increased to 186 miles (300 kilometres) and the inclination for all six increased to 57 degrees.

For ATLAS-1's two-pallet/igloo combo, 12 instruments would support 14 experiments—six devoted to atmospheric science, four to solar science, three to space plasma physics and one to astrophysics—from the United States, France, Germany, Belgium, Switzerland, the Netherlands and Japan. Several were also expected to fly on the single-pallet follow-on ATLAS missions. "Reuse of these facilities," explained NASA's ATLAS-1 press kit, "also will allow scientists to expand their base of knowledge to provide a more accurate, long-term picture of Planet Earth and its environment."

A COMPLEX PAYLOAD

The forward pallet housed four instruments: Atmospheric Lyman-Alpha Emissions (ALAE), provided by the Service d'aéronomie of France's Centre National de la Recherche Scientifique, JPL's Atmospheric Trace Molecule

Spectroscopy (ATMOS), the Millimetre-Wave Atmospheric Sounder (MAS), supplied by Germany's Max-Planck Institute for Aeronomy and the Space Experiments with Particle Accelerators (SEPAC), developed by Japan's Institute of Space and Astronautical Science (ISAS) and overseen for ATLAS-1 by the Southwest Research Institute (SwRI) in San Antonio, Texas. All four had flown previously: ALAE, MAS and SEPAC as pallet-based payloads on Spacelab-1 and ATMOS atop the MPESS truss on Spacelab-3. Of this suite of experiments, ALAE sought to measure the abundance of 'common' hydrogen and deuterium ('heavy hydrogen') by observing ultraviolet Lyman-alpha emissions, which both radiated at slightly different wavelengths to understand atmospheric turbulence in the lower thermosphere and the rate of water evolution. ATMOS mapped trace molecules, including carbon dioxide and ozone, in the middle atmosphere during orbital sunrises and sunsets via infrared absorption for comparison with global models of worldwide, seasonal and long-term change. During ATLAS-1, it would play a particularly important role in studying aerosols emitted from 1991's eruption of Mount Pinatubo in the Philippines. And MAS examined the strength of millimetre-level waves radiating at specific frequencies of water vapour, chlorine monoxide and ozone to better model their distribution in Earth's upper atmosphere.

But perhaps the most visible of the quartet was SEPAC, whose three large black spheres—two on opposing sides of Pallet One, a third affixed to the starboard payload bay wall—explored ionospheric charged-particle dynamics by emitting 'streams' of xenon plasma to create very low-frequency to very high-frequency waves, 'clamping' the Shuttle's electrical potential to that of neighbouring plasmas. This was hoped to yield insights into the evolution of aurorae, as well as Earth's magnetic field and the effects upon the spacecraft. During Spacelab-1, SEPAC's electron beam assembly, capable of operating at voltages between 500 volts and 7.5 kilovolts at 1.6 amps, failed to function in its high-power mode, although it returned pleasing results.

"We were all pretty jazzed up about this," Sullivan said of SEPAC. "The idea was to have the orbiter oriented so that the aperture of the instrument would inject these electrons roughly along the magnetic field-line down towards the atmosphere, near the polar regions." She described it as a 'dose-response' type of experiment: just like in medicine, SEPAC would inject known energy 'doses' into the atmosphere, then measure the brightness of the resultant auroral glow. "If I know I put this many kilowatts of energy and I measured that luminosity," she said later, "maybe I can start to get a clearer understanding of how the energy of the incoming solar particles couples into the atmosphere and creates auroral luminescence." In Sullivan's mind, SEPAC was nothing less than "the biggie" among ATLAS-1's payload. Its failure on

Fig. 8.2: The payload crew for ATLAS-1. Payload commander Kathy Sullivan (seated, second from left) replaced EOM-1/2's Owen Garriott. Also assigned to the mission were Mike Foale (standing, right), who went on to serve as payload commander on ATLAS-2, and payload specialists Byron Lichtenberg (standing, left) and Dirk Frimout. The latter replaced Mike Lampton, diagnosed with kidney cancer a few months before launch.

Spacelab-1 had turned it into a prime driver for a reflight opportunity, which Sullivan described as "one of scientific life's tragic little ironies" (Fig. 8.2).

The aft pallet contained eight additional experiments. Of these, a notable holdover from Spacelab-1 was the University of California at Berkeley's Far Ultraviolet Space Telescope (FAUST), whose wide field of view was put to work on ATLAS-1 to acquire spectra of high-temperature celestial sources at far ultraviolet wavelengths between 1,300-1,800 ångströms to better understand the lifecycles of distant hot stars, faint diffuse galactic features, large nearby galaxies, quasars and stellar nebulae. Fogged film on Spacelab-1 ruined many of FAUST's images and on its second flight it recorded incoming photons electronically. However, on ATLAS-1 the telescope found itself hamstrung by a blown fuse which rendered it without power and left its aperture door stuck open, precluding any possibility of reactivating it in flight.

Abutting FAUST on the aft pallet were JPL's Active Cavity Radiometer (ACR), the Measurement of Solar Constant (SOLCON) from Belgium's Institut Royal Météorologique de Belgique and the Measurement of Solar Spectrum (SOLSPEC) from the Service d'aéronomie of France's CNRS, all three of which flew on Spacelab-1. Individually and as a unit, they tracked the total amount of solar energy received by Earth's atmosphere and its impact upon our planet's environment to further investigate solar-terrestrial relationships. SOLCON's self-calibrating radiometer worked in close conjunction with SOLSPEC's three spectrometers to observe the variability of solar irradiance with an accuracy of better than 0.1 percent and a sensitivity above 0.05 percent.

The final four ATLAS-1 experiments, also on the aft pallet, were the five integrated spectrometers of Utah State University's Imaging Spectrometric Observatory (ISO), the Atmospheric Emissions Photometric Imager (AEPI) from Lockheed's Palo Alto Research Laboratories in Palo Alto, California, the European Grille Spectrometer—a joint effort between Belgium's Institut d'Aeronomie Spatiale de Belgique and France's Office National d'Etudes et de Recherches Aerospatiales—and the Solar Ultraviolet Spectral Irradiance Monitor (SUSIM), provided by the Naval Research Laboratory. All four had flown on earlier Spacelab missions. During ATLAS-1, Grille data revealed clear aerosol 'bands' from the Pinatubo eruption and AEPI would function in close alignment with ISO to observe the optical properties of SEPAC's electron beams, as well as natural aurorae and atmospheric 'airglow' and Shuttle-generated emissions.

In addition to these 12 primary instruments, a further two investigations were also aboard ATLAS-1: Energetic Neutral Atom Precipitation (ENAP), provided by the University of Texas at Dallas, which used ISO to observe faint emissions in the thermosphere at visible and ultraviolet wavelengths during periods of orbital darkness, and GSFC's 720-pound (350-kilogram) Shuttle Solar Backscatter Ultraviolet (SSBUV), mounted on the Shuttle's payload bay wall in a pair of dustbin-sized Getaway Special (GAS) canisters. The first canister (equipped with a motorised 'door') contained SSBUV's spectrometer, five optical sensors and an in-flight calibration system, whilst the second housed command, data and power systems. Although it sat apart from the pallets, SSBUV formed an integral part of the payload and, according to NASA, was "equal in priority to the ATLAS-1 experiment science requirements". It observed atmospheric ozone levels and its data from ATLAS-1 helped to confirm a ten-percent ozone depletion in the northern hemisphere, and at mid-latitudes, most likely triggered by residual aerosols lingering in the upper atmosphere in the months following the Pinatubo eruption. Additional targets of interest included forest fires and the results of 1991 oil-fires, ignited by retreating Iraqi troops following Saddam Hussein's invasion of Kuwait. On later ATLAS flights, it would examine cold stratospheric temperatures in

1992-1993. All told, the ATLAS-1 payload, fully loaded with instruments and utilities, totalled 15,100 pounds (6,850 kilograms).

PERSONALITIES AND RESPONSIBILITIES

Late in September 1989, the ATLAS-1 science crew was announced for a nine-day mission redesignated STS-45 and slated to launch aboard Columbia in March 1991. Veteran astronaut Kathy Sullivan, a geologist, and British-born astrophysicist Mike Foale would serve as mission specialists, teamed with payload specialists Mike Lampton and Byron Lichtenberg, who had worked on ATLAS ever since their assignment to EOM-1/2. Rick Chappell and Dirk Frimout reprised their earlier roles on EOM-1/2 as ATLAS-1's alternate payload specialists. At the time, Sullivan was training for another Shuttle flight and it would be several months before she joined the ATLAS-1 crew in a full-time capacity. "We were rolling in pretty fast, because it was a multi-payload Spacelab flight," she said. "I probably joined it for real, fully, in late April, early May of '90 and we flew in March '92. That's a pretty quick turn-around for bringing a Spacelab crew, all the international linkages and science teams back up to speed."

By the time Sullivan began training, she had also picked up another title: 'payload commander', the mission specialist responsible for planning, integration, on-orbit co-ordination and long-range leadership of payload-related issues. "Payload commander," she said, "was the NASA mission specialist who would oversee and organise the typically two mission specialists and two payload specialists who work back-to-back shifts operating complex, multi-experiment Spacelab flights." Training for so many experiments, Sullivan explained, required a 'responsible' mission specialist to be assigned before the commander and pilot, to build relationships with scientific and payload teams and visit factories and laboratories for hands-on interaction with actual flight hardware. "To provide a long lead-time for the mission crew, to be sure that one of the NASA mission specialists is considered and recognised as authoritative in all those early planning decisions, you want that group to be able to make effective decisions and move the flight preparations forward," she said. "Naming a mission specialist as payload commander gave that authority."

That balance shifted in May 1990, when the orbiter crew—Commander Charlie Bolden, Pilot Brian Duffy and flight engineer Dave Leestma—were named. "When you combine the payload crew with the Shuttle crew, that balance shifts around," she said. "The Shuttle commander is the Shuttle commander; there's no two ways around that. You help the commander in that sense because you know the mission teams. You know the experiment teams. You have a little more insight about the personalities, cultures, background, mindsets that the payload team brings to bear and can help jumpstart the overall crew's understanding of that by the time investment that you've made." Bolden accepted an offer from the NASA psychologists to have his crew participate in a personality evaluation, based upon the Myers-Briggs psychometric questionnaire, to ensure that they could function at their peak in terms of performance and cohesion. "We were going to fly two out of four guys who'd been sitting around for a decade, waiting," said Sullivan of Lampton and Lichtenberg. "Quite a different mix of folks. I think Charlie knew he wanted to look at everybody to have a sense of how best to move them and drive them, support, encourage and propel them."

The tests revolved around six personality types, two of which described 'focused, goal-oriented' individuals, which Sullivan recognised as a trait representative of 15 percent of the general population…and about 98 percent of the astronaut population. Such personality characteristics for high-achieving astronauts are unsurprising, but Bolden and Sullivan considered the tests enlightening in that they uncovered each crewmembers' strategies for handling periods of calm and periods of extreme stress. "If strain goes up, pressure goes up, or anxiety goes up," Sullivan said. "Everyone has got sort of a home-base or default response pattern that they tend to move back to when things get strained. That's just normal human nature." Years later, Bolden rationalised his own thinking. "Although nine days is not a long time," he said, "I had learned from my first two flights that things do happen to crews once they get on orbit and I wanted to optimise our chances of being very successful and not having any problems." And Duffy added, tellingly, that relationships between the NASA crewmembers and the payload specialists was not entirely untroubled. "It wasn't friction-free by any means and there were sometimes when Charlie had to make sure things were smooth and that feathers weren't ruffled," he reflected. "For the most part, it was fine. But there was the occasional incident or two where people get a little crosswise with each other."

The hydrogen leaks which plagued the Shuttle fleet in the summer of 1990 (see Chapter Seven) rippled into the downstream manifest and with Columbia set for a lengthy period of maintenance and refurbishment, beginning in mid-1991, two of her planned Spacelab missions were correspondingly shifted onto her sisters, with Atlantis picking up STS-45 and ATLAS-1, now rescheduled for March 1992.

Six months prior to launch, in early September 1991, Lampton was suddenly removed from the crew "for medical reasons", and replaced by Frimout, a co-investigator for the Grille Spectrometer. Years later, Bolden remembered that Lampton had complained of back problems and was diagnosed with kidney cancer, from which he recovered. Sullivan recalled that Lampton's sickness began around Christmas 1990, but as he "got life-threateningly ill" over the succeeding months it was left to NASA Associate Administrator for Space Science and Applications Lennard Fisk to announce his replacement, with Chappell and Frimout both in the running. "The cancer," Lampton told Croft and Youskauskas, "was the sad thing." After so many years training for a chance to fly, Lampton's last opportunity was snatched from him. In the meantime, Chappell and Frimout were wrung through the Shuttle simulators—"Everyone take a look at them, bring them up the curve a little bit and see if we find any grounds in performance that argue one way or the other," Sullivan said—but Chappell and Frimout revealed no cracks or flaws. "Both very competent," she said of them both. "There wasn't really any high-level distinguishing factor there" (Fig. 8.3).

Sullivan's perspective on payload specialists harked back to her pre-NASA days working on oceanographic ships. "I was scientific party; I wasn't ship's party," she recalled. "I knew lots about my investigations and the equipment I was bringing aboard and because I always liked to learn those things, and it's who I am, I always tended to know a reasonable to fair amount about the ship." The relationship between 'professional' NASA crewmembers and payload

Fig. 8.3: Technicians install hardware into ATLAS-1's igloo, prior to integration. The twin pallets and instrument suite are visible in the background.

specialists followed this analogy: "a really good pairing of someone whose primary expertise was the instruments and had learned a little to a fair amount about the orbiter with someone whose primary expertise is the orbiter and learned a little to a fair amount about the payloads," she said. "You put one of each of those guys together as a pair on a shift and you've got a pretty good package of great depth and the ability to work really effectively at that interface."

Like several earlier Spacelab flights, the ATLAS-1 crew would work in dual shifts: a red team of Foale and Lichtenberg, led by Leestma, and a blue team of Sullivan and Frimout, led by Duffy. "I got to do a lot of things that, as a mission specialist, wouldn't normally get to do," remembered Leestma. "I got to fly the orbiter, do the manoeuvres. I got to manually fly it a couple of times, because the payload desired a manual slough through things by the orbiter, so I got to do a lot of neat things." As the flight's commander, Bolden anchored his schedule across both teams, but tended to align his work pattern with the blues, since it factored suitably into STS-45's planned re-entry and landing timelines. The planning was logical, Sullivan reasoned, as it enabled a proper spread of expertise across the two teams; for she, Bolden, Leestma and Lichtenberg had flown before, whilst Duffy, Foale and Frimout were 'rookies'. Asked years later if any criteria were used to separate the teams, Duffy deferred to the military chain of command. "You did what the commander told you to do," he laughed. "Charlie Bolden was the criteria!" And although many ATLAS-1 experiments could run autonomously, a shift system was still needed. "There was not enough automated interface," said Sullivan, "to operate all the experiments from the ground if you put the whole crew to sleep at the same time."

By March 1992, a mission almost a decade in the making was zeroing-in on a launch date, with Atlantis aiming for a 2.5-hour 'window' which opened at 8:01 a.m. EDT on the 23rd. Shifting from Columbia's long-duration capabilities, whose additional cryogenic reserves comfortably pledged a nine-day mission, with the potential for an additional 24 hours if consumables permitted, onto her sister Atlantis came with a corresponding decrease in flight duration. According to the STS-45 press kit, the mission was scheduled for a couple of hours shy of eight days, touching down at 6:08 a.m. EDT on 31 March after 126 orbits.

"LOOK AT THAT!"

Atlantis' opening launch attempt on 23 March was scrubbed when concentrations of liquid hydrogen and oxygen in the aft fuselage peaked above maximum allowable 'redlines'. Efforts to troubleshoot the problem failed to reproduce the exact cause and engineers concluded that the incident resulted from plumbing in the Shuttle's main propulsion system being improperly conditioned to the propellants. A second attempt the following morning saw liquid oxygen concentrations again peak above their redline, but they rapidly recovered in a situation that NASA described as "anticipated and acceptable". Without further ado, Atlantis took flight at 8:13 a.m. EDT. Within minutes of reaching their 186-mile (300-kilometre) orbit, the crew divided themselves into their shifts to commence activation of ATLAS-1, with Leestma's red shift slated to bed down for their first sleep period at 3.5 hours after launch to permit Duffy's blues to bring the laboratory alive.

Inclined 57 degrees to the equator, Atlantis' orbit carried her as far north at Juneau in Alaska and further south than Tierra del Fuego in Argentina, enabling atmospheric scientists to gather data from the tropics to the auroral regions and over diverse geographical areas, from rainforests to deserts and oceans to landmasses. "The challenge of the blue shift was to quickly jump out of the suits and grab some clothes and the books that we needed to begin operating," remembered Sullivan, "and by about two hours after liftoff, be up on the flight deck, with the doors open, and ready to power up the laboratory." The process was complicated, as the red team busied themselves on the middeck with their pre-sleep rituals, whilst Bolden, Duffy, Sullivan and Frimout laboured on the flight deck. A little past two hours into the mission, Sullivan activated the high-rate multiplexers and the DDUs that she and Frimout needed to command the ATLAS-1 instruments and completed initial activation of the Spacelab command and data-management subsystem, its three computers respectively supporting the experiments, subsystems and one in reserve as a backup. By the time Leestma's red team bedded down for the 'night', three hours after launch, ATLAS-1 was alive and humming, ready for eight days of continuous science. But the 12-minute launch delay, though meagre in the grand scheme of things, had shifted 'shadow' times of targets on Earth's surface by about one degree and 34 minutes, requiring several experiments' observation timelines to be tweaked.

Neither of the two shifts regarded themselves as 'competing' with the other. "Charlie didn't ever really use a device like that to drive performance," said Sullivan. "Commitment to each other, commitment to the mission [were] the intrinsic factors that he exemplified and reinforced. He wouldn't have needed to set up some fake game for me to make me do anything better." Leestma broadly agreed, although there was some light-hearted one-upmanship during training simulations. "Maybe one day the red team has their sim and the blue team has it the next day," he told the NASA oral historian. "When you're done, you ask your training team: 'Well, how did we do compared to those guys?' There's always that natural competitiveness."

The rear of Atlantis' flight deck proved an incessant hive of activity during both shifts, its relatively small volume packed to capacity with hand-held cameras, photomultipliers, filters, diffraction gratings and shades at the windows to reduce Sun-glare and allow the astronauts to adapt their eyes to the darkness. And from the front seats, the pilots would conduct over 220 manoeuvres. Twenty-four hours after launch, the first firing from SEPAC's coffee-can-sized electron beam generator was scheduled to occur in the early stages of the red shift's turn on duty. "Us blue guys were all down below" on the middeck, remembered Sullivan, "mucking around with dinner and starting to get changed for sleep. We knew it was coming and we were eager to see it, but we thought we should get out of their way, let them get set up for this and get into this."

All at once came a shout from the flight deck: "Oh, my God. Look at *that*!"

The cry broke a cardinal rule of spaceflight: astronauts ought never to utter sentences which ended in *that*—such as "What the hell was *that*?"—because it risked terrifying the other crewmembers on the middeck who could see nothing. In response to the red crewman's shout, the blues quickly swam up from the middeck, through the small access hatch to the flight deck to witness the cause of the commotion.

It was SEPAC, coming alive.

"Man, it takes about three seconds," said Leestma, "and everybody's up there…wondering what's going on." Seven noses were soon pasted to Atlantis' windows.

The spectacle reminded Sullivan of a sci-fi movie, as an oscillating blob of energy, like "some luminescent blue creature…about to ooze out", lingered momentarily atop the SEPAC canister, then abruptly shot into the atmospheric plasmas. "It goes by quickly," she said. "It was starting to curve away. You could see the curvature of the magnetic field-line. You could just see it begin to spiral along. All this material you drilled into your head in college physics…and now you're seeing, in front of your eyes, the curvature of the magnetic field-lines, the electron gyro-radius, as this thing spirals around it."

The red team's exclamation was surely understandable when faced with a sight of such captivating splendour. The eerie luminescence of SEPAC's beam reminded Leestma and Lichtenberg of a phaser, or some futuristic space 'gun' out of *Star Wars*. Interestingly, the STS-45 crew participated in the presentation of the Irving G. Thalberg Memorial Award to *Star Wars* creator George Lucas whilst in orbit. They also carried a real Oscar statuette aboard Atlantis.

Fig. 8.4: Glorious view of ATLAS-1 in Atlantis' payload bay, backdropped by the grandeur of Earth beyond. Note the twin canisters of the Shuttle Solar Backscatter Ultraviolet (SSBUV) instrument on the payload bay wall in left foreground.

But though Sullivan was unashamedly a fan of Lucas' films, "*our* photon torpedoes", from SEPAC, "were *much* better!" (Fig. 8.4).

Sadly, they were not nearly as long-lived.

SEPAC's demise was not unanticipated and the crew had foreseen just such an eventuality. "An experiment like this takes electric power from the orbiter into the pallet and then distributes it to the different experiments with an electrical bus on the pallet," Sullivan said later. "There's a fuse, of course, between the orbiter primary payload bus and the pallet. If really dumb stuff happens on the pallet, the fuse blows and the orbiter is protected. There was also some fusing within the distribution bus on the pallet, in fact, which probably came from all the littler experiments, saying *If this sucker fries, it'll kill all of us*." Privately, Sullivan was convinced that the ATLAS-1 principal investigators insisted on fusing SEPAC in order that any power outage would not take down the other instruments on the pallet. "I think it was like a 20-amp fuse between SEPAC and the other," she remembered. "It was 50 amps between the orbiter primary payload bus and the pallet and I'm sure it was a lower rating, like 20 amps, into SEPAC."

The STS-45 crew had squinted in surprise when they learned this fact.

"Isn't that awfully low?" one of them asked.

The SEPAC team agreed that, yes, the fuse rating was low, but it was buried so deep inside the experiment's housing unit that nothing could be done to rectify it. In Sullivan's mind, it was a problem waiting to rear its head. "Hold that thought," she told the NASA oral historian. "It comes back later."

And *later* for SEPAC came very quickly indeed. Without warning, on only its second or third firing, SEPAC's electron beam assembly abruptly arced and shorted out. "The fuse died," Sullivan concluded, grimly. "They got two or three doses off. We were distraught. We *told* them it was the wrong fuse." However, the SEPAC team announced that the data already gained was sufficient for their research needs. Years later, Sullivan looked back on STS-45 as the only Shuttle mission to have fired photon torpedoes.

Despite the disappointing loss of SEPAC, the remainder of ATLAS-1 proceeded without incident and produced spectacular results. ATMOS obtained measurements over latitudes between 30 degrees north and 55 degrees south, revealing a substantial uptick in the quantities of released industrial chloroflurocarbons since Spacelab-3: levels of inorganic chlorine had increased from 2.77 to 3.44 parts per billion and fluorine had risen from 0.76 to 1.23 parts per billion. Its data pinpointed stark atmospheric change following the Pinatubo eruption, including a thick layer of sulphuric acid droplets formed from sulphur dioxide. MAS observed a large day-to-night ozone variation at altitudes above 43 miles (70 kilometres), with far greater quantities on the 'night' side of the Shuttle's orbit. SOLCON and SOLSPEC produced values near to predicted levels, helping to further validate models of interaction between sunlight and the atmosphere. And SUSIM collected more than a hundred solar ultraviolet radiation measurements.

Atmospheric science stations as far afield as India, Indonesia, Japan and New Zealand made joint observations with the Shuttle. And Sullivan found time to marvel at the view of the aurora, a brilliant red and purple mass, extending from Africa to Australia. It reminded her of "a huge, richly brocaded theatre curtain", albeit that it hung *upwards* from Earth into space. The scientific part of her brain was momentarily overtaken by an unmistakable sense of childlike wonder.

Exceptionally economical use of cryogenic consumables meant that on 29 March the mission management team added an extra day to the flight, extending STS-45 from 190 hours and 126 orbits to 214 hours and 143 orbits, with landing moved to 2 April. Early that morning, as Sullivan had done in the opening hours of the flight, so Foale did the reverse in ATLAS-1's closing hours: deactivating the high-rate multiplexers, switching off the electrical power distribution boxes and the experiment computer and finally shutting down the Spacelab command and data-management subsystem. At 6:23 a.m. EDT, less than two hours shy of nine full days since launch, Atlantis landed at KSC, the first dedicated Spacelab mission to touch down back to her launch site.

The weeks following ATLAS-1 were consumed by traditional post-flight tours and, having transported Frimout—Belgium's first astronaut—into space, one notable focus was a journey to the Kingdom of Belgium. The astronauts had spoken to Prince Philippe in orbit and Duffy recalled the trip with fondness, for two reasons. "We were treated like royalty," he reflected, but the second reason was far more poignant. "It was nice for the spouses, because they don't get the rewards that we get when we fly."

ASTRONAUT ROYALTY

Notions of royalty and regal reward in the astronaut corps were not about meeting foreign dignitaries or lavish overseas trips. Rather, the mark of astronaut royalty was assignment to a space mission and the reward came when the SRBs' hold-down bolts detonated at T-0 to release the Shuttle from the launch pad and into flight. In short, astronaut royals were the ones who got to fly often. And for Mike Foale, that sense of 'oftenness' became apparent on 16 March 1992, a week before he launched on STS-45, when he was assigned as payload commander for STS-56, the ATLAS-2 mission, then scheduled for the following spring. In NASA's January 1992 manifest, the mission was listed aboard Endeavour in May 1993, but by the time of Foale's assignment another flight had moved further downstream, allowing ATLAS-2 to be reassigned to Discovery and brought forward to late March. Joining Foale were Commander Ken Cameron, Pilot Steve Oswald, flight engineer Ken Cockrell and electrical engineer Ellen Ochoa. Like the first mission, the astronauts would be divided into two shifts, labouring around the clock: a blue team of Cameron, Oswald and Ochoa and a red team of Cockrell and Foale.

Targeting an altitude of 186 miles (300 kilometres), highly inclined at 57 degrees, STS-56 was baselined for 198 hours and 132 orbits, although NASA noted that "an additional day is highly desirable and may be added if consumables allow", with "planning [to] accommodate the longer duration wherever appropriate". The mission was loaded with close to a hundred rolls of film, with expectations that the crew might acquire up to 5,500 photographs. And the high inclination orbit allowed for photography of southern areas rarely seen by the Shuttle: the Chatham, Bounty and Antipodes islands of the South Pacific Ocean, Iles Crozet and Heard Island in the Indian Ocean and Bouvetoya and the South Sandwich Islands in the South Atlantic Ocean. "Crew training for this task has consisted primarily of reviewing maps, since in the majority of cases there are no representative photos in the Space Shuttle Earth Observations Project database," NASA explained before the flight. "In addition, many of the islands are small and will be difficult to spot from the orbiter."

But with Shuttle delays having already impacted another Spacelab mission (see Chapter Six), STS-56 found itself pushed into early April, furnishing ATLAS investigators with an opportunity to "observe changes in Earth's ozone," according to Shuttle program director Tom Utsman, "during the seasonal transition between spring and summer in the northern hemisphere". Unlike ATLAS-1, the bulk of whose observations emphasised the southern hemisphere, this second mission focused upon the north and would make detailed measurements of stratospheric ozone over the Arctic. In order to view orbital sunrises at high latitudes, a night launch was highly desirable. Provisionally targeted to fly on 6-7 April, Discovery's countdown proved uneventful; that is, until its final seconds. During a pre-planned 'hold' at T-9 minutes, discussions among launch controllers about higher-than-expected temperatures on the anti-flood valve of the Shuttle's upper main engine were still unfinished, pushing the resumption of the countdown back by almost an hour. Eventually, at 1:23 a.m. EDT on the 7th, the count resumed. At T-31 seconds, right on the timeline, Discovery's computers assumed primary command of all critical vehicle parameters. The safe and arm devices of the SRBs were armed and at T-16 seconds the sound suppression water system came online, flooding Pad 39B with water to reduce the reflected energy at liftoff. Inside the cockpit, Cameron *et. al.* braced themselves for the immense jolt of Main Engine Start in the milliseconds after T-6.6 seconds.

All at once, at T-11 seconds, the clock stopped. The launch was scrubbed (Fig. 8.5).

It later became apparent that the hydrogen high-point bleed valve's closed indicator was not present at T-21 seconds. Although its absence was later verified to be little more than an instrumentation problem—and not an actual failure—it generated a breach of the launch commit criteria and mandated an abort. Engineers later checked the valve and found nothing amiss. Liftoff was rescheduled a day later on the 8th and Discovery and her quintet of astronauts speared into the darkened Florida sky at 1:29 a.m. EDT.

Several hours later, Discovery's payload bay doors were opened to reveal STS-56's twin primary cargoes: ATLAS-2, led by Mission Scientist Tim Miller of MSFC, and a unique retrievable satellite which the finest acronym-makers within NASA had carved into the name 'Spartan'. Built by GSFC, the 2,840-pound (1,300-kilogram) 'Shuttle Pointed Autonomous Research Tool for Astronomy' was a cube-shaped facility to accommodate instrumentation for

Fig. 8.5: Flying as a single-pallet configuration, and as a co-manifested payload, for its second and third missions, ATLAS-2 is readied for launch.

astrophysics or solar physics research. ATLAS-2 was demonstrably smaller than ATLAS-1, weighing about 8,360 pounds (3,790 kilograms) and housing half of the instruments—ATMOS, MAS, SOLSPEC, ACR, SUSIM and SOLCON—on a single pallet, rather than two, situated at the midpoint of the payload bay. An igloo sat at the front of the pallet, facilitating command and control functionalities. As such, ATLAS-2 emphasised two of the four scientific disciplines of ATLAS-1 (atmospheric science and solar science) but omitted its predecessor's space plasma physics and astrophysics focuses. With MAS provided by Germany, SOLSPEC a French-led instrument, SOLCON from Belgium and the others from the United States, and with direct mission participation from the Netherlands and Switzerland, ATLAS-2 represented six sovereign nations. Also reprising its role from STS-45 was SSBUV, its twin GAS canisters located on Discovery's starboard payload bay wall.

Activation of these payloads was complete four hours after launch and ATLAS-2 operations continued virtually uninterrupted until a dozen hours prior to landing. 'Virtually' uninterrupted, that is, because the instruments were temporarily paused during the Spartan deployment and retrieval, since Shuttle manoeuvres imposed constraints upon their pointing parameters. Deployment occurred at 12:11 p.m. EDT on 11 April, after which Cameron and Oswald manoeuvred the Shuttle to a separation distance about 170 miles

(275 kilometres) 'behind' the satellite, leaving it alone to conduct its solar observations. Two days later, Discovery drew close to Spartan, passing directly 'beneath' it and taking up a position for capture. The satellite was grappled by the RMS at 1:20 p.m. EDT on the 13th. A half-hour later, Spartan was secured onto its berth in the payload bay.

In the meantime, ATLAS-2's instruments went on to gather atmospheric data over 94 percent of the home planet. Although ATMOS suffered a problem with its high-data-rate transmitting capability, a new downlink format was developed to prevent any science from being lost. Meanwhile, the investigators responsible for MAS—whose three-feet-wide (0.9-metre) reflector dish rhythmically moved back and forth on the pallet during the mission—faced their own trials when their experiment suffered a glitch with its pointing system; this was eventually rectified through real-time software modifications. It successfully observed ozone at high latitudes in both the southern and northern hemispheres and measured the prevalence of water vapour, an excellent indicator of atmospheric motion. MAS data also confirmed that 'upward' atmospheric movement occurred close to the equatorial regions and 'downward' motion nearer the poles, at least during the springtime equinox. And although not strictly a member of ATLAS-2, SSBUV suggested a marked decrease in atmospheric ozone levels at northern mid-latitudes and performed the first observations of tropospheric sulphur dioxide concentrations over urban and industrial areas of the eastern United States, Europe and eastern Asia.

Atmospheric aurorae were also one of the mission's many spectacular sights, although on at least one occasion none of the crew were on the flight deck to witness it. Watching an automated payload bay camera from the ground, the CAPCOM in Mission Control was clearly impressed as the aurora's snakish fingers curled around the planet's pole.

"Discovery, Houston," he began. "We're seeing a beautiful view of what looks like the aurora out there. How's it look out the window?"

A few seconds of silence ensued. Then came Cameron's voice.

"We're missing it," the commander said, regretfully. "We're all downstairs in the middeck, reconfiguring things and getting the laptops done before the red shift goes to sleep. We'll have to take your word for it."

"Roger," replied the CAPCOM with a chuckle. "We'll see you on the other side."

Originally scheduled to return to Earth after eight days, the STS-56 crew were waved off by 24 hours, due to unacceptable weather conditions in Florida. Discovery's payload bay doors were reopened and some additional ATLAS-2 research was undertaken before Cameron and Oswald guided their ship back to KSC at 5:37 p.m. EDT after nine days in space and 148 orbits.

ENDS AND TRENDS

By the April 1993 Shuttle manifest, released just days after STS-56 landed, the future of ATLAS remained nominally healthy. ATLAS-3—co-manifested with a deployable satellite carrying infrared spectrometers and telescopes for atmospheric observations—was assigned to STS-66, scheduled for September 1994. The final missions in the series, ATLAS-4, ATLAS-5 and ATLAS-6, were listed as 'required' for launch in January 1995, January 1996 and January 1997, although NASA cautiously footnoted their assignments with an advisory that none were currently budgeted and their inclusion on the manifest was "for planning purposes only". In fact, the space agency had determined that U.S. taxpayer dollars could be better spent building free-flying spacecraft and launching them via expandable rockets, rather than the Shuttle, was cheaper in the long term. As such, funding lined up to finance the latter ATLAS missions was funnelled into NASA's Earth Observing System (EOS) to investigate the home planet's climate, atmosphere, weather and landmasses. The inevitable consequence was that ATLAS-3 marked the final mission of this highly successful program.

It was anticipated that ATLAS-3 would also employ a single pallet and igloo and refly the six instruments from ATLAS-2, together with SSBUV on the payload bay wall. "This reduces the cost of this space-based research," NASA noted, "and demonstrates the capability to return sophisticated equipment from space to Earth for refurbishment and reuse." All told, the payload totalled 8,287 pounds (3,758 kilograms) (Fig. 8.6).

Astronaut assignments for STS-66 began in August 1993, when ATLAS-2's Ellen Ochoa was named as payload commander, a rotation from one flight to the next that apparently was not intentional but offered important crew continuity. Five months later, on 10 January 1994, the rest of the crew was announced. Commanding Atlantis for the projected 259-hour, 172-orbit flight was Don McMonagle, with Curt Brown as pilot. Rounding out the six-member crew were flight engineer Joe Tanner, NASA physician Scott Parazynski and ESA astronaut Jean-François Clervoy, a French aeronautical engineer. With substantial European involvement in the program (and specifically, the presence of France's SOLSPEC) and it perhaps unsurprising that a French citizen was aboard. As the crew came together, launch was rescheduled for 27 October, but an additional week-long delay to 3 November later became necessary when an earlier flight's main engines (see Chapter Nine) had to be substituted with those originally earmarked for STS-66.

Fig. 8.6: The ATLAS-3 pallet is readied for installation aboard Atlantis.

Like their predecessors, the ATLAS-3 crew worked in shifts: a red team of Ochoa and Tanner, led by McMonagle, and a blue team of Clervoy and Parazynski, led by Brown. With launch targeted close to midday, part of the crew benefitted from what McMonagle described as a "very comfortable wake-up" at seven that morning, although the other half had sleep-shifted their schedule and had been awake since ten o'clock the previous evening. Atlantis roared uphill at 11:59 a.m. EST on 3 November, following a three-minute delay in her countdown, due to unfavourable wind speeds at one of the Transoceanic Abort Landing (TAL) sites. Weather at Zaragoza and Moron, both in Spain, were initially unacceptable, with Ben Guerir in Morocco also characterised as "marginal" in terms of crosswinds. Fortunately, the situation brightened by T-5 minutes and launch proceeded. This proved fortuitous, for if STS-66 had been scrubbed on the 3rd launch teams would have been required to stand down until no sooner than the 14th for another try, due to time limitations associated with cryogens aboard the CRISTA-SPAS payload.

As was the case with Spartan on STS-56, CRISTA-SPAS was not part of the ATLAS payload or the Spacelab program and is briefly mentioned here to establish context for the whole mission. However, unlike Spartan, it was classified as a 'primary' (rather than 'secondary') payload, co-manifested with the ATLAS-3 pallet. CRISTA—a tongue-twister of an acronym, denoting the Cryogenic Infrared Spectrometers and Telescopes for the Atmosphere—was led by Germany's University of Wuppental and had been proposed for development a decade earlier. It included three telescopes and four spectrometers to measure atmospheric emissions in the near-infrared and far-infrared portions of the electromagnetic spectrum. It sought to gather the first global measurements of medium-scale and small-scale disturbances in the middle atmosphere, including winds, wave interactions, turbulence and other physical processes. A second experiment, the Naval Research Laboratory's Middle Atmosphere High Resolution Spectrograph Investigation (MAHRSI), scanned the horizon to observe ultraviolet emissions from nitric oxide and hydroxyl to understand the processes responsible for destroying atmospheric ozone. "In short," Parazynski later wrote, "we'll be flying an alphabet soup of instruments studying climate change." Mounted atop a German-built Shuttle Pallet Satellite (SPAS), and deployed and retrieved using the RMS mechanical arm, CRISTA/MAHRSI spent eight days in free flight, gathering about 200 hours of atmospheric data.

STS-66 marked the first time that ATLAS had flown in the autumn/winter months and the instruments took measurements of the home planet's northern hemisphere at a time when atmospheric factors began to 'shift' away from relatively quiescent summertime conditions to the more active wintertime. It was known, for example, that chemical processes associated with the much-publicised Antarctic ozone 'hole' tended to peak in early October, as increased springtime sunlight struck air cooled during the southern hemisphere's winter season and solar ultraviolet radiation triggered chemical reactions which both created and destroyed ozone. (An approximately ten-percent decrease in total ozone had been noted at mid-latitudes in the northern hemisphere in the year between ATLAS-1 and ATLAS-2.) By late November, ozone-rich air from mid-latitudes mixed with Antarctic air to fill the lost ozone and chemicals such as nitrogen oxides began to consume 'free' chlorine, serving to 'repair' the lost ozone. As such, ATLAS-3 occurred on the cusp of this intermediate period of the atmospheric year, when the ozone hole had commenced recovery but had not yet closed.

As such, ozone science was a key thrust of ATLAS-3's research. "Observations of both areas," noted NASA's pre-flight press kit, "should provide valuable data for comparison with the spring data of ATLAS-1 and 2." After STS-66, preliminary analysis in the post-mission report indicated that the Antarctic hole was a self-contained region. Although it had seen an increase freon-22—a chemical "used as a replacement for chlorofluorocarbons (CFCs)" in the stratosphere, it was not as great a threat as CFCs. "It is, however, still a growing source of stratospheric chlorine," cautioned the report. "Also of note is the lack of a direct link between the Antarctic ozone hole and ozone depletion in the mid-latitudes, indicating that there are atmospheric processes that are still not well understood."

In a similar vein to the two missions which preceded it, ATLAS-3 sought to measure global temperatures in the middle atmosphere, together with trace-gas concentrations, and to provide this to the scientific community for comparison with data from other spacecraft. "The ATLAS-3 mission is the most complete global health check on the atmosphere that has ever been done," explained Mission Scientist Tim Miller, "measuring more trace gases that are important in ozone chemistry than any previous research effort." All of its instruments had flown before, but several had been extensively modified in readiness for STS-66 and their precise objectives shifted slightly. ATMOS, for example, benefitted from an improved recorder controller to provide ground-based scientists with more data about its status and performance. The solar science experiments were also prepared for quite different results, compared to their predecessors; for during ATLAS-1 and 2 the Sun had been at a period of near-maximum activity in 1992-1993 and on ATLAS-3 the orbiter's payload bay was directed toward our parent star on four occasions to investigate its behaviour at a time when its cycle of activity was wearing down toward a 'minimum' in 1996-1997.

The mission's high inclination also afforded spectacular views through Atlantis' overhead flight deck windows, including Mount Everest. One day, Parazynski decided to spot and photograph the fabled Himalayan peak. "I studied maps that showed our anticipated flight tracks over the ground, in particular the prominent landmarks in Tibet to the west of the mountain, including aptly nicknamed Bowtie Lake and Champagne Glass Lake," he wrote. "Even though we'll be zooming by at orbital velocity, they should guide me to major Himalayan glacial features, including the Rongbuk Glacier, which in turn will point me right to the mountain's summit. Sure enough, I'm able to spot it through my telephoto lens on the first opportunity and I feverishly snap some of the very best Everest photos ever acquired on a perfectly cloud-free day. These include some amazing stereo pairs, two images taken a few seconds apart, creating almost a topographic view of the various routes to the top" (Fig. 8.7).

With the successful retrieval of CRISTA-SPAS, the mission of STS-66 drew inexorably toward its close. Early on 14 November, ATLAS-3 was deactivated and the crew prepared for their return to Earth. That return was hampered, in Florida, at least, by the ravages of Tropical Storm Gordon, which battered the East Coast and the Caribbean with high winds, rain and cloud from the 8th until the 21st. Both scheduled landing opportunities at KSC were thus called off and McMonagle guided his ship smoothly into Edwards,

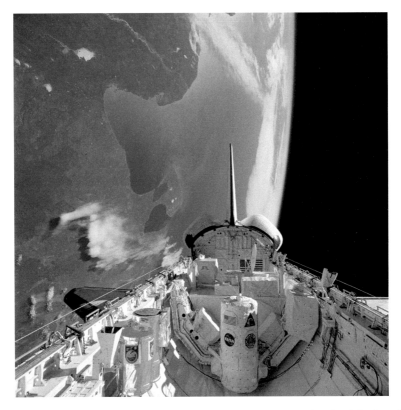

Fig. 8.7: Backdropped by the home planet, ATLAS-3 marked the final flight of this remarkable suite of Earth-watching instruments.

touching down three hours later than intended at 7:33 a.m. PST after 162 hours and 144 Earth orbits.

And despite ATLAS' tremendous raft of successes and a contribution to the annals of Earth and solar science which continues to this day, it remains disappointing that only half of the number of missions planned for this promising program were ultimately flown.

"Wheels Stop, Houston," radioed McMonagle, as Atlantis rolled to a halt.

"Copy, wheels stop, Atlantis," replied CAPCOM Kevin Chilton from his console in Mission Control. "Welcome home. It's a great way to end '94. Beautiful mission."

"Absolutely agree," McMonagle said. Then, keenly aware that with his picture-perfect touchdown the curtain had finally fallen on ATLAS, and perhaps speaking for the entire program and the half-dozen sovereigns it represented, he added with a measure of poignancy: "We appreciate all the international support we've had on this Mission to Planet Earth."

9

X-SAR-Crossed Lovers

OF WORLDS UNSEEN

The Empty Quarter, or *Rub' al Khali*, in southern Oman, is a barren and desolate place. Rich in scorpions and rodent life, it is the largest sand desert in the world, covering much of the southernmost third of the Arabian Peninsula, with a total extent of a quarter-million square miles (650,000 square kilometres), measuring 700 miles (1,100 kilometres) from its southwestern to northwestern extremities and 380 miles (500 kilometres) from north to south. An endless sea of 'hyper-arid' dunes, burnished a fiery reddish-orange by the ubiquitous presence of feldspar, together with plains of gravel and gypsum and brackish salt-flats, render this blasted landscape one of the most inhospitable on Earth. Thousands of years ago, several shallow lakes existed here, particularly in the Empty Quarter's southwestern corner, supporting a multitude of plant, animal and even human life.

Human habitation, that is, with the notable exception that none of our remains have been found here; only the nascent traces of our ancestors' chipped tools, more than two millennia old. Nowadays, the Empty Quarter is the most oil-rich site in the world, but in remote antiquity it was a caravan route for the frankincense trade through the interior of Arabia. One biblical city, supposedly involved in this highly lucrative trade, was Ubar (the 'Atlantis of the Sands'), whose residents became so corrupted by their new-found wealth that the Qur'an states that God obliterated them. The legend of the lost city endued for centuries and many explorers and adventurers (including Britain's T.E. Lawrence) vowed unsuccessfully to locate it. Then, in 1983, Los Angeles filmmaker Nicholas Clapp drew upon the possibility of using the

Shuttle Imaging Radar (SIR), to find Ubar from orbit. The radar first flew aboard Columbia on STS-2 in November 1981 (see Chapter Two) and mapped more than 15 million square miles (40 million square kilometres) at resolutions finer than 130 feet (40 metres). SIR-A penetrated dry sand dunes in northern Sudan and revealed the presence of long-dried-up river channels, ancient oases and Palaeolithic human settlements.

The radar's full capabilities were largely undemonstrated on STS-2 because the planned five-day mission was cut short by a fuel cell malfunction and Columbia returned to Earth after only 54 hours. Three years later, it flew again as SIR-B and its success provoked astonishment. Over eight days in October 1984, it found ancient caravan trails, enabled geologists to construct three-dimensional maps of subtle features on California's Mount Shasta, permitted contour modelling of parts of eastern and southern Africa and examined intricate structural features, including fault-lines, folds, fractures, dunes and rock layers. The mythical lost city of Ubar, though, remained magnetic in its siren-like appeal. "I was surprised to find that we were able to readily detect ancient caravan tracks in the enhanced Shuttle images," admitted JPL geologist Ronald Blom. "One can easily separate many modern and ancient tracks on the computer-enhanced images, because older tracks often go directly under very large sand dunes. We could never have surveyed the vast area where Ubar may have been, nor could we be confident of its location without the advantage of computer-enhanced images from space." Excavation began in the summer of 1990 and found a remote Bedouin well at Shisr, together with towers, rooms and human artefacts, all dating from at least two millennia before Christ.

The first Gulf War, which erupted early the following year, prevented further exploration, but in November 1991 archaeologists returned to the Omani desert, this time laden with sensitive ground-penetrating radar. Suspicion was acute that the lost city, which the Qur'an described as having literally been swallowed by the desert, may have collapsed into an underground cavity. As digging progressed, the remains of the site closely mirrored the description posited by Islam's holy book: an octagonal fortification, surrounded by stout towers and thick walls, within which resided storage rooms, frankincense burners and pottery sherds. In February 1992, the *New York Times* exulted that archaeologists were "virtually sure" that the site was Ubar. More recently, scientific scepticism has fallen upon the site's actual provenance as the mythical Arabian citadel, but even today the discoveries continue to whet archaeological appetites.

MAPPING PLANET EARTH

Who says 'no' when offered a ride into space?

Bob Crippen certainly did not, when George Abbey, JSC's director of flight crew operations, approached him late in 1983 with an offer to command the Shuttle's first dedicated Earth sciences mission the following August. STS-17 (later redesignated 'STS-41G') would deploy the Earth Radiation Budget Satellite (ERBS) and operate an expansive payload, sponsored by NASA's Office of Space and Terrestrial Applications and labelled 'OSTA-3', whose centrepiece was the SIR-B radar, mounted atop a single Spacelab pallet. It was an exciting offer, but triggered an acute problem, for Crippen was midway through training for another mission, targeted for April 1984 to repair the Solar Max satellite. Abbey felt that an anticipated ramp-up of the Shuttle launch rate to two dozen missions per year by 1987 made it important to understand how quickly individual astronauts (and particularly commanders) could be 'turned around' between flights. Crippen accepted Abbey's offer, but his Solar Max duties rendered him unavailable to join his new crew until the late spring of 1984. It was decided that his flight engineer on STS-41G, Sally Ride, who flew with Crippen on an earlier mission, would oversee crew training until the commander became available. Crippen and Ride were joined by pilot Jon McBride and mission specialists Dave Leestma and Kathy Sullivan, with the latter pair also expected to perform a spacewalk to evaluate in-flight satellite refuelling methods. By the time Challenger launched at 7:03 a.m. EDT on 5 October 1984, STS-41G had also picked up two payload specialists: Australian-born oceanographer Paul Scully-Power and Canada's first man in space, Marc Garneau. This gave the mission the largest crew ever to fly into space at that time.

ERBS and OSTA-3 were characterised as 'unique' payloads and as several other missions slipped on the manifest, due to problems with their boosters (see Chapter Three), STS-41G stayed firmly where it was and flew almost on time. Both payloads' Earth-observing commitments demanded a high-inclination orbit of 57 degrees and a higher-than-normal altitude of 218 miles (350 kilometres). "I don't know if that's good, bad or indifferent, but it was sure exciting to see our flight kind of hold its position while others slipped, because we figured, as they slipped, then we'd slip," he said later. "But because of the requirements of our payload, we had to fly at a certain time of the year and a certain inclination."

When Crippen, McBride, Sullivan, Ride and Leestma were named to the mission in November 1983, their launch was targeted for the following

Fig. 9.1: Technicians work on the SIR-B radar instrument, mounted atop its Spacelab pallet in KSC's Operations and Checkout Building.

August. But for a while, their orbiter remained a question mark. "When we first got assigned, it was on Columbia," said Leestma. "We ended up flying on Challenger. NASA always told us when we got selected for a spaceflight, not to fall in love with our orbiter or our payload, because they were liable to change." But with Columbia's additional cryogenic reserves, she could comfortably remain in space for ten days, longer than her sisters. And the Shuttle manifests of November 1983 and January 1984 baselined STS-41G as a ten-day flight on Columbia. Unfortunately, the gremlins very quickly reared their heads (Fig. 9.1).

Following Spacelab-1, Columbia entered a protracted period of refurbishment and maintenance and it soon became apparent that the amount of work needed—removing disarmed ejection seats, installing new instruments to gather aerodynamic re-entry data, fitting a Heads-Up Display (HUD) for the pilots and tackling general wear and tear on the Shuttle's airframe—would consume more time than expected. Challenger, as a newer vehicle, was roomier for STS-41G's anticipated large crew, although the astronauts preferred Columbia's ability to stay longer in orbit. But with STS-41G also scheduled to land on KSC's narrow, swamp-fringed runway, it was essential that the orbiter carried an updated instrument suite, including a HUD. (Indeed, *Flight International* reported in May 1984 that if Columbia did fly STS-41G, she would have to land at Edwards, on account of her lack of a HUD.) Not only were landings in Florida highly desirable as NASA sought to shorten Shuttle turnaround times, but Crippen himself wanted to land there, having been thwarted by bad weather on his two previous missions, which both diverted to Edwards. In any case, by the time NASA published a revised Shuttle manifest in May 1984, STS-41G was reassigned formally to fly Challenger, although her lack of sufficient cryogenic consumables meant the mission's duration was shortened from ten to eight days.

Still, even Challenger needed substantial refurbishment before she could fly again. Her previous landing in April suffered damaged and contaminated brakes and tarnished thermal protection materials, which pushed launch back to 5 October. It is important at this juncture to note that STS-41G was not a dedicated Spacelab mission, but rather the 4,254-pound (1,929-kilogram) OSTA-3 was co-manifested as a pallet-based payload alongside ERBS. The reader will recall that OSTA-1 flew on an engineering version of the pallet on STS-2 in November 1981 (see Chapter Two). Contracts worth $3.77 million were awarded by JSC in May 1982 to Rockwell International to integrate OSTA-3, initially "into Shuttle Orbiter Columbia". At that time, OSTA-3 was slated to fly STS-16, although as the manifest contorted in response to ongoing programmatic change it eventually moved one mission to the right.

Sullivan had been working on OSTA-3 mission development for several months before she was named to its crew, only learning of her assignment from Ride after returning from a backpacking trip. "I was the early consultant to a cargo group that's putting together a flight and I brought the crew perspective into all their planning efforts before they had an exact flight crew chosen," said Sullivan. "You may or may not end up flying that payload, so some crew is going to be assigned about 12 months before a launch. You'd like for them to be able to pick up a stack of checklists and systems documentation and know that there's been an informed crew viewpoint shaping them up to that point, so they're getting a fairly mature product to start doing the final detail and honing their own flight procedures."

Many of OSTA-3's payloads were OSTA-1 reflights: JPL's second-generation SIR-B radar stood proudly dominant, its large dining-table-like bulk overhanging the pallet when fully unfurled. Modifications incorporated since its first flight included an antenna tilt mechanism, which enabled it to move between 15 and 57 degrees, permitting target areas on Earth's surface to be viewed from several angles during successive orbital passes. This allowed SIR-B to image ground, water or vegetation, each of which exhibited greater levels of detail when seen from steeper or more shallow angular projections. During STS-41G, it was scheduled to collect 50 hours of data—including 42 hours of digital data and eight hours of optical data—at resolutions as fine as 80 feet (25 metres), covering 18 million square miles (46.6 million square kilometres), with particular emphasis upon southwestern Africa, India, the western coastline of Peru, China, central Australia, Saudi Arabia and California's Mojave Desert. The success of SIR-A in identifying traces of ancient human habitation beneath the Sahara Desert made this region another area of focus for the radar's second mission. SIR-B was also targeted at the Swedish island of Oland in the Baltic Sea to survey ancient Nordic ruins. Scientific tasks included studies of tectonic features in eastern Africa, Canada, the U.S. Midwest and Turkey, coastal landforms in the Netherlands, meteor craters in Canada, vegetation in North and South America, New Zealand, Australia and Europe, tropical deforestation in Brazil, the effects of acid rain in Germany and crop monitoring in Japan. Oceanographic experiments included the propagation of extreme waves in the southern Atlantic Ocean, sea ice in the Southern Ocean, icebergs off the Labrador coastline and oil spills in the Pacific Ocean and the North Sea.

Abutting SIR-B were the Feature Identification and Location Experiment (FILE) and Measurement of Air Pollution from Satellites (MAPS), both provided by NASA's LaRC, together with the JSC-managed Large Format Camera (LFC). The latter was built by the Itek Optical Systems Division of Litton Industries in Boston, Massachusetts, on behalf of JSC, and utilised 70 pounds (32 kilograms) of photographic film to acquire 2,400 negatives in colour and black and white. The 900-pound (400-kilogram) LFC sat on a truss-like MPESS and its 12-inch (305-millimetre) focal length sought to acquire more stable images with greater accuracy than earlier cameras, thanks to improved optics and electronics. Its unique lens combined high resolution with a wide field of view for precise stereo photography and a ground-level resolution as fine as 65 feet (20 metres). Specific targets on STS-41G included oil and mineral resource monitoring and mapping. Elsewhere, MAPS used an electro-optical sensor, digital tape recorder and aerial camera to measure distributions of carbon monoxide in the troposphere on a global scale. And FILE worked to classify surface features, such as water, vegetation, bare ground and snow cover, clouds and ice to prioritise timings for future Earth-watching instruments and reduce unnecessary downlinked data. Its hardware on OSTA-3 was identical to OSTA-1, save for the absence of its pyramidal sunrise sensor, deleted from STS-41G to afford "greater flexibility…for ground commands and alternative operation by the crew".

Targeting a 197-hour, 132-orbit mission, Challenger roared aloft at sunrise on 5 October 1984. Original plans called for ERBS to be deployed at 218 miles (350 kilometres) on Challenger's sixth orbit, eight hours after launch. Crippen and McBride would then lower their altitude to 170 miles (273 kilometres) on the second day of the flight, then 139 miles (222 kilometres) on the fifth day to support SIR-B observations. Before launch, some SIR-B investigators pressed for the radar to begin operations at the ERBS deployment altitude. But engineers determined that the Shuttle thruster firings to lower Challenger's orbit might impart excessive shocks on SIR-B's delicate carbon-fibre panels. Very quickly, STS-41G's carefully choreographed timeline fell apart. One of ERBS' two solar arrays took longer than predicted to unfold and the satellite was finally deployed three hours behind schedule.

But if ERBS' woes had already hamstrung the flight, worse was to come when SIR-B suffered difficulties transmitting data through Challenger's Ku-band antenna. Towards the end of their first day in space, the crew deployed the radar and for about two minutes it began transmitting data smoothly, then stopped. Engineers determined that SIR-B had lost its 'lock' on the geostationary-orbiting TDRS, due to a failed motor in the antenna's beta gimbal. With one of the antenna's movement axes effectively disabled and the other swinging listlessly backwards and forwards, a repair was needed. Ride and Leestma were tasked to unplug a wire that routed power to the antenna's motors; if they pulled it at the right moment—just as the Ku-band dish swung out at right-angles to Challenger's cockpit—the crew could reorient the Shuttle to reacquire TDRS. The wire sat behind a row of lockers and, after painstakingly removing them, Crippen watched the antenna's motions out of the aft flight deck window, then yelled down to Ride on the middeck to pull the wire at the right instant. The fix worked and the antenna ceased its erratic motions (Fig. 9.2).

Meanwhile, Sullivan retracted SIR-B to prepare for the manoeuvre to a lower altitude. But when she commanded the radar's two outer 'leaves' to fold onto the central section, it was obvious from their clunky motions that something was awry. She tried closing it with backup controls; again, to no avail. A third option was to fire pyrotechnics to slam the leaves shut, but such drastic action would render SIR-B unusable. Flight rules dictated that the radar *had* to be closed before any Shuttle manoeuvre could occur, for fear that it might suffer structural damage. When she first deployed SIR-B, Sullivan noticed that it wiggled and writhed in what Henry Cooper described as "a classic case of dynamic instability". Ride's dexterous handling of the RMS saved the day, when she gently pushed the antenna leaves into place. Mission Control was unhappy with this technique because there was no way of accurately gauging

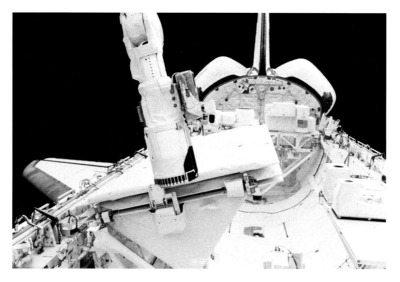

Fig. 9.2: Challenger's Remote Manipulator System (RMS) mechanical arm, deftly controlled by Sally Ride, gently pushes one of the SIR-B antenna 'leaves' into place.

the amount of force imposed on the fragile panels, but neither the arm, nor the radar, appeared dented or scraped. Ride also earned brownie points with Leestma and Sullivan, whose spacewalk—scheduled for 9 October, but later postponed until the 11th—might have been cancelled if SIR-B had not latched back into place.

The remainder of the mission was spent in the lower orbit, where the radar and two other Earth monitoring instruments could acquire their best results. Although MAPS encountered thermal fluctuations in its coolant loop, the experiment gathered 80 hours of data and clearly confirmed that burning in South America and southern Africa were a major source of atmospheric carbon monoxide. FILE acquired 240 images across a wide range of differing environments and the LFC shot 2,280 photographs, including high-priority targets of Mount Everest and the Dead Sea and oblique-angle views of aircraft contrails. Some of its images were of such extreme resolution that they were immediately classified by the Department of Defense. Other perspectives of Australia's Great Barrier Reef and Maine's national forests allowed for the development of improved maps and more accurately plotted topography. Fossil-fuel deposits were found in the Middle East, water sources pinpointed in southern Egypt and Ethiopia and geological evidence unearthed to reveal that blocks of land in China were forcing their way into the Pacific Ocean along the Kunlan fault-line.

Elsewhere, SIR-B serendipitously examined Hurricane Josephine, which had arisen in the western Atlantic Ocean on 7 October, observing wave patterns associated with its motion and speed. In other regions, soil moisture was measured to identify new water sources, support agricultural monitoring and crop forecasting and, during a pass over Bangladesh, hidden breeding grounds of malaria-carrying mosquitoes were found. Plant types in Florida and South America were successfully discerned, ocean waves over 65 feet (20 metres) high were recorded and polar ice flows and evidence of oil spills were detected. Despite the importance of this work for 'real-world' applications, SIR-B's claim to fame in the eyes of the public on STS-41G were its observations of Oman's Empty Quarter and the Shisr region's potential candidacy for the lost citadel of Ubar. "I was surprised to find that we were able to readily detect ancient caravan tracks in the enhanced Shuttle images," admitted Ronald Blom. "One can easily separate many modern and ancient tracks on the computer-enhanced images, because older tracks often go directly under very large sand dunes. We could never have surveyed the vast area where Ubar may have been, nor could we be confident of its location without the advantage of computer enhanced images from space."

Aside from opening and closing SIR-B with the RMS, the astronauts had little interaction with the OSTA-3 instruments. "We turned them on and off, changed the parameters and settings and did some fine-tuning," said Ride. "We changed data tapes for SIR-B. Our direct involvement was not really as scientists, but as operators." Yet the crew's modesty belied an undeniable truth: the failure of Challenger's Ku-band antenna would have wholly ruined SIR-B, were it not for their inputs.

With the dish now rigidly fixed in one place, Challenger had to firstly point the radar towards the ground, record as much data as possible on tape, then reorient herself to point the dish at TDRS for playback. This repetitive process slowed down how much radar imagery SIR-B could acquire. Seven tapes were aboard Challenger—each capable of storing 20 minutes' worth of data—but playback took just as long as recording. Although some steps were taken to maximise scientific return, such as 'dumping' data through TDRS whilst over an ocean so that the radar could be faced back to Earth when the Shuttle approached a landmass, it proved a laborious process. "We'd take data and then do data 'dumps' and point the orbiter at the TDRS, then we'd go back and do data 'writes', rather than being able to take data the whole time and point the antenna and dump it," recalled Leestma. "The SIR-B scientists didn't get all the data they wanted, but the mission was not a loss and they got almost everything." At one stage, the TDRS lost attitude control for almost 16 hours, then lost its lock on the White Sands ground terminal in New

Mexico, preventing it from being commanded. Given these problems, it is quite remarkable that so much valuable data was successfully returned from SIR-B.

Towards the end of her spacewalk with Leestma on 11 October, Sullivan took a few minutes to look at the radar at close range, with her own eyes, to discover why it had proven so stubbornly irksome to latch into place. "The insulation is billowing enough," Sullivan said, "to interfere with a single motor closing and you don't need to miss by much to keep the latch from shutting." This pure white blanketing had frustrated all previous efforts to close the antenna leaves. "It looked like the insulation was just a little bit too thick in the fold areas," added Leestma. "And so she could squeeze that down and she could tell that was…it was just the insulation, there was nothing else binding" (Fig. 9.3).

Despite the maelstrom of problems which so impaired STS-41G, a moment of fun came in the moments after Challenger's landing on 13 October, when Crippen finally made landfall at KSC. Sitting at the CAPCOM's console in Mission Control was astronaut Dave Hilmers, keenly aware that Crippen's three previous Shuttle landings ended at Edwards in California. As Challenger slowed to a halt, Crippen announced "Wheels Stop".

"You outfoxed us again," Hilmers said, dryly. "You landed at KSC, but the beer's been sent to Edwards!"

Fig. 9.3: Unusual perspective of the SIR-B radar, atop its Spacelab pallet, backdropped by the grandeur of Earth. The 50-foot-long (15-metre) RMS mechanical arm appears to curl, snake-like, in the foreground.

The tension of eight difficult days in space now behind him, Crippen guffawed at Hilmers' wit. "I don't believe it," he laughed. "Don't believe it!"

UNDER THE RADAR'S STEELY GLARE

As Challenger returned to Earth with SIR-B safely tucked away in her payload bay, at least two more missions of the radar (and potentially three) were planned but met with substantial delay and a marked change in launch site. On 19 October 1984, less than a week after Crippen discovered that his crew's welcome-home beer had been inadvertently dispatched to Edwards, a Shuttle Radar Laboratory (SRL) was booked as a payload for a future mission. Like OSTA-1 and OSTA-3 before it, this new facility comprised a single pallet/MPESS to accommodate the large radar and adjunct instrumentation.

Although it did not appear in NASA's March 1985 manifest, either as a designated Shuttle mission number or a 'requested' payload, it did turn up three months later. In the June 1985 manifest, it was identified not as 'SRL-1', but as 'SRL-2' (for reasons never made explicitly clear) and was expected to mark one of the first missions out of Vandenberg Air Force Base, a secondary Shuttle launch site in California. Assigned to STS-72A aboard Discovery, it was targeted for February 1987 as a five-day mission with a crew of five. Notably, the Vandenberg location in Santa Barbara County—whose unique geography permitted Shuttles to fly almost due-south and pass over virtually no landmasses until Antarctica—would permit SRL-2 to reach polar orbit. STS-72A was targeted for 90 degrees of inclination, the highest ever planned for any Shuttle mission, with an operational altitude of 186 miles (300 kilometres). By the November 1985 manifest, it had slipped slightly to mid-March 1987, retaining its duration and crew size, but its inclination was tweaked slightly to 88 degrees and its altitude raised to 210 miles (340 kilometres). "Polar orbit would be spectacular for looking at the Earth," wrote Rhea Seddon in *Go for Orbit*. "Launching due south out of Vandenberg toward the South Pole, the Shuttle would orbit across the pole, then head northward on the other side, crossing the North Pole. As it orbited, the Earth would revolve eastward, so that on the next orbit another swath of Earth would be visible."

In the aftermath of Challenger, Vandenberg as the Shuttle's notional second home base was indefinitely stood down and mothballed for good in November 1989. New filament-wound SRB motor casings, flimsy in construction but necessary for the weight savings to achieve polar orbit, were cancelled. Risks that hydrogen gas could become entrapped in the exhaust duct beneath the Shuttle's main engines during the ignition sequence were another worry. Surrounded by mountainous terrain, Vandenberg's acoustic, blast and thermal excesses threatened to upset computers and cause structural damage to the Shuttle. Added to that unpalatable list of safety hazards, the site's propensity to unpredictable weather, including ice, heavy rain and thick fog, made Vandenberg unattractive to fly from. By the time NASA's next definitive Shuttle manifest emerged in August 1988, the newly redesignated 'SRL-1' was 'requested' for July 1991 and formally assigned to fly on STS-54 in February 1992. Launching from KSC furnished no option to reach polar orbit and SRL-1's inclination was set at 57 degrees, although its altitude remained 186 miles (300 kilometres). It was expected that SRL-1 would feature seven crewmembers and last seven days.

Under this new manifest, two follow-on SRL missions were also baselined. SRL-2 was requested for November 1992 and provisionally pencilled-in for STS-69 in March 1993, with SRL-3 requested for April 1995 but its inclusion was footnoted "for NASA planning purposes only". Both missions were expected to include the same number of crew, the same flight duration, the same orbital altitude and inclination as SRL-1. A measure of surety had emerged by January 1990, with SRL-1 now requested by June 1992 and scheduled to fly a month later. Now assigned to Discovery on STS-57, its altitude was reduced to 150 miles (240 kilometres), its inclination remained 57 degrees and its crew of five would spend nine days aloft. SRL-2 and SRL-3, respectively targeted to fly aboard Columbia on STS-75 in December 1993 and Atlantis on STS-97 in December 1995, would follow, the only difference being that the final mission was scheduled for seven days rather than nine. Shuttle delays throughout 1990 and 1991 eventually pushed SRL-1 into late 1993.

In August 1991, crew names for this first mission took form. Physicist Linda Godwin was assigned as payload commander for STS-60, which was not expected to fly before October 1993. Six months later, planetary scientist Tom Jones was named as a second mission specialist. In his memoir, *Skywalking*, Jones recounted the electrifying news of his first flight assignment, delivered by Don Puddy, JSC's director of flight crew operations. Sitting at the conference table with Puddy was Godwin. "I vaguely remembered that Linda had been assigned earlier to work on long-range planning for one of next year's Shuttle missions," Jones wrote, "but I couldn't recall which one."

That forgetfulness was quickly assuaged.

Operating from 57 degrees, SRL-1—now known as the 'Space' Radar Laboratory, neatly enabling it to retain its pre-Challenger acronym—would map a substantial portion of Earth, including regions as far north as Juneau in Alaska to a tad further south of Tierra del Fuego in Argentina. By now, SRL-1's mission designation had changed to STS-59 and in the January 1992 manifest was baselined aboard Endeavour in September 1993, with a crew of seven. The follow-on SRL-2 and SRL-3 missions were listed as 'required' for the first quarter of Fiscal Year 1995 and the third quarter of Fiscal Year 1996. In early March 1993, the remainder of the crew, which by now had moved on the manifest to become STS-59, were announced. Commander Sid Gutierrez, Pilot Kevin Chilton, Army helicopter pilot Rich Clifford and physicist Jay Apt rounded out the crew. Jones recalled a joyful telephone call from Clifford: "We're flying together on '59, T.J." Gutierrez divided the crew into two shifts, leading the red team with Godwin and Chilton, and having Apt lead the blue team with Clifford and Jones. "He needed the two pilots on the same sleep cycle, both fresh and wide awake for launch and landing days," Jones wrote. "Linda would join them, enabling Sid to get the payload commander's input on any experiment or orbiter problems without having to wake the opposite shift" (Fig. 9.4).

Fig. 9.4: SRL-1's red shift, pictured in their coffin-like sleep stations on Endeavour's middeck. From top to bottom are Sid Gutierrez, Linda Godwin and Kevin Chilton.

Whereas Gutierrez, Chilton and Godwin had flown previously to low-inclination orbits, only Apt and Clifford had experience of 57 degrees and they were keenly aware of its usefulness for Earth observations. In his book *Orbit*, Apt drew parallels between his first and second Shuttle flights: the first entered a 28.5-degree orbit, the second (see Chapter Ten) attained 57 degrees. "On my first flight, I flew no farther north than the glorious Himalaya," he wrote. "I shot photo after photo of Tibet, with the Sun low in the sky and the shadows long. Central and northern Asia were a mystery to me. On my second and third flights, I was on the flight deck for a 12-hour shift, when it was 'night' in Houston and 'day' in Asia. We flew over almost the entire continent."

From 57 degrees, Apt saw smoke rising from cellulose plants at the southern shore of Lake Baikal, irrigation channels cutting across the Taklimakan Desert, multi-coloured soil tones in Kamchatka (the latter of which, he wrote, was "worth the trip up north") and the puzzling landscape of the Korean peninsula. In the northern Americas, he beheld drifting volcanic ash from Mount Spurr in Alaska, together with the breathtaking grandeur of Yellowstone National Park and the geological variety of the western states. South America was most memorable for Apt, as he was able to see smoke and fires in Tierra del Fuego and Patagonia, presented to his eyes as far-off points of light. Even though he had flown to 57 degrees on his second mission, Apt found that lighting conditions were not good—"Our windows were pointed into the Sun most of the time"—but his third flight on SRL-1 benefitted from much better conditions. In preparation for the immense amount of photography which would be conducted, alongside the radar observations, all six crewmembers became proficient in the use of appropriate camera lenses with consistent shutter speeds and light meters to determine proper exposure times.

To understand how SRL-1 evolved, it is important to comprehend the accomplishments of SIR-A and SIR-B. The first mission was restricted to recording the ground-track directly 'beneath' the orbiter, but in preparation for SIR-B the radar was engineered to 'tilt' at angles of between 15 and 57 degrees to the side and its imaging resolution was enhanced from 130 feet (40 metres) to 82 feet (25 metres). By varying its 'look' angle in this fashion, it became possible to assemble 'mosaics' of adjacent surface features, collected over periods of several days. The third mission, SIR-C—renamed the 'Spaceborne' Imaging Radar, rather than the 'Shuttle' Imaging Radar, which, like SRL, neatly enabled it to keep its pre-Challenger acronym—offered multi-frequency, multi-polarisation imaging and represented the first spaceborne radar with the ability to transmit and receive horizontally and vertically polarised waves at both L-band and C-band wavelengths.

Unlike previous SIR missions, its radar beam was formed from hundreds of transmitters, embedded in the antenna's surface. By properly adjusting the energy from these transmitters, the beam could be electronically 'steered' and, when combined with Shuttle manoeuvres, offered the scope to acquire images from various directions. STS-59 would conduct over 460 manoeuvres, requiring thousands of keystrokes by the crew, more than any other Shuttle mission at that time. All told, inclusive of support equipment and utilities, SRL-1 tipped the scales at 23,800 pounds (10,800 kilograms). But SIR-C was not the only radar aboard SRL-1. Running like a strip along its upper edge was 'X-SAR', a 39.4-foot-long (12-metre) synthetic-aperture radar. It was built by the German Dornier and Italian Alenia Spazio industrial firms, together with the Deutsche Agentur für Raumfahrtangelegenheiten (DARA, the German Agency for Space Flight Affairs) and the Agenzia Spaziale Italiana (ASI, the Italian national space agency). It provided a single-polarisation radar, operating at X-band wavelengths. X-SAR was a follow-on project from Germany's Microwave Remote Sensing Experiment (MRSE) flown on Spacelab-1 (see Chapter Three). Its 'slotted waveguide antenna' was finely tuned to produce a narrow energy beam and X-SAR was mechanically aligned with the L-band and C-band beams of SIR-C. Before launch, NASA's SRL-1 press kit touted resolutions as fine as 33 feet (11 metres). Throughout a projected nine-day mission, the combined SIR-C/X-SAR suite was expected to conduct around 50 hours of observations, cover more than 19.3 million square miles (50 million square kilometres), acquire 32 terabytes of raw data and store it all on 180 cassettes, using three high-density, digital, rotary-head tape recorders.

Both radars would work in tandem. The shorter-wavelength X-SAR was useful for determining snow typology, with the L-band and C-band capabilities of SIR-C estimating snow volumes. Although the United States, Germany and Italy were involved in the development of the payload, 49 scientific investigators and three associates from 13 nations—Australia, Austria, Brazil, Canada, China, the United Kingdom, France, Germany, Italy, Japan, Mexico, Saudi Arabia and the United States—constituted the science team. Also operating on the payload was MAPS, the radar's old friend from STS-2 and STS-41G. Mounted on an MPESS truss at the forward end of the payload bay, it was elevated on a tilted support structure in order that its sensors were pointed in a nadir-viewing direction.

Led by three project scientists—JPL's Diane Evans for the United States, DLR's Herwig Ottl for Germany and Mario Calamia of Italy's University of Florence—and overseen by Program Scientist Miriam Baltuck of NASA Headquarters, SRL-1 did not scan the planet indiscriminately. Data was acquired from 400 worldwide sites, including 19 'super-sites' (and 15 backup super-sites) of particular geological, hydrological or ecological import to be observed during daily radar 'passes' in conjunction with intensive fieldwork carried out before, during and after the mission on the ground. Experiments were set up on every continent, save Antarctica, and particular focal regions included Mammoth Lakes in California for remote-sensing of the water content of the Sierra Nevada ice-pack, the Austrian Alps and southern Chile's Patagonia for snow-cover analysis, Chichkasha in Oklahoma, Ötztal in Austria, Bebedouro in Brazil and Montespertoli in Italy for soil-moisture studies, Michigan's Upper Peninsula to investigate forest biomass, together with temperate forests in North America and Central Europe and tropical forests in South America's Amazon basin, surface and internal waves in the Mid-Atlantic Gulf Stream and Hawaii's Volcanoes National Park. During SRL-1, 'ground-truth' teams at these sites measured vegetation, soil moisture, sea state, snow cover and weather conditions to correlate with the Shuttle data.

ROUND TWO

With such an enormous workload, it came as a measure of relief in mid-1992 when NASA opted to shift STS-59 into the spring of 1994 and move the Shuttle's critical repair of the Hubble Space Telescope forward to December 1993; since both missions used Shuttle Endeavour, their order on the manifest was correspondingly switched. According to the STS-59 press kit, launch was scheduled for 8:07 a.m. EDT on 7 April 1994, although teams were reportedly "protecting an option in the countdown timeline" to fly up to an hour sooner at 7:07 a.m. EDT to permit managers to "evaluate predicted climatological and atmospheric conditions" in the KSC area. But neither option proved propitious for Endeavour. An inspection of a Shuttle main engine at prime contractor Rocketdyne's facility in Canoga Park, California, found that the critical dimensions of nickel-alloy liquid oxygen guide-vanes in the turbopump preburner were out of tolerance. Inspectors discovered that two components had sharp (rather than rounded) tips, posing a risk of fragmentation…and, in flight, a premature engine shutdown. "A thinned or deformed vane," wrote Jones, "could break off in the oxidiser flow, shattering the turbopump blades downstream."

Liftoff was delayed to 8 April to inspect Endeavour's vanes, which turned out to be of the correct configuration. High winds put paid to STS-59's first attempt to go airborne and early on the 9th the astronauts departed their quarters—"Round Two, Round Two," chirped Clifford, confidently—and headed for Pad 39A. This time, the weather gods were with them. For Jones,

Fig. 9.5: Tom Jones looks through Endeavour's aft flight deck windows into the payload bay. Notice the large-format Linhof camera in an overhead window at far left and the Velcro patches, recommended by Godwin, to hold cameras and lenses securely in place.

the 7:05 a.m. EDT liftoff yielded a "nasty shaking" and a peculiar sense that the entire cabin was whipsawing around him. Thoughts of his father, who died a year earlier, flooded into his brain. Eight and a half minutes later, he was in orbit (Fig. 9.5).

But the clock was forever their enemy. As time ticked down toward the first sleep period for the blue team, Godwin's red shift activated SRL-1. In her recollections, there was "no doubt" that Endeavour's aft flight deck was the focus of activity. Before launch, the payload commander requested extra covers to be made for the aft flight deck panels, containing Velcro squares to hold cameras, filters, lenses and spot-meters securely in place. Downstairs on the middeck's starboard wall was a four-berth sleep station: three bunks for the astronauts and a lower tier for storage. Despite initial problems with the power-up of the X-SAR amplifier—caused by an overly sensitive protection circuit—the German-built radar entered operations on 10 April. Although the mission had always been baselined for nine days, it was considered "highly desirable" for a tenth day to be eked out of Endeavour. The decision to extend STS-59 came late on the first day of the flight, setting the mission on its path of science-gathering discovery in fine fashion.

Overall, SRL-1 was a remarkable scientific triumph. In its post-flight mission report, NASA announced that SIR-C/X-SAR accomplished 97 percent of their required data takes from 400 primary science targets and 99 percent from the 19 critical super-sites. The crew also handled additional requests, including imaging Germany's Rugen Island in the Baltic Sea and Japanese rice fields. Ninety-four hours of radar data, taken over 44 discrete nations and covering an area in excess of 43.75 million square miles (70 million square kilometres), were stored on 166 digital tapes. MAPS functioned flawlessly, acquiring data on the regrowth of forests in fire-scarred areas of China. Jones also relayed a verbal report on thunderstorms over Taiwan, the Philippines and New Guinea to augment the MAPS data.

On 19 April—after Godwin took one last, wistful glance at the laboratory whose development she had led for almost three years—the payload bay doors were closed and sealed, but cloudy conditions at KSC forced the first landing attempt to be waved off. "Unfavourable and dynamic" weather later that afternoon put paid to a second attempt to bring STS-59 home. This forced mission managers to reschedule landing for the following day, the 20th. But weather in Florida remained unacceptable and Gutierrez and his crew were diverted to Edwards.

"We be home," yelled Chilton, as the Shuttle rolled to a halt.

"All *riiiggghhhttt*!" whooped the rest of the crew.

"How'd it feel, bud?" Jones asked Gutierrez of his first landing as a Shuttle commander.

"Felt good," Gutierrez replied. "Little fast."

LURE OF THE UNEXPECTED

A little fast, indeed, was an apt turn of phrase for a mission which passed from start to finish like a whirlwind, but might also serve well to encapsulate Jones' meteoric astronaut career in 1994. "No longer a rookie," Chilton said. "Ready for his next flight."

And that next flight would not be long coming.

With the ink barely dry on their STS-59 flight report, Jones returned to his Houston office two weeks later to begin his next assignment. More than two years of his professional life had been devoted to SRL and in August 1993 he was assigned as payload commander for SRL-2. JPL requested that two of its experts ought to serve as payload specialists on the flight but NASA rebuffed this, countering that its mission specialists could handle the work. Next, JPL insisted that at least one crewmember should fly both flights. This followed typical NASA practice of 'carrying-over' an experienced crewmember from one payload to the next on important science flights. But Jones' transition from SRL-1 to SRL-2 proved somewhat different. Originally, the two flights sat a year apart, as shown by NASA's February 1991 and January 1992 manifests, which anticipated a 15-month gap between them. However, when SRL-1 slipped behind Endeavour's Hubble repair, the radar missions drew closer together. By the April 1993 manifest, they were only eight months apart: with SRL-1 on STS-59 in April 1994 and SRL-2 on STS-68 the following December. "JPL had always planned to fly the new instrument at least two and preferably three times," wrote Jones, "testing the radars' ability to monitor seasonal variations in rain forests, croplands, wetlands, ocean currents, sea ice, soil moisture, glaciers and snow cover." And MAPS sought to track ongoing carbon monoxide changes. "JPL would have preferred six months between missions," Jones said, "to capture a full seasonal swing, but competing demands on the Shuttle schedule moved SRL-2 up to late summer."

So it was that when Jones was named as SRL-2's payload commander, he was still eight months away from flying SRL-1. By then, the manifest had shifted yet again with SRL-2 now retargeted to fly Endeavour in August 1994, narrowing the gap between the pair to only four months. In October 1993, NASA named the rest of the crew. Commanding STS-68 was Mike Baker, with Terry Wilcutt as pilot and mission specialists Steve Smith, Dan Bursch and Jeff Wisoff. They mercilessly pestered Jones during his SRL-1 training, even emailing him questions whilst STS-59 was in space.

"Don't forget, you start sims with us next week," read one note from Baker. He signed off with simplicity: "Your STS-68 Associates."

Early on 18 August, Jones found himself sitting next to Wisoff on Endeavour's darkened middeck, ready to set a new record of only 120 days between two missions by one astronaut. Like SRL-1 before it, this mission would ascend to 139 miles (222 kilometres), inclined 57 degrees to the equator, and run for 244 hours and 163 orbits. Liftoff was targeted for 6:54 a.m. EDT. After strapping in, Jones and Wisoff killed time, playing rock, paper, scissors. Endeavour was primed and ready to go. Or so it seemed.

"Go for Autosequence Start. Endeavour's on-board computers now have primary control of all the vehicle's critical functions..."

As the countdown ticked towards the all important T-31 seconds, control of the final stages was handed off from the ground launch sequencer to Endeavour's on-board computers. The disembodied voice of the launch announcer echoed from loudspeakers, crisply acknowledging the seconds as they worked their way backwards towards Main Engine Start and the ignition of the twin SRBs.

"Twelve, eleven, ten, nine, eight, seven..."

At ten seconds, the pre-dawn darkness was punctuated by a bright cascade of sparks from swirling hydrogen burn igniters, as they fired off to disperse unburnt gas in the vicinity of the main engine nozzles.

"We have a Go for main engine start..."

With a familiar rumble and a sheet of translucent orange flame, Endeavour's engines thundered alive. From the roof of the Launch Control Center, Jones' wife, Liz, together with their two children, the other crew families and a handful of astronaut escorts—including STS-59's Rich Clifford—steeled themselves for the upcoming wave of sound. "In the growing light of dawn," Jones wrote, "she saw the gout of orange exhaust flare beneath the orbiter and saw the steam billow from the flame trench as the engines spooled up to full power."

"We have three main engines running..."

Then, everything in the world fell still.

"Three, two, one...and...we have main engine cutoff. GLS safing is in progress."

As the Ground Launch Sequencer automatically safed the vehicle, the three blazing engines fell silent. A Redundant Set Launch Sequencer (RSLS) abort had been commanded, forcing an engine shutdown. However, whereas previous RSLS aborts occurred at around T-3 seconds, STS-68's count had gotten down to T-1.9 seconds. Almost immediately, cooling water was sprayed onto the hot engines and the attention of launch controllers was riveted upon the Main Propulsion System (MPS) fire detectors. If any had tripped, the abort carried the prospects of turning into a bad day, for a fire at the base of the engines could easily spread and trigger an explosion. And that would require the astronauts to evacuate.

"We have a cutoff of the main engines. The countdown clock has stopped."

In the seconds thereafter, an urgent flurry of acronym-laden calls flowed between launch controllers and the crew. *"We have main engine cutoff...RSLS safing is in progress...All three main engines are in post-shutdown standby...GLS is Go for orbiter APU shutdown..."* (Fig. 9.6).

The gathered spectators watched in alarm as the famous countdown clock starkly read T-00:00:00, yet no Shuttle ascended into the heavens and a smudge of grey cloud rose ominously above Pad 39A. Then came the call which calmed the proceedings: *"No MPS fire detectors tripped."* There was no evidence of fire on the pad, meaning an early evacuation from the Shuttle would go unneeded. The 'white room' was moved back into position alongside Endeavour's crew access hatch, to facilitate the departure of the crew. Wilcutt shut down the three APUs. And through Endeavour's tiny side hatch window, Jones saw Pad 39A's gantry visibly swaying backwards and forwards; the vehicle was still rocking from the 'twang' effect induced by the momentary ignition of her main engines.

Fig. 9.6: The SRL-2 payload is lowered onto the Spacelab pallet during STS-68 preflight preparations.

After the abort, Endeavour rolled back to the VAB for an engine replacement and a revised launch attempt was established for 30 September. Early that morning, STS-68's blue team of Jones and Smith, led by Bursch, were already awake, with the red shift of Baker, Wilcutt and Wisoff having arisen from their slumbers five hours before launch. Launch occurred without incident at 7:16 a.m. EDT. Upon arrival in orbit, Jones expressed satisfaction that he had taken anti-nausea medication before launch, but Smith soon fell victim to space sickness. Bursch offered a Phenergan shot, which did the trick. It was fortuitous that Smith's blue team bedded down for their first abbreviated sleep period, shortly after reaching space. Activation of SRL-2 was conducted by the red shift, with Wisoff taking charge of the radar instruments.

Endeavour flew over many of the same sites that SRL-1 did, enabling the SIR-C and X-SAR teams to examine change between seasons. In addition to its enormous scientific yield, the first mission demonstrated that it could acquire high-resolution data and endure adjustments to its timeline to cater for evolving events on the ground. Notably, SRL-1 observed severe floodwaters in the U.S. Midwest and near Thoringen in Germany, as well as taking three different views of Tropical Cyclone Odille as it formed in the Pacific Ocean. Snow and ice classification maps were assembled from data over a supersite at Ötztal in Austria and the summertime flight of STS-68 offered a chance to observe the Patagonian district of southern Chile, home to the largest glaciers and ice fields in South America. Volcanic sites were a key objective. On SRL-1, the radar observed Mount Pinatubo in the Philippines and several locations in the Galapagos Islands. Imagery of Pinatubo during the summer monsoon season, when new mud flows were predicted to occur, was high on SRL-2's list of priorities…but, early in the flight, a serendipitous observation of another eruption, in the Russian Far East, was made. Klyuchevskaya Sopka, the highest mountain on the Kamchatka Peninsula and the highest active volcano in the whole of Eurasia, burst into fiery life eight hours after Endeavour entered orbit, giving the STS-68 crew a surprise. They saw its tremendous black plume on the horizon, rising to 50,000 feet (15,000 metres), dead ahead of their flight path. At first, it looked like a vast thunderhead. Baker was first to recognise it as a volcano. Wisoff was amazed. "It shows how much change nature can produce in a short period of time," he recalled. "You had this huge eruption, but then by the end of the flight it had largely stopped erupting. There was still a small smoke trail, but it had re-snowed on top of the soot. In the span of ten days, it was almost white again!"

Jones remembered the event lucidly in *Skywalking*. It was the red team, he wrote, who summoned the blues up to the flight deck to see the eruption. Like Wisoff, Jones thought at first it was an anvil-shaped thunderhead or a clump of dust lofted by high winds, but the smoke-free nature of the surrounding terrain (and the location in Kamchatka) quickly assured them that it was a volcano. Jones had seen it before, during SRL-1, but less than six months earlier Klyuchevskaya had lain silently under a blanket of April snow. Now, at the end of September, it was in full fury. "We soon had every camera…zeroed-in on the eruption," Jones wrote, "as Endeavour gave us a dramatic, down the throat view of this impressive geology lesson." In another example of responding to unforeseen events, the SRL-2 investigators reprogrammed several radar passes over the coming week to scan the eruption site up to three times daily and Jones hoped that their work might lead to the implementation of permanent, Earth-orbiting satellites to watch for volcanic events.

By the sixth day of STS-68, consumables remained at a level sufficient to enable mission managers to formally extend the flight an additional 24 hours. Endeavour would now land on 11 October. (She would actually touch down at Edwards at 10:02 a.m. PDT, after thick cloud at KSC forced NASA to switch landing sites.) The radar's filing-cabinet-sized payload recorders performed well, although one had to be removed and replaced after it failed to play back properly. Rerouting the data stream between the remaining machines, a repair by Smith and Wisoff occurred over a comparatively 'empty' Pacific orbital pass. "With the radar inactive over the ocean," wrote Jones, "Steve and Jeff were well into the repair before we made landfall again. An hour later, the swap was complete and the pair had stowed their wrenches and screwdrivers with almost no loss in science data." More than 110 hours of radar observations were acquired across a cumulative 950 data takes, recorded on 199 digital tapes and covering an area of 32 million square miles (83 million square kilometres). Fourteen on-board cameras captured 14,000 photographs and around 22,000 keystrokes were inputted in support of 409 manoeuvres.

Photography, indeed, was one aspect of STS-68 which really stood out years later for Wilcutt. "We took an incredible amount of photos," he told the NASA oral historian. "One of the things that everybody enjoys when they go to space is looking out the window. We were being *paid* to do that." As far as Wilcutt was concerned, in the days of film-based photography, "if you've got it, we felt we should burn it up…I'd be surprised if we left any film that was left unexposed" (Fig. 9.7).

One of the bonuses of flying a second time was the pioneering use of 'interferometry' during the final three days of STS-68 and SIR-C/X-SAR data was used to record topographical changes between April and October in California's Long Valley Caldera and Hawaii's Kilauea. Interferometry is analogous to stereo photography, although to achieve it between STS-59 and STS-68 demanded that their repeat orbital paths be closely aligned. "Mission Control and our crew combined to perform the most precise orbital manoeuvres ever seen in the Shuttle program," explained Jones, "putting Endeavour in an orbit for the first six days that nearly matched our SRL-1 flight path." In support of the interferometry exercise, Bursch worked with the flight dynamics team to 'trim' 14 manoeuvres down to an accuracy of 0.05 feet per second (0.01 metres per second). At times, the respective flight paths of the two missions differed by less than 33 feet (11 metres).

Fig. 9.7: An orbital sunrise glints over SRL-2.

Also flawless were operations with MAPS, which gathered 256 hours of data and achieved a full success rate. However, MAPS' results reinforced the STS-59 consensus that although carbon monoxide concentrations in the southern hemisphere were relatively low, with exceptionally clean air, the situation worsened in the northern hemisphere, with the highest concentrations to the north of the 40-degree latitudinal band. MAPS also observed intentionally-set fires, monitored by scientists from the University of Iowa and the Canadian Forest Service, to assess their wind fields, thermal evolution and carbon monoxide emissions and calibrate SRL-2 infrared data. "These fires," noted NASA, "were planned in advance of the mission and would have been set for forest-management purposes, even if the Shuttle mission were not in progress." Other 'controlled' observations included an experimental 'spill' of 105 gallons (477 litres) of diesel oil and 26 gallons (117 litres) of algae products in the North Sea to test SRL-2's capacity to differentiate between the spill and naturally-produced film caused by the products of fish and plankton.

LASER-EYED CYCLOPS

Another Spacelab mission sat between the first and second SRL flights. Known as the Lidar In-Space Technology Experiment (LITE), developed by NASA's Office of Advanced Concepts and Technology and executed under the Mission to Planet Earth umbrella, it sought to transmit narrow pulses of laser light into Earth's upper atmosphere and used a telescope, mounted atop a single Spacelab pallet, to measure the reflectance qualities of clouds, suspended aerosol particulates and the surface. The usefulness of lidar as a remote-sensing tool stemmed from the fact that it could obtain high vertical and horizontal resolutions and LITE marked the first time that such technologies had been trialled in space for atmospheric observations. This 5,920-pound (2,680-kilogram) payload first made an appearance in NASA's August 1988 manifest, requested for a Shuttle flight in February 1993, with no definitive mission number. By January 1990, it was scheduled to fly—albeit as a 'secondary' payload, rather than a 'primary' one—aboard Columbia on STS-67 in May 1993. Targeted as a seven-day mission, with a five-member crew, LITE would conduct observations from an altitude of 186 miles (300 kilometres), inclined 28.5 degrees to the equator.

When NASA released its January 1992 manifest, LITE-I had moved to STS-63 in March 1994 and a second mission, LITE-II, was added for the second quarter of Fiscal Year 1995. Circumstances changed, however, and by the April 1993 manifest LITE-I had realigned to STS-64 aboard Discovery, scheduled for September 1994, and LITE-II vanished entirely from NASA's planning charts. Interestingly, the April 1993 manifest recategorised LITE as a 'primary' payload. Its mission duration was increased to nine days and its orbital inclination and altitude also changed to 57 degrees and 160 miles (260 kilometres).

In November 1993, Commander Dick Richards, Pilot Blaine Hammond and Mission Specialists Susan Helms, Carl Meade and Mark Lee were assigned to STS-64. Years later, Richards described the flight as "hodge-podge", on account of its multitude of payloads: in addition to the pallet-based LITE, the astronauts would deploy and recover a Spartan solar physics satellite, support a wide range of technological experiments and Lee and Meade performed a spacewalk. For Richards, the Spartan rendezvous commitment appealed to him for his next mission as a commander. In February 1994, the crew changed with another mission specialist, physician Jerry Linenger, assigned "to more efficiently distribute the crew workload for this complex flight".

For the purposes of this survey, only STS-64's Spacelab element—the pallet-mounted LITE—will be discussed here; the mission's other payloads are explored elsewhere. Launch was targeted for a 2.5-hour 'window', opening at 4:30 p.m. EDT on 9 September 1994, for a mission timelined for 212 hours and 143 orbits. A late afternoon launch was necessary to allow LITE to conduct 'nighttime' observations during the early stages of the flight, but weather violations at KSC forced mission managers to hold the countdown clock for almost two hours. Eventually, with the closure of the window drawing uncomfortably close, the weather suddenly brightened. Discovery sprang from the launch pad at 6:22 p.m. EDT and upon arrival in orbit, the six astronauts set to work. With so many laptops needed on the flight deck for STS-64's tasks, the payload general support computer for LITE was set up by Meade on the middeck, with Lee activating the Spacelab pallet and its single experiment. LITE would operate almost continuously throughout the flight, its operations halted for the deployment and retrieval of Spartan and during Lee and Meade's spacewalk. The astronauts supported LITE from the flight deck, performing 'cross-tracks', taking a certain 'spot' on the landmark track to determine differences in the surface reflectance of the laser.

Over the next ten days, 43 hours of data were collected and Discovery's relatively low altitude of 160 miles (260 kilometres) and high-inclination orbit facilitated observations over a broad area of the home planet. On the ground and in the air, investigators at 50 locations in 20 discrete countries made their own observations. Five international aircraft flew directly underneath portions of the Shuttle's flight path, covering parts of Europe, the southwestern United States, the Caribbean Sea, South America and the South Atlantic. LITE's data was gathered during ten sessions, each lasting about 4.5 hours, together with a handful of 15-minute 'snapshots' of selected sites which establish 'ground-truth' comparison points. The returning lidar signals were converted to digital data and stored on tapes aboard Discovery for transmission to the ground. With pulses transmitted ten times per second, and each lasting less than 30 billionths of a second, at precisely-known wavelengths corresponding to ultraviolet, infrared and visible green light, it was possible to achieve ground resolutions of about 50 feet (15 metres). By the end of the mission's first day, the LITE team was reporting "terrific-looking returns" from the instrument (Fig. 9.8).

The lidar investigated the organisation of cloud structures in the Western Pacific Ocean, together with cloud decks off the coasts of California and Peru, smoke plumes from biomass fires in South America and Africa and the transport of dust from the Sahara Desert. Low-atmosphere aerosols were examined over the Amazon rainforest and gravity waves (a mechanism that transfers momentum from the troposphere to the stratosphere) over the Andes mountain-chain and the reflection characteristics of desert surfaces in the United States, Africa and China. LITE also dramatically outlined the structure of Super Typhoon Melissa, providing unprecedented views of the storm's eye. It was expected that the technology demonstration could lead to future dedicated satellites to perform year-round lidar observations of clouds, urban smog or dust storms. The instrument took the form of a 3.2-foot-wide (0.9-metre) telescope and optics package with photomultiplier tubes to provide visible green and ultraviolet sensitivity and a silicon avalanche photodiode for

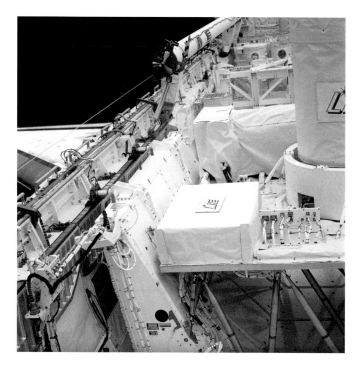

Fig. 9.8: The tall barrel of the LITE instrument (far right) and its adjunct instrumentation affixed to the Spacelab pallet in Discovery's payload bay during STS-64.

infrared capability. In spite of periodic laser shutdowns, caused by coolant-loop problems, the system "performed more effectively than originally expected, demonstrating the ability of a lidar...to penetrate multiple clouds and aerosol layers down to the Earth's surface".

Having already been extended from nine to ten days, Discovery's landing at KSC in Florida was called off on 19 September, due to thunderstorms and low clouds. The mission was extended an additional 24 hours, but with rainstorms continuing to plague Florida, Richards was directed to land at Edwards. Under picture-perfect California skies, Discovery touched down at 2:12 p.m. EDT after 11 days, completing the first—and only—flight of a dedicated lidar on the Shuttle.

BRAVE NEW WORLD

It is one of the ironies of history that at the dawn of the millennium, more precise radar-generated maps existed of Venus than of our own world. Although some areas of Earth, including the United States, much of Europe, Australia and New Zealand, had digital maps at a roughly hundred-foot

(30-metre) resolution level, the vast majority of Earth lacked such reliable data, primarily due to cloud cover in equatorial regions, which precluded imaging by optical satellites or aircraft. "Humans always want to know about their environment and for the first time we will generate a coherent data-set, a picture of the Earth," said Germany's Gerhard Thiele. "We don't have that yet, which is pretty surprising." On the cusp of a new millennium, Shuttle Endeavour's STS-99 mission, with Thiele as one of its six crewmembers, sought to redress the balance.

Building upon the success of SRL-1 and SRL-2, the new flight progressed one step further as the ambitious Shuttle Radar Topography Mission (SRTM). Its crew utilised an innovative radar-mapping methodology called 'single-pass interferometry' to assemble the most comprehensive, three-dimensional maps ever acquired. SRTM was spearheaded by NASA and the Department of Defense's National Imagery and Mapping Agency (NIMA) and supported by the German and Italian space agencies, DARA and ASI. Operating from a high-inclination orbit of 57 degrees, it would generate maps from 60 degrees North to 56 degrees South latitude, constituting 80 percent of Earth's surface and representing home to 95 percent of humanity. More than a thousand radar data-takes would be acquired, with the STS-99 crew operating around the clock in two 12-hour shifts.

The two SRTM instruments—dubbed the 'main antenna' and mounted on a single Spacelab pallet and MPESS—were SIR-C and X-SAR, both of which flew on SRL-1 and SRL-2. A second receiver, the 'outboard antenna', sat at the end of a 200-foot-long (60-metre) mast, deployed sideways into space over the Shuttle's port-side payload bay wall. Radar echoes from both instruments were captured, amplified and their data recorded onto digital tapes for analysis, the distance between them affording the maps a degree of three-dimensionality. Radar interferometry was not entirely new ground for the Shuttle, having been trialled on SRL-2. But since that flight only carried SIR-C/X-SAR in the payload bay, and not an outboard antenna, the Shuttle had to fly over the same patch of ground track and acquire dual radar images on separate orbital passes. It was therefore only possible to achieve 'multiple-pass interferometry', rather than acquiring the data on a single orbital pass.

Nevertheless, the seed of an idea was planted for a fixed-baseline, single-pass interferometer and in July 1996 SRTM crystallised into a formal mission proposal, targeted for launch in May 2000. A third SRL had been under consideration for some time. But selling a quarter-billion-dollar reflight was not easy, said Charles Kennel, associate administrator for the Mission to Planet Earth Enterprise at NASA Headquarters in 1994-1996. He recalled discussions with JPL's Charles Elachi, principal investigator for SIR, who proposed reflying SRL with few technical changes. Kennel felt that a straightforward SIR-C/X-SAR reflight was unwise. However, a substantial technology demonstration, he told Elachi, might swing the proposal towards acceptance. "My objection was that the science-per-unit-dollar on that Shuttle mission, just as a reflight, was not worth it," said Kennel. "The demonstration? Yes, because that then set the way for new science. But just to repeat it: No, sir."

He passed the proposal to the National Academy of Sciences, which created a proposal to fly the mission in collaboration not only with Germany and Italy, but in partnership with the U.S. military. "JPL got very creative and understood that there was a big truss that was down at [JSC] that they could stick on the Shuttle," Kennel explained. "You could use this as a three-dimensional mapper and get resolved altimetry; a three-dimensional image because of the binocular vision. It turns out that the military had been wanting to do one of these digital topography maps for a long time."

SRTM thus carried the elements of the first single-pass interferometer ever carried aboard the Shuttle, capable of taking two simultaneous images of Earth from slightly different locations and at differing frequencies, to be combined into three-dimensional topographical maps for civilian, military and scientific purposes. The presence of the outboard antenna on SRTM allowed this interferometric data to be gathered with greater efficiency at C- and X-band frequencies. Since it would be possible to acquire both images in the same orbital pass, the potential data return was expected to be higher (Fig. 9.9).

The outboard antenna and its mast were deployed from a nine-foot-tall (2.9-metre) canister, mounted onto the Spacelab pallet, next the main antenna. Its design was based upon the same technology used for the truss segments destined for the ISS and it was fabricated by AEC-ABLE Engineering of Goleta, California. It consisted of carbon-fibre-reinforced plastic, stainless steel, alpha titanium and an iron-nickel alloy, known as 'Invar'. Officially designated the ABLE Deployable Articulated Mast (ADAM), when fully extended the 640-pound (290-kilogram) device was the longest rigid structure ever deployed from the Shuttle. During launch and re-entry, the mast was folded up in a manner not dissimilar to a giant accordion in its canister. It was deployed via a rotating nut, which grasped the edges and corners and pushed them outwards into a succession of 87 almost-perfect 'cubes'. It was designed to deploy without twisting, which allowed it to house 485 pounds (220 kilograms) of electrical wiring, fibre-optic cables and gas thruster lines along its length to connect the outboard antenna with the instruments in the payload bay. AEC-ABLE guaranteed that the mast would remain stable to within 0.8 inches (a mere two centimetres) of tip displacement, which was

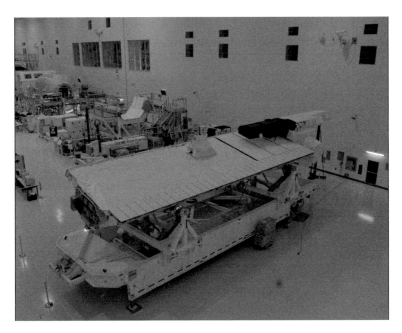

Fig. 9.9: The Shuttle Radar Topography Mission (SRTM) payload is pictured during final integration in the Operations and Checkout Building. The SIR-C radar dominates the payload, with the X-SAR and mast canister visible at far right.

considered essential if the radar-mapping standards demanded of SRTM were to be met. At the mast's tip was the 800-pound (360-kilogram) outboard antenna itself, assembled from spares left over from the two SRL missions, which included C-band and X-band passive receiver panels to collect radar echoes from the SIR-C/X-SAR transmissions. All told, SRTM weighed 29,900 pounds (13,600 kilograms).

In spite of its exciting potential, STS-99 was a long time coming. When its six-strong crew—Commander Kevin Kregel, Pilot Dom Gorie and Mission Specialists Janet Kavandi, Janice Voss, Japan's Mamoru Mohri and Germany's Thiele—were announced by NASA in October 1998, they anticipated launching the following September for their 11-day flight. But the Shuttle fleet fell foul to extensive wiring defects, causing STS-99 to slip firstly to October 1999 and eventually January 2000.

An opening launch attempt on the 31st was scrubbed due to low clouds and rain, together with a technical issue pertaining to one of the Shuttle's master events controllers. A second try on 9 February also came to nought and eventually—hopeful of third-time-lucky good fortune—Kregel led his crew out to the launch pad on the morning of the 11th. This time, Endeavour rocketed into orbit at 12:43 p.m. EST, "on a 21st-century mission," exulted the launch announcer, "placing Earth back on the map".

Within hours of establishing themselves in a 145-mile-high (233-kilometre) orbit, the crew split into their two teams. The blue shift of Voss and Mohri, led by Gorie, set up a network of laptop computers, then headed to bed for an abbreviated, six-hour sleep period. Meanwhile, the red shift of Kavandi and Thiele, led by Kregel, deployed the outboard antenna. Since Kavandi and Thiele had trained to perform a contingency spacewalk to manually unfurl or retrieve the mast in the event of problems, it made sense for Kregel to assign them to the same shift, as well as to the shift which initiated the deployment process.

The red shift initiated the extension of the mast at 6:27 p.m. EST and all 87 cube-shaped bays smoothly unfurled from their canister within 17 minutes, reaching a length of 199.96 feet (60.95 metres). The astronauts manoeuvred Endeavour into the correct mapping attitude, with the orbiter's tail facing into the direction of travel and executed a series of thruster firings to test the ability of dampers to absorb the force of manoeuvres and keep both the main and outboard antennas correctly aligned with their targets. By the time Kregel, Kavandi and Thiele retired to bed about ten hours after launch, they had executed STS-99's most complex task. "Everything has to go like clockwork and the red shift will deploy the mast, calibrate the radar, verify that both antennas...look at the same point on Earth," Thiele explained. "We will calibrate it with the help of the ground and then at the end of our shift, the mission will be rolling." If the mast-based antenna did not deploy correctly, the mission would be significantly impaired, "not so much that we could not live with a shorter [radar] baseline", said Thiele, "but the key thing here is safety". The mast needed to be firmly locked in place in order to compensate for Shuttle motions and it could not be firmly locked into place if it was improperly deployed.

However, it soon became evident that the mast's damping system, which acted as a shock absorber, was not behaving as expected. Mission Control opted to leave the dampers in their 'locked' position, since calculations indicated that the mast was at no risk in this configuration and did not pose any threat in terms of science capability. By the morning of 12 February, the SRTM radars had begun their first 'swath' over southern Asia and by the end of their first full day in orbit had mapped a total of 1.7 million square miles (4.5 million square kilometres), roughly equivalent to an area half the size of the United States. X-band imagery of the White Sands region in New Mexico was met with delight from investigators at its level of detail and quality. The breadth of their radar mapping covered all of Earth's southern areas, save

Antarctica, and all northern areas to the south of the southern tip of Greenland, southern Alaska and central Russia. During his second shift, Kregel performed a series of thruster pules to measure the movement of the mast and noted that the tip moved, as predicted, by less than 11 inches (28 centimetres).

Later, on 13 February, the crew also performed their first 'flycast' manoeuvre, pulsing Endeavour's thrusters to maintain optimum conditions for mapping. It was named in honour of the casting technique adopted by fly fishermen and sought to limit the loads on the mast during frequent Shuttle attitude-control manoeuvres. "If we did this without the flycast manoeuvre, we would have a deflection at the mast tip and we would approach the mast limit loads at the base," said Gorie. "So we pulse for just a short time—one-sixth of the natural frequency of the mast—and the mast will deflect 'back' slightly. Right when it starts springing backwards, we will pulse again and, in essence, 'catch' it in its deflected state." During each flycast manoeuvre, Endeavour was reoriented into a nose-first attitude, with the mast extending away into space, then her thrusters were fired to deflect the mast backwards and rebound it forwards. When it reached 'vertical', a stronger thrust was applied to arrest its motion and increase the orbiter's velocity, thereby keeping Endeavour in the most optimum attitude for radar mapping.

Overall, SRTM operations ran with exceptional smoothness, with flight controllers' attention drawn only to a minor propellant consumption issue associated with the nitrogen gas thruster at the mast's tip. This was designed to keep the mast from 'righting' itself in response to Earth's gravity, removing the need for additional thruster firings to keep the antenna in its data-collection position. Although propellant was flowing to the thruster, it appeared that little or no thrust was being produced, in spite of the crew periodically cycling its valve open and closed. In the meantime, Kregel and Gorie were required to use orbiter thruster firings to maintain the fine position, which increased propellant expenditure by about 0.07 to 0.15 percent per hour. Efforts to conserve propellant were explored over the following days, including a relaxation of the requirement to maintain the mast's attitude, which was already showing its stability to be far better than predicted. Other measures included changing the manner in which wastewater was dumped overboard and limiting usage of non-critical equipment. On the 17th, Gorie cycled the nitrogen line and Voss reported seeing the dislocation of a small, white object—presumably a small piece of ice—and flight controllers noticed that the thruster began to generate some thrust (Fig. 9.10).

Gradually, the propellant conservation measures paid off and the crew was advised that it would be granted a ninth day of science, as originally intended in the mission plan. "That's great news," exulted Gorie. "They're getting some

Fig. 9.10: Gerhard Thiele and another STS-99 crewmember inspect the SRTM payload during pre-flight preparations.

fantastic data on this mission!" Fantastic, indeed, for SRTM was uncovering the home planet as never seen before and Project Scientist Michael Kobrick of JPL noted that the radar was mapping 40,000 square miles (100,000 square kilometres) every minute, more than tripling the world's supply of digital terrain elevation data. By the fifth day of the flight, 15 February, more than half of the targeted land surface had been mapped and around 20 percent had been mapped twice, covering regions the size of Africa, North America and Australia combined. Total mapping coverage of the world had reached 29 million square miles (75 million square kilometres) by this stage and continued to grow: 34.9 million square miles (90.6 million square kilometres) by the 16th, then 36.4 million square miles (94.3 million square kilometres) by the 17th, then 42 million square miles (108.8 million square kilometres) by the 18th. Dozens of sites were mapped two and three times and images of Brazil, South Africa and New Zealand's South Island were considered so spectacular in topographical detail that enthusiastic scientists gushed that the data would be used for decades to come. Although 'quick-look' data was available to researchers, the bulk of SRTM's results were stored on high-density tapes.

Additional flycast manoeuvres fine-tuned Endeavour's orientation. The sixth such manoeuvre came early on 18 February and lasted slightly longer than previous burns, in order to conserve propellant by eliminating another flycast planned for the 20th. By now, the rate of mapping coverage allowed SRTM to cover the territorial equivalent of the state of Alaska in just 15 minutes and Rhode Island in under two seconds. Images of the San Andreas Fault and the Rose Bowl area of southern California, together with Kamchatka and the Hawaiian island of Oahu led to new insights into the nature and causes of wildfires, lava flows, tsunamis and localised flooding.

With the additional hours of radar time, Endeavour's mapping ended early on 21 February, a day ahead of the scheduled return to Earth. The SRTM instruments and the STS-99 crew had imaged almost all of their targeted areas, save small portions of the United States, which were already topographically well characterised. The scientists who oversaw the mission were united in their praise. After 222 hours of uninterrupted data-gathering, the process of retracting the mast began at 8:17 a.m. EST. The procedure ran crisply, until the mast got to within 0.9 inches (2.5 centimetres) of the canister…then stopped.

"Janet's and my hopes were soaring very high," Thiele said. "Unfortunately, the hopes were soaring high, but not the reality." One of the three latches on the canister's lid failed to engage, although Kregel saw no obstruction and was convinced there would be no difficulty in closing the payload bay doors. Nor did he consider it necessary to awaken Kavandi and Thiele for a spacewalk. Mission Control agreed. Suspicion on the cause centred on the possibility that the mast had endured exceptionally cold ambient temperatures whilst fully deployed, reducing its flexibility. The astronauts waited for the Sun to warm up any stiffened components inside the canister, then drove open the latches and turned on the retraction motors, hopeful that the mast would retract and the latches re-engage.

They did not.

"We saw perhaps just a tiny movement at initial power-up, but then nothing else," reported Kregel, glumly. Even switching the retraction motors to maximum torque served only to move the mast an additional 0.2 inches (0.6 centimetres), before stalling (Fig. 9.11).

Finally, after three failed attempts, the mast successfully folded into its canister. That 222 hours of mapping was acquired was quite remarkable. Original plans called for 250 hours, but in late January 2000 mission managers opted to halt operations 24 hours sooner, giving the SRTM teams only nine full days for science. Concern was raised about how well the mast might retract at the flight's end and thus preserve options for a contingency spacewalk. If Kavandi and Thiele needed to manually crank the mast closed, the additional time would be necessary for them to prepare their space suits, pre-breathe

Fig. 9.11: The SRTM mast pictured fully extended over Endeavour's payload bay wall during STS-99.

pure oxygen and go outside. Leaving the retraction until the final day before landing offered no time to handle unforeseen issues. And since the radar consumed so much energy (about 900 kilowatt-hours), a spacewalk would not have been possible after 250 hours of mapping.

"This is a device we've never deployed before," explained STS-99 Lead Flight Director Paul Dye. "It's the longest space structure anybody's ever put out and we would like to be able to get it back. If we had a problem on the stow day, we'd have had no opportunity to try to fix that problem. We had stretched the mission as long as we possibly could." Dye compared the situation to the difference between "a glass that's ten-percent empty or 90-percent full". With the extension of mapping activities by an additional nine hours, then another ten minutes on top of that, most of SRTM's radar requirements were met. When judging the success of SRTM, the numbers spoke for themselves: eight terabytes of data gathered, 332 high-density data tapes filled and 99.98 percent of targeted sites mapped on at least one occasion and 94.6 percent on at least two occasions. In fact, several sites were radar-surveyed as many as four times.

Landing day, 22 February 2000, brought news of grim weather conditions, with high winds and potential heavy cloud cover seriously threatening a return to KSC. On the United States' west coast, Edwards had been activated and placed on standby. The first KSC opportunity was waved off, due to ongoing concerns about the weather, but mission managers eventually opted for the second opportunity. Kregel and Gorie guided the Shuttle to a smooth touchdown in Florida at 6:22 p.m. EST, wrapping up a spectacular mission which uncovered the home planet as never before and became the first to fly in the new millennium. STS-99 gained the honour of being first purely by happenstance, but there surely could have been no better mission than one which cast humanity's gaze back to the cradle from which we came and enhanced our understanding of our home's beauty, majesty and mystery.

10

Mission for Japan

FAREWELL TRADITIONS

Traditions frequently arise from the unlikeliest of places. One tradition which had become entrenched in NASA's astronaut psyche since the early Shuttle era was the practice of a crew's spouses to organise and host farewell parties. Of course, astronauts became as close as family with their crewmates, to such an extent that invitations to dinner or excursions with each other's children were commonplace. In the early days of Mercury, Gemini and Apollo, when space missions were male-dominated, parties were organised by the astronauts' wives, but as women began flying aboard the Shuttle, party-making responsibilities fell to both genders. It made little difference, of course, and much celebration was planned and executed both before and after missions, with few obvious problems.

Then there was STS-47.

In spite of its numerical designator, Endeavour's second space voyage was actually the 50th Shuttle mission and proved historic for other reasons, too. Among the seven-strong crew was the first (and only) married couple ever to fly into orbit together. Air Force mechanical engineer Mark Lee and former Texaco mechanical engineer Jan Davis were named as mission specialists for STS-47 in September 1989 and married during training, but only revealed this fact to NASA management shortly before the flight. At first, managers considered removing either Lee or Davis from the crew; although there were no children in the marriage, the risk of losing both astronauts in a disaster would prove doubly traumatic for an entire family. "It was difficult to decide who to take off the flight, if they were going to do that, and so they let them fly together," said astronaut Rhea Seddon, wife of STS-47 Commander Robert 'Hoot' Gibson. "They did tell us it was an unwritten rule, that they really didn't want to do that." The word crept down from on high: "We're making an exception for Jan and Mark, but we don't want any other married couples to fly together."

Their marriage was noteworthy not just for its novelty, but also for the fact that STS-47 included only three crew spouses. With Lee and Davis married, and pilot Curt Brown and mission specialist Mae Jemison both single, it was left to Seddon and Eleanor 'E.B.' Apt (wife of mission specialist Jay Apt) and Akiko Mohri (wife of Japanese payload specialist Mamoru Mohri) to handle the partying arrangements.

"THE WAY SCIENCE IS DONE"

STS-47 would carry Spacelab-J, a collaborative life and microgravity science flight, organised by NASA and the National Space Development Agency of Japan (NASDA). Japan was a major player in the Spacelab-1 mission (see Chapter Three). As early as October 1976, during the 27th Congress of the International Astronautical Federation, held in Anaheim, California, the East Asian island-nation expressed enthusiasm to participate "in another Spacelab in the 1980s", as well as "co-operation with ESA", according to Shigebumi Saito, commissioner of the Japan's space activities committee, and a joint ESA/NASDA summit in Paris in October 1979 considered "using Spacelab for research in materials and life sciences". A dedicated 'Japanese Spacelab' mission, conceived by NASDA in 1979 as the First Materials Processing Test (FMPT), went on to receive 103 experiment proposals from the Japanese scientific community and in March 1980 a total of 62 finalists were selected for consideration. Thirty-four experiments—22 in materials science and 12 in the life sciences—were chosen by NASDA and principal investigators were named and an IWG assembled in July 1984. Materials research emphasised semiconductor crystal growth, composites casting, drop and bubble dynamics, solidification, behaviour of glasses under extreme temperatures and convective processes in weightlessness, whilst the mission's life sciences thrust included studies of endocrine and metabolic change in the Japanese payload specialist, the motions and behaviour of fish, the functionality of mammalian cells, the effect of microgravity upon calcium metabolism and bone formation and the impact of cosmic radiation on biological organisms. However, it was found that the 34 Japanese-sponsored experiments did not fill an entire Spacelab module and NASA developed seven of its own investigations, as well as two joint studies with NASDA, to complement the life and microgravity science themes.

Spacelab-J was requested as a potential future Shuttle payload in June 1981 and a launch services agreement with the United States was signed in March 1984. Japan reportedly contributed $90 million to the mission's $363 million price-tag. "Missions such as Spacelab-J mirror the way science is done on Earth," said Program Scientist Robert Sokolowski of NASA's Office of Space Science and Applications. "Astronauts aboard the orbiting laboratory will conduct experiments around the clock. These experiments will add to basic knowledge about the behaviour of everything from crystals, fluids and even humans, when exposed to the near-weightless environment of spaceflight." When NASA released its March 1985 Shuttle planning chart, Spacelab-J was listed as 'required' for January 1988; the subsequent June manifest saw it assigned to STS-81F on Columbia, as a seven-day flight, with a crew of seven, destined for an altitude of 230 miles (370 kilometres) and an orbital inclination of 57 degrees. Experiments would be housed inside a Spacelab long module and atop an MPESS truss. By November 1985, it had shifted onto Challenger on STS-81G in February 1988. But following the loss of Challenger early in 1986, the mission was stood down. By the August 1988 manifest, it reappeared aboard Columbia on STS-49 in July 1991, baselined as a long module (without MPESS), targeting an altitude of 186 miles (300 kilometres), inclined 44 degrees to the equator. And when the first Spacelab-J crewmembers were assigned in September 1989, the mission had moved to STS-47 in June 1991, its inclination adjusted again to 57 degrees. All told, the payload tipped the scales at 21,860 pounds (9,900 kilograms).

Twenty-four of Spacelab-J's experiments focused upon materials and processes in the microgravity environment, with emphasis upon the production and analysis of protein crystals, electronic components, fluid dynamics, glasses and ceramics, metals and alloys. The Protein Crystal Growth investigation sought to yield crystals by the vapour-diffusion and liquid-to-liquid diffusion processes, whilst four high-temperature furnaces—the Gradient Heating Furnace, the Image Furnace, the Crystal Growth Furnace and the Continuous Heating Furnace—melt and solidified a variety of materials. Of these, the gradient furnace facilitated the exploration of crystal formation processes in semiconductors, ceramics and alloys and featured three temperature zones, which allowed gradients to be 'moved' up to a maximum 1,100 degrees Celsius (2,000 degrees Fahrenheit) along the length of a sample. During Spacelab-J, the furnace was employed to process specimens of lead-tin-tellurium for potential electronic applications, including infrared detectors for fire security and imaging systems. The image furnace, meanwhile,

Fig. 10.1: Mark Lee and Jan Davis, the first married couple to fly together in space, examine the interior components of the Spacelab-J module.

supported investigations which used the 'floating-zone' growth procedure, whereby a liquid 'zone' was moved through a material during the crystal-formation process, and its samples included indium-antimonide, a compound whose Earth-bound applications span the infrared technology arena, ranging from thermal imaging cameras to missile guidance systems and telescopes (Fig. 10.1).

"We used those to make the world's largest single crystal of indium-antimonide," recalled Gibson. "The reason the scientists wanted to do single crystals in orbit is that down here in the gravity field you start progressing a melt-zone through a block of indium-antimonide to make a single crystal out of it [and] gravity will cause the melt-zone to collapse, so you can't make a very big crystal. In weightlessness, we don't have any gravity, so we made the largest single crystal of indium-antimonide that had ever been built."

Other experiments sought to understand flow processes in a viscous, gold-laced glass sphere. Finally, the continuous heating furnace provided high temperatures of up to 1,300 degrees Celsius (2,370 degrees Fahrenheit), together with a rapid-cooling capability, to two samples in tandem. Specific materials heated and cooled in the furnace included compounds of aluminium-lead-bismuth, silver-copper and silver-yttrium-barium-copper. The choice of the continuous heating furnace for these samples meant that two could be heated as two others were being cooled, thereby achieving increased processing efficiency. A Large Isothermal Furnace uniformly heated large samples, including tungsten, to maximum temperatures of 1,600 degrees Celsius (2,900 degrees Fahrenheit), then rapidly cooled them through helium purging. Protein crystal growth formed an additional thrust of the research conducted in the Spacelab module and middeck. One investigation from Shunji Takekawa of Japan's National Institute for Research in Inorganic Materials, grew crystals of a rare-earth mineral compound called samarskite—"I have no clue what samarskite is or what it's used for," admitted Gibson—and even the experiment itself proved 'experimental' in nature, seeking simply "to better understand its properties and possible usefulness".

The remaining 19 experiments emphasised cellular biology and life sciences and employed a variety of organisms, including the Spacelab-J astronauts, four female South African clawed frogs in a specialised box, the osteoblastic (bone-forming) cells of rats, kidney cells, fungi, chicken embryos and a pair of Japanese koi carp—"big goldfish," according to Gibson—one of which had its gravity-sensing organ, the otolith, removed before flight to enable comparisons of space adaptation processes with its twin. And the mutative effects of cosmic radiation upon the larvae of fruit flies were examined.

Notably, the frogs' eggs—each a spherical cell, no bigger than 0.8 inches or a couple of millimetres in size—were fertilised and incubated in orbit, then examined at various developmental stages, from embryonic level to tadpoles to adulthood. "This is the first time we can fertilise them in space and watch their development through hatching," said principal investigator Kenneth Souza. "The stage most sensitive to gravity changes and the stage at which the symmetry—left, right and head-tail location—of the frog is established occurs shortly after fertilisation. This critical stage was missed by previous spaceflight studies." Early in the flight, Spacelab-J crewmembers injected the frogs with hormones to stimulate them to shed their eggs, after which a sperm solution was added for fertilisation.

More than a hundred live tadpoles conceived and born in space were brought back to Earth alive, revealing that reproduction and maturation (at least with amphibians' eggs) could occur 'normally' in space. And the behaviour of the adult frogs themselves induced intrigue both from investigators on the ground and the astronauts in space. "We had in some of our video Jan Davis taking one of the frogs in the glovebox and turning it loose in weightlessness to see what it would do," said Gibson. "Of course, it tried to swim. It kicked its hind legs and it was trying to swim in weightlessness, in the air. That was kind of cute to watch." As for the tadpoles, their behaviour was more

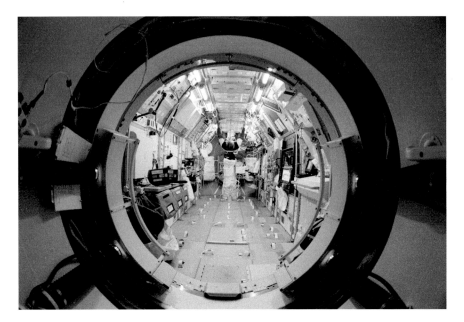

Fig. 10.2: Impressive view of the interior of the Spacelab-J module, as seen from a crewmember translating through the tunnel adapter.

anomalous: rather than swimming in straight lines, they flew in continuous loops, although their development was characterised as normal (Fig. 10.2).

As the United States and Japan moved together as partners in the Space Station Freedom development effort—soon to become the ISS—a key investigatory component of Spacelab-J was to identify the underlying cause of space sickness, known to affect a quarter of all astronauts and cosmonauts. Aboard Endeavour was the Autogenic Feedback Training (AFT) experiment, previously flown on Spacelab-3, which included belt-worn electronic instrumentation to record physiological data such as sweat, pulse, heart and respiration rates. The experiment offered clear insights into the effects of crew workload and behavioural responses to environmental stress; 'baseline' information which proved important when planning future long-duration missions.

A UNIQUE CREW

For NASDA, the presence of a Japanese payload specialist aboard Spacelab-J was of paramount import in these studies. During the flight, Mamoru Mohri, a chemist by profession—and, as Gibson's recalled years later "an absolute rock star in Japan"—received scans of his spine and legs before and after the mission, from which comparisons would be drawn about changes in the muscle volume of his calves and thighs, changes in fat and water content in his spinal bone marrow and changes in the volume, shape and water content of his spinal vertebrae. Mohri subjected himself to blood and urine samples to assess his stress levels, the extent of bone-muscle atrophy and the impact of the microgravity adaptation process on his normal body biochemistry. He also participated in advanced studies of space sickness, including tracking flickering light targets whilst anchored in different orientations within the Spacelab module.

At the time of the naming of principal investigators and IWG for Spacelab-J in July 1984, a call was issued for NASDA astronaut candidates. Mohri applied and in August 1985 he and aerospace engineer Takao Doi and physician Chiaki Naito (later Mukai) were selected from 533 qualified candidates reviewed by NASDA. Following the Challenger disaster, Spacelab-J stood down and in 1987 Mohri was appointed an adjunct professor of physics at the University of Alabama in Huntsville. For the next two years, he worked on microgravity experiments in alloy solidification and liquid behaviour, using MSFC's drop-tower facility. Finally, in April 1990 he resumed training as Spacelab-J's primary payload specialist, with Mukai as his backup. In *Go for Orbit*, Seddon remembered that Mohri and Mukai both spoke fluent English…almost. "One word that they had trouble with was Hoot's name," she wrote. "There is no *h*-sound in Japanese and Hoot got used to being called *Foot*."

By this stage, the NASA members of the science crew had been announced. In September 1989, veteran astronaut Lee, together with Davis and physician Jemison were named as mission specialists. Four months later, Lee was assigned as payload commander, his leadership responsibilities including oversight of payload issues, direct interface with MSFC, co-ordinator of the mission's training plan and timeline and the singular crewmember detailed to represent the crew at meetings, documentation reviews and hardware/software change boards. Of note, Jemison became the first African-American woman to fly into space and filled a unique role as a 'science mission specialist'. "Under new NASA guidelines for missions requiring a payload specialist not provided by the customer," it was noted, "NASA selects a mission specialist to fill those duties…Since the chosen NASA astronaut will be performing payload specialist duties and is a trained mission specialist, the term 'science mission specialist' has been developed." Jemison's backup in this unique role—a role never seen again in the Spacelab era—was a University of

California at Riverside biochemist and protein crystal growth specialist named Stan Koszelak. He was selected for Spacelab-J training in October 1989.

At the time of the science crew's selection, Spacelab-J was definitively pointed at STS-47 in June 1991, originally assigned to Shuttle Discovery but switched to Atlantis by January 1990, then Endeavour by March 1991. But as the delays impacted the Shuttle fleet during 1990 (see Chapter Seven), the mission found itself pushed backwards in the pecking order and by the time the final three members of the STS-47 crew—Gibson, Brown and Apt—were named in August 1991 launch had slipped to no sooner than the summer of the following year. It was Gibson's fourth Shuttle flight and his third mission in command and he followed a personal ritual at his first crew meeting. In precis, they were going to make it look easy and fun. "This is going to be challenging," he told them. "Anybody can go do this and make it look difficult. We're going to go make it look easy. And at any point, if we're not having fun, we're doing something wrong. This is the most exciting thing in the world that we're getting to do. It ought to be fun, and if it isn't fun, we're not doing it right" (Fig. 10.3).

During those dozen months of training as a full crew, the astronauts travelled to Japan and on one occasion rode the famous bullet train to a traditional spa in Kobe. Whilst there, clad in sandals and kimonos, they had the opportunity to try Japanese delicacies, which left Brown (a committed meat-and-potatoes person) frequently hungry. Even Gibson, who could eat anything, found himself thinking twice. On one occasion, the commander tucked

Fig. 10.3: Mae Jemison inspects the amphibians' eggs during Spacelab-J.

into a bowl of what he thought were noodles, but which, upon closer inspection, turned out to be black-eyed baby eels. Worse was to come. During a visit to Ishijima Heavy Industries to inspect Spacelab-J hardware under construction, the crew ate a beef dish for lunch. Gibson enjoyed its tenderness and unique flavour.

He asked Mohri if it was Kobe beef. Mohri asked one of the waiters. After a short verbal exchange in Japanese, he turned back to Gibson.

"Yes, it's Kobe beef," he said, then paused. "It's tongue."

Suddenly, Jan Davis' face froze, her lips curled into a scowl and she stopped eating.

If Davis and Brown were turned off by some of these culinary delights, Mark Lee revelled in them. Owner of a voracious appetite, he encouraged his crewmates to try *takoyaki*—dough-balls, filled with chunks of octopus meat—cooked and purchased from Japanese street vendors. "With Mark on the crew, it was always a contest to see who could eat the most," said Gibson. "Usually, astronauts would come back and they'd have lost six or eight pounds in the course of a flight. Mark put *on* weight." During food tasting sessions on the ground, in which the astronauts picked their menu options, the crew challenged each other to produce the highest scores. "Mark always had higher scores than any of us," Gibson chuckled. "I eat everything…but you couldn't beat Mark Lee."

"PUTTING ON A SHOW"

Like several previous Spacelab flights, the STS-47 crew was divided into two teams, each working 12-hour shifts to operate the experiments around-the-clock. The 'red' team consisted of Lee and Mohri, led by Brown, whilst the 'blue' team comprised Davis and Jemison, led by Apt, with Gibson anchoring his schedule across both. The biological nature of many of the experiments meant that from around 30 hours before launch many of the 'time-sensitive' items—the frogs and koi carp, for example—had to be carefully loaded into the Spacelab module, by now in a vertical orientation on KSC's Pad 39B. Consequently, a technician was lowered in a sling-like chair (see Chapter Five) down the connecting tunnel between Endeavour's middeck and the Spacelab module. Other samples, including the seed cultures for the protein crystal growth investigations, were loaded even later, barely 14 hours ahead of the mission's 12 September 1992 liftoff.

"You are aware that you are sitting on a controlled explosion, but you also realise that you've taken all the precautions," Jemison remembered. "You trust the people you have been working with and you know they have worked to try to keep things safe. After that, you have to leave it alone. If you keep worrying about it, then you're not going to be able to do your job." Endeavour rose perfectly from Earth at 10:23 a.m. EDT that morning and entered a circular orbit of 186 miles (300 kilometres), inclined 57 degrees to the equator, whereupon Gibson and Brown set to work readying their ship for a scheduled 164 hours in space. Led by Lee, the activation of Spacelab-J was begun a little over two hours after launch—"like taking the subway to work," he quipped as he navigated the tunnel into the module—and experiments began "almost immediately," about 3.5 hours into the flight, NASA's mission report noted, "to ensure maximum exposure to the microgravity environment".

As Spacelab-J came alive, Apt's blue shift retired to bed, with a four-tier sleep station mounted against the starboard wall of the middeck. Gibson, whose shift overlapped the reds and blues, opted for the lowermost tier. "I let the rest of the crew have the good ones," he said. "The one that was down on the very bottom, right by the floor, was the smallest one, so I said 'I'll take the crummy one'." Sleeping in such tight, coffin-like confines proved claustrophobic. "It had a little light, so if you wanted to read, you could read during your sleep period," Gibson said. "Usually, you didn't. Usually, you had been busy enough that when it was time to get in the sleep station, it was time to go to sleep." From time to time, he would wake, momentarily forgetful of where he was, a panic also experienced by Guy Bluford on Spacelab-D1 (see Chapter Six). "You could get a little bit of claustrophobia before you either found the door and could open it and see outside or you found where the light was and you could turn the light on," Gibson added. "I had that happen to me once or twice. They weren't the most comfortable things to have" (Fig. 10.4).

One of the earliest experiments to begin running was the Space Acceleration Measurement System (SAMS), an instrument designed to record low-level vehicle motions during orbital operations, to help to characterise the environment for microgravity experiments. Its main unit was situated near the rear of the Spacelab module, with sensor heads located close to major experiment facilities. As STS-47's flight engineer, Apt was in charge of the flight deck when 'his' shift was on duty and his geographical training enabled him to occasionally look up from his work and instantly recognise where he was over the home planet. From 57 degrees of inclination, there was something that Apt wanted to see...something which eluded him, but which remains a popular misconception to this very day. "We spent several passes looking for the Great Wall of China with no luck," he wrote in *Orbit*. "Although we can see things as small as airport runways, the Great Wall seems to be made largely of

Fig. 10.4: Commander Robert 'Hoot' Gibson emerges from the tunnel adapter and enters the brightly lit Spacelab-J module.

materials that have the same colour as the surrounding soil. Despite persistent stories that it can be seen from the Moon, the Great Wall is almost invisible from only 180 miles up!"

For Mohri, each day of STS-47 felt like he was an extra for *Alice in Wonderland*. "Everything I saw no longer fit within its known parameters," he explained later, "so as a scientist I could explain what was occurring around me and yet I was very much entranced." The Japanese emphasis of the mission was reflected in the foods carried aboard Endeavour, with the crew eating many of their meals with chopsticks and curried rice taking centre-stage as a main menu staple. "One of the delicacies that he brought was a thing called pickled pears," remembered Gibson. "It was an actual pear, but they were bite-size. We had not tasted these before the mission." But exotic food was not Brown's forte. "Anyway, I tasted one of these pickled pears, and I only ate one of them, because it was the world's biggest salt-bomb," said Gibson. "A huge dose of salt is what it tasted like." Unsurprisingly, Brown tasted it, winced, then cast it aside.

Economical expenditure of consumables enabled mission managers to authorise an additional (eighth) day to the flight and it was late on 19 September that the final deactivation of Spacelab-J got underway, under Davis' and Jemison's supervision. As Gibson remembered, another factor played into the mission extension. "It was given to us because we had a problem, right after we launched, and that was we had a water leak in the coolant system for the furnace back in the Spacelab," he said. Lee and Mohri were trained for In-Flight Maintenance (IFM). "What they had to do was remove some panels and get down inside and repair this water leak. Because we couldn't power up the furnace the first day, having to do that IFM, they extended the mission a day so that we'd get everything out of the furnace that we had planned to get. We knew right away, on the first day, that we were extended from seven days into eight." Lee recalled that the repair was straightforward—not unlike fixing a leaky water faucet at home—which allowed the investigators to regain functionality in the loop.

So it was that at 8:53 a.m. EDT on the 20th, the pilots landed Endeavour safely at KSC, Gibson executing a smooth touchdown and Brown deploying the landing gear and drag chute. After the Shuttle's wheels came to a halt, the pair did what Gibson called "our final red-shift tag-up", playfully punching each other in the arm, to celebrate a flight well flown. Years later, Gibson—who had never flown Spacelab before—was full of praise for Europe's laboratory, likening it to an entirely separate spacecraft. All told, it doubled STS-47's habitable area from 2,500 cubic feet (70 cubic metres) available on the flight deck and middeck to 5,000 cubic feet (140 cubic metres) with the long module. "It did things a little differently than what we did on the Shuttle," he said. "It took a bunch of learning to really learn the Spacelab. But golly, what a capability it gives you."

11

Mission for Italy

"WEIRD SCIENCE"

One day in 1987, Claude Nicollier came to Jeff Hoffman with an invite to an interesting meeting. It was an Italian project called the Tethered Satellite System (TSS).

"Jeff," Nicollier said, "you're a physicist. This might intrigue you."

A year had passed since the loss of Challenger and several scientist-astronauts were taking sabbaticals or pursuing advanced degrees. Hoffman was enrolled at Rice University on a master's course in materials science, when Nicollier invited him to the TSS meeting. Nicollier was about to be sent to the Empire Test Pilots' School in Boscombe Down in the United Kingdom and when the pair attended the meeting Hoffman was quickly hooked. So began a nine-year involvement with one of the strangest Shuttle experiments of all time, an experiment which astronaut Marsha Ivins once described as "weird science".

Weird is certainly an apt choice of descriptor for the TSS, which constituted a satellite on a string, trawled through the electrically-charged ionosphere to demonstrate the electrodynamics of a conducting tether to convert kinetic energy into electrical energy. Originally proposed to NASA in the early 1970s by Guiseppe Colombo of Padua University and Mario Grossi of the Smithsonian Astrophysical Observatory, it was hoped that the concept might ultimately lead to systems featuring electricity-generating tethers, using Earth's magnetic field as a power source. Moreover, by 'reversing' the direction of current in the tether, the force created by its interaction with Earth's magnetic field could potentially place objects into motion, boosting the velocity of a spacecraft without propellant, thereby counteracting the effects of atmospheric drag. Following Colombo and Grossi's initial proposal, the Facilities Requirements Definition Team met in 1979 to consider the applications of tethered satellites and its report, published the following year, strongly endorsed a Shuttle mission. A memorandum of understanding was signed in 1984, which called for NASA to develop the deployment mechanism and Italy to build the satellite. A science advisory team provided guidance in preparation to a formal announcement of opportunity, in April of that year, for experiments. Twelve experiments were selected, featuring participants from NASA, ASI and the Air Force's Phillips Laboratory.

The TSS Deployer Core Equipment and Satellite Core Equipment (DCORE/SCORE), provided by ASI, controlled electrical current between the satellite and the Shuttle. It featured an electron accelerator with two electron-beam emitters, each capable of ejecting up to half an amp of current from the system. The Research on Electrodynamic Tether Effects (RETE), supplied by the Italian National Research Council in Rome, employed a pair of instrumented booms to measure the electrical potential in the plasma 'sheath' formed around the satellite during deployment. The Magnetic Field Experiment for TSS Missions (TEMAG), built at the Second University of Rome, used two magnetometers to map fluctuations in magnetic fields around the satellite. Elsewhere, the University of Genoa sought to investigate the extent to which TSS could 'broadcast' from space, with magnetometer emplacements around the world and extreme-low-frequency receivers at Arecibo in Puerto Rico primed to track emissions and plasma wave directions. Finally, the University of Padua supplied a pair of experiments to analyse the satellite's oscillations over a range of frequencies in real time.

The United States suite of experiments included MSFC's Research on Orbital Plasma Electrodynamics (ROPE), which studied ambient charged particles in the ionosphere and ionised neutral particles in the vicinity of the satellite. The Shuttle Electrodynamic Tether System (SETS), provided by the University of Michigan at Ann Arbor, collected electrons by determining current and voltage and measured the resistance to current flow in the tether. The Smithsonian Astrophysical Observatory conducted research into electromagnetic emissions and dynamic noise from TSS, whilst the Tether Optical Phenomena (TOP) investigation gathered visual data via hand-held, low-light-level television cameras. Other experiments provided theoretical electrodynamic assistance and measured the levels of the Shuttle's own electrical potential in comparison to the ambient space plasma.

The satellite itself was a 5.2-foot-diameter (1.6-metre) sphere, weighing 1,140 pounds (517 kilograms), boasting an outer skin of aluminium alloy, coated with an electrically-conducting layer of white paint. Piercing its shell were apertures for charged-particle sensors, a connector for the umbilical tether and access doors for its on-board batteries. Extending from one side was a fixed instrument boom, whilst a shorter antenna sprouted from the opposite face. To assist with thermal control, the interior of the spherical shell was painted black. Inside the shell were a payload module for the scientific experiments and a service module for subsystems. At the shell's centre was a tank of pressurised nitrogen, which provided propellant for 12 cold-gas manoeuvring thrusters. If the satellite represented a technological marvel, then the 0.08-inch-thick (2.1-millimetre) tether which connected it via a supporting mast to a Spacelab pallet was equally impressive. Surrounding its Nomex core was electrically-conducting copper wire, insulated with Teflon and coated with ultra-strong braided Kevlar-29 and an outer 'jacket' of braided Nomex to protect it from abrasion and the corrosive effects of atomic oxygen in low-Earth orbit. During deployment, the tether was unreeled from a 4,470-pound (2,030-kilogram) mechanism affixed to the pallet and an MPESS. This regulated the length, tension and rate of deployment and could unreel the tether at a maximum speed of about 4.3 mph (7.1 km/h).

The structure took the form of a four-sided 'tower', not unlike a small broadcasting pylon, which unfolded slowly from a storage canister using rollers. As the canister rotated, fibreglass batons popped out of their stowed positions to form cross-members which supported the vertical segments. The tower rose to 38.7 feet (11.8 metres) above the payload bay to minimise the risk of impacting the Shuttle during deployment and retrieval. "The complexity of the experiment is extreme," said astronaut Andy Allen, who piloted the first TSS mission and commanded the second. Its sheer audacity was amply illustrated by the numbers: when the tether was unreeled to its full length of 12.8 miles (20.6 kilometres), the TSS/orbiter combination would be a hundred times longer than any spacecraft in human history, its electrical potential anticipated around 5,000 volts and its maximum current output approximately one ampere (Fig. 11.1).

At the time of the Challenger disaster, NASA's Shuttle planning charts ended in August 1988 and TSS-1 was not attached to a formal mission. According to the March and June 1985 manifests, the payload was booked in February 1985 and 'required' for launch in September 1988. Following the standing-down of the Shuttle fleet, the August 1988 manifest listed TSS-1 aboard Atlantis on STS-45 in January, a seven-day flight baselined with a seven-member crew, targeting a 210-mile-high (340-kilometre) orbit, inclined 28.5 degrees. A pair of follow-on missions (TSS-2 and TSS-3) were identified "for NASA planning purposes", requested for October 1992 and October

Fig. 11.1: Franco Malerba poses with the Tethered Satellite System (TSS), loaded aboard Atlantis' payload bay.

1994. By the time crewmembers were assigned to TSS-1 in September 1989, launch was scheduled aboard STS-46 in May 1991.

But the crew had a traumatic road ahead of them.

A CHANGING CREW

Named in September 1991 were mission specialists Hoffman, Nicollier and Costa Rica-born plasma physicist Franklin Chang-Díaz. A few months later, Hoffman was assigned as payload commander. In addition to TSS-1, their mission was co-manifested to deploy the European Retrievable Carrier (EURECA), a free-flying scientific and technology satellite. But by March 1990, Shuttle delays had pushed STS-46's launch from May to September 1991. Hoffman, the reader will recall from Chapter Seven, was also training to fly ASTRO-1 and was unable to dedicate his time to STS-46 until after December 1990. "It was like learning to go to the Moon," he said of the complexity of TSS-1. "Nobody knew how to control a tethered satellite. It was going to be completely automatic and all you would do is push a button. It would go 'up' and then you'd push a button and it would come back." The satellite did have an attitude-control mechanism, but there was no pitch or roll functionality, which reared concerns in Hoffman's mind. What if the tether went 'slack', he pondered, or a malfunction obliged a halt at mid-deployment. Would there be a loss of control?

"The problem is when the thing really gets going, it's coming out at several metres per second," he said. "That's pretty fast. If you just slam the brakes on, it's going to go wildly unstable. They basically had never designed for all these contingencies. As we did more and more simulations, and we learned more and more about the system, we came up with more and more scenarios where you need these manual capabilities." Hoffman worked with TSS program managers, who proved very accommodating in organising the financial resources to install extra capabilities on the satellite.

In addition to Hoffman, Nicollier and Chang-Díaz, NASA also assigned veteran Shuttle commander Robert 'Hoot' Gibson to lead STS-46. Although payload crews were typically named ahead of the flight crew, it was highly unusual for a commander to be appointed at such an early date. "There was going to be a lot of procedure development to handle some of the modes of the tethered satellite and handle some of the flight control issues that were going to come up," Gibson recalled. But his time on the crew was short. In July 1990, he was removed from the mission and grounded from flight status following "violation of a policy which restricts high-risk recreational activities for astronauts named to flight crews". Gibson was involved in an air racing incident in which one pilot was killed. "Our high-risk activity policy defines plain and simple guidelines for astronauts assigned to flight crews," NASA explained. "They are intended to preserve our crews as assigned and apply regardless of the time prior to launch." Chief Astronaut Dan Brandenstein pressed for Gibson to be fired, but he was saved by deputy chief Mike Coats and grounded for a year. Gibson received legal advice—"They can't do this to you," he was told—but fearful of winning a battle and losing a war, he took his punishment and went on to command two Spacelab flights, including the first docking with Russia's Mir orbital station (see Chapter Thirteen). In December 1990, Loren Shriver took Gibson's place on STS-46, joined by pilot Jim Wetherbee and flight engineer Andy Allen. By this point, launch had slipped to March 1992. But following the retirements of two veteran Shuttle commanders the following summer, STS-46's crew line-up changed again. Wetherbee was promoted to command his own mission and Allen moved into the pilot's seat, with Marsha Ivins selected as the new flight engineer.

Shriver shared Hoffman's concerns about TSS controllability. He knew the dangers of tether slackening was an acute worry: although gravity gradient forces would keep it tight during part of its deployment, the risk was very real in the early stages. "The dynamics," said Shriver, "could do most anything and it could go frontwards and then backwards and side to side." The obvious implication was that the Shuttle itself could be endangered. "The trouble was finding any computer system that NASA had that could model that," he continued. "Within the last couple of months of going to fly, [we] had some stand-alone trainers that began to show some of that in a reasonable manner, but we never did have really good training set-ups in the Shuttle mission simulator. There were some approximations, but they were never that complete or that good."

Rounding out the STS-46 crew was a single payload specialist, destined to become the first Italian citizen in space. In November 1984, ASI picked physicists Cristiano Batalli-Cosmovici and Franco Rossito, together with Italian Air Force officer Andrea Lorenzoni, as candidates, but all three were stood down after the Challenger disaster. Following the resumption of Shuttle flight operations, in May 1989 Batalli-Cosmovici and physicists Franco Malerba—an early candidate for Spacelab-1 (see Chapter Three)—and Umberto Guidoni were selected for TSS-1 training. In September 1991, Malerba was selected as the prime payload specialist, with Guidoni as alternate. Together with Switzerland's Nicollier and five Americans, STS-46 became one of few Shuttle missions to include three nations on its crew, although they got a kick out of counting Hoffman's English wife, Barbara, on the roster to make it four. "It never ceases to amaze me how quickly crews coalesce into a highly functional unit," remembered Shriver. "Once you get worked out who is going to be doing what, then the plans start to fall in place and go off and start training."

TETHER JAM

The TSS deployment required the crew to work in two shifts—red and blue—for around-the-clock monitoring activities. On the red team were Ivins, Hoffman and Chang-Díaz, with Allen, Nicollier and Malerba forming the blues and Shriver anchoring his schedule across both. Their 166 hours in space were tightly packed: activation and checkout of the TSS was scheduled to begin soon after orbital insertion, followed by work with the TOP investigation and deployment on the fourth day. Within hours of deployment, TSS-1 would reach its maximum distance of 12.8 miles (20.6 kilometres), whereupon it would be reeled back in to about 1.5 miles (2.5 kilometres) for further experiments, then retrieved and docked back onto the top of the mast. Prior to the TSS-1 operations, EURECA was set to be deployed. With two ambitious satellites tucked inside her cavernous payload bay, Atlantis rocketed into space at 9:56 a.m. EDT on 31 July 1992. But the intricate flight plan

Fig. 11.2: Deftly manipulated by Claude Nicollier, Atlantis' RMS mechanical arm deploys EURECA. In the payload bay, the spherical TSS is clearly visible atop its Spacelab pallet.

quickly ran astray when EURECA suffered intermittent data glitches, delaying its deployment by 24 hours. Even after the satellite had been released, Shriver and Allen had to station-keep close by until its checkout was completed (Fig. 11.2).

EURECA's late departure obliged mission managers to add an extra (eighth) day to STS-46, with TSS-1 deployment having been correspondingly pushed back. Original plans called for deployment to occur with Atlantis' payload bay facing away from Earth—tail slanted 'upward' and nose pitched slightly 'down'—such that the satellite would unreel at an angle of 40 degrees behind the Shuttle's flight path. Departing at a relatively slowpoke pace, it would be halted at a distance of 0.9 miles (1.6 kilometres) to adjust the deployment angle to five degrees, putting the satellite in the same plane, directly overhead. Deployment would then resume and upon reaching a distance of 3.4 miles (5.5 kilometres), a quarter-revolution-per-minute spin would be imparted via the TSS's attitude thrusters, thus kicking off a lengthy series of scientific investigations.

The next stage would see the satellite extended to a peak velocity of around 4.1 mph (8.1 km/h), reaching a distance of nine miles (15 kilometres), after which the rate of tether unreeling would be slowed. Five and a half hours after first motion from the top of the mast, by now having reached its maximum unreeled distance, measurements of tether dynamics would be taken. Of specific interest were explorations of the validity of theoretical concepts that tethers became more stable with increasing length, together with analysing the effects of induced disturbances, such as 'bobbing', pendulum-like 'librations' or skipping-rope-type motions. One particular test—the 'Ten-Degree Deadband'—involved adjusting the Shuttle's autopilot to induce a slight 'wobble' of up to ten degrees in any direction, then damping it out and assessing disturbances upon Atlantis herself. Working on the theoretical prediction that payload instability reached its most acute whilst the tether was at its shortest deployed extent, the retrieval speed of the TSS was planned to decrease at a distance of about 4.5 miles (7.2 kilometres). At length, it would be halted again at 1.5 miles (2.4 kilometres), preparatory to several hours of science observations, then the retrieval back onto the mast. Since many of the dynamic features of the tether/satellite combination were unknown, a guillotine feature provided for the cutting of the payload at any stage.

Many of these plans were cast into disarray when problems were encountered with the TSS. Four days into the mission, at 12:12 p.m. EDT on 4 August, the mast was deployed without incident and—despite a troublesome umbilical which initially refused to separate— the first attempt to release the satellite started five hours later. It was quickly aborted by the crew, due to excessive side-to-side motion in the tether. Following a lengthy checkout of the reel mechanism and vernier motors, a second attempt at 6:51 p.m. proved successful and the tether deployed smoothly to 587 feet (179 metres). Motions stopped again at 7:47 p.m. and the tether was reeled in a short distance, then reeled back out at a slightly higher rate to 840 feet (256 metres).

From Atlantis' aft flight deck windows, the view was spectacular. "When the Sun set," recalled Hoffman, "and everything turned red, it was just glorious." Ninety minutes later, deployment resumed, but stalled again after two minutes at 843 feet (257 metres) and was powered down to survival levels to maintain its battery lifetime. Next morning, the tether was retracted to 735 feet (224 metres).

And at that point, it stopped and refused to budge in either direction.

"I remember clearly looking through the camera," Hoffman said. "All of a sudden, it started to get all these wiggles in it." Wiggles were worrisome, implying slackness in the tether and suggestive of a break or jam. "The tether hadn't broken," he continued. "We could see that. The tether had jammed. The satellite had a jet of nitrogen gas to pull it away, and that was still on, but now it had bounced back." The nitrogen burst was pushing the satellite in a sideways direction and Hoffman and Nicollier struggled to regain control. Shriver knew that its current extent "was right in the middle of the so-called unstable zone" and flight rules forbade the tether from exceeding a 45-degree 'redline' angle, beyond which it would have to be cut loose for safety reasons. Shriver was determined to prevent that from happening. "Every once in a while, I listen to the audio from that," Hoffman reminisced. "It was certainly the wildest time that I've ever been in space. We were really up against the wall. We got very close to the redline, where we would've had to cut the tether" (Fig. 11.3).

A contingency spacewalk now became a realistic possibility—"We had that in our hip-pocket," remembered flight controller Jeff Hanley—and the Shuttle's cabin pressure was reduced to enable Hoffman and Chang-Díaz to begin pre-breathing protocols before donning their space suits. If a spacewalk

Fig. 11.3: The TSS tether pictured during deployment on STS-46.

had been ordered, Hoffman would have climbed the TSS mast "basically pull it in, hand over hand, and Franklin was going to wrap up the tether". Thankfully, their efforts went unneeded, as the satellite's motions gradually calmed. The location of the jam was believed to reside in the upper or lower tether control mechanism and, mindful of the problems associated with the actual deployment, a consensus was reached to clear the glitch and bring the satellite back into the payload bay. "The motor that extends the boom was actually more powerful than the motor which reels in the tether," said Hoffman, "so maybe if there was a kink in there, by extending the boom, that would be able to pull the kink free."

Firstly, the mast was retracted by one panel, to allow the crew to visually check for tether slackness, but nothing was found. This indicated that the problem existed in the upper tether control mechanism. The astronauts then re-extended the mast with its reel brakes engaged, which cleared the jam and enabled them to retrieve the satellite. "It was probably my longest shift I ever worked in Mission Control," said Hanley. "It was an 18-hour shift, but we worked the problem." The maximum deployed distance had been 843 feet (257 metres) and although the predicted 45-volt electromagnetic flux was developed, the induced voltage proved insufficient to excite the physical process for many of the mission's primary (Category I and II) scientific objectives. Some Category III tasks—including electron-beam propagation, beam-gas cloud interactions and Shuttle surface glow phenomena—were achieved and the basic concept of the tethered satellite was successfully demonstrated.

After eight intensive days, Atlantis swept into KSC on 8 August, touching down at 9:11 a.m. EDT. As Shriver sat in crew quarters, along came Gibson, who had by August 1992 long since finished his year's purgatory and was training to lead the very next Shuttle mission (see Chapter Ten). Perhaps losing command of the troubled STS-46 worked in his favour after all. Gibson warmly shook Shriver by the hand, a mischievous glint in his eye.

"Hey, Loren, thanks for taking that mission for me."

Shriver only half-smiled. "Well, you're welcome, Hoot."

BAD SIMULATION RUN

The limited data return from TSS-1 proved sufficient to warrant calls for another mission, but in the days after landing an investigative board, chaired by LaRC chief engineer Darrell Branscome, convened to assess the problems. Later in August, the board concluded that the tether had snagged on a 0.2-inch (6.1-millimetre) bolt and corrective action should not impair the prospect of a repeat flight. "This was an engineering change that happened just three or four months before launch," said Hanley. "They didn't check the clearance with this level-wind mechanism [which] is supposed to hit the end of the stop and then start the other direction, but it hit this bolt that's sticking out too far." By April 1993, TSS-2 and TSS-3 were listed as 'requested' for October 1994 and October 1996, although still footnoted "for NASA planning purposes". In March 1994, NASA Administrator Dan Goldin and ASI Special Administrator Giampietro Puppi confirmed that the second mission—identified not as 'TSS-2', but as 'TSS-1R', a 'reflight' of the first—would occur in February 1996. "NASA and ASI long have planned this reflight, but a formal commitment awaited U.S. Congressional approval for NASA to spend Fiscal Year 1994 funds," it was reported. "NASA and ASI completed a study last year of the jointly-developed TSS and confirmed their judgement of its usefulness as a unique Shuttle-based experiment carrier."

Most of the original STS-46 crew was kept together. In September 1994, Chang-Díaz was appointed as payload commander for STS-75, with Hoffman and Nicollier reprising their roles as mission specialists and Allen named as commander in January of the following year. Rounding out the crew were pilot Scott 'Doc' Horowitz, Italian mission specialist Maurizio Cheli and payload specialist Umberto Guidoni, who served as an alternate on STS-46. "We were elated," said Hanley, "that we were going to get another crack at it." The mission aboard Columbia was baselined for 13 days to enable additional time for the deployment and several days of ionospheric research thereafter. "It was unbelievably embarrassing when we discovered the cause of the tether jam," Hoffman reflected, "so there was a significant amount of redesign." Having said this, he believed there was a "strong scientific case" to refly TSS and felt that the inclusion of himself, Allen, Chang-Díaz, Nicollier and Guidoni was critical to mission success. "The astronaut office decided that the core of the payload crew…should refly," he said, "which was great, because we were all good friends." NASA also carried STS-46 Lead Flight Director Chuck Shaw onto the second mission (Fig. 11.4).

Unusually, the crew was split into three teams for the TSS-1R deployment, before reverting to a standard dual-shift system after the tether activities were complete. The red team comprised Horowitz, Cheli and Guidoni, the blues were Chang-Díaz and Nicollier and the 'whites' were Allen and Hoffman, allowing them to operate a broad-based payload. After the deployment and retrieval of the satellite, it was intended that Allen would rejoin the blues and Hoffman the reds.

Although deployment of the mast and satellite had been done on STS-46, the task for Allen's men was to get the tether unreeled to its full extent.

Fig. 11.4: A playful portrait of the STS-46 crew, seemingly entangled in the tether. Front row, left to right, are Franco Malerba, Jeff Hoffman and Loren Shriver. Back row, left to right, are Claude Nicollier, Marsha Ivins, Franklin Chang-Díaz and Andy Allen. More than half of the crew went on to fly STS-75.

"Arrivederci, au revoir, auf wiedersehen and adios," Allen cheerily radioed from Columbia's flight deck on 22 February 1996 as he and his crewmates lowered their visors for launch. "We'll see you in a couple of weeks." After a perfect countdown, STS-75 roared aloft at 3:18 p.m. EST. But the first portion of their ascent did not prove quite as perfect. Four seconds after liftoff, data indicated that one of the Shuttle's main engines was running at 40 percent of its rated performance. Mission Control confirmed that telemetry indicated that all engines were performing satisfactorily at full power and Columbia continued safely into orbit. It turned out to be an instrumentation error, but if the engine problem had been for real Allen and Horowitz would have been forced to make an emergency landing. "We had a couple of moments there that we got a little adrenaline rush," Allen said later. "This looks like a bad simulation run."

Safely in orbit, the crew divided into their shifts and blue crewmen Chang-Díaz and Nicollier set to work activating TSS-1R in readiness for deployment on 24 February. Before turning in at the end of his first shift, Hoffman tested the reel motor and latching mechanism which secured the satellite onto its docking ring. By the 23rd, with less than 24 hours to go before deployment, the mission had encountered its first spate of problems when the Smartflex relay experienced an electrical overload. Next, a laptop computer encountered difficulties and was exchanged for a spare, which itself behaved sluggishly. Nevertheless, in the early hours of 24 February, the methodical procedure to activate the experiments got underway. Firstly, the experiments were switched on individually, then together, in an effort to isolate the computer problems.

Mission Control elected to postpone deployment until the 25th to gain additional confidence with the Smartflex. At 3:45 p.m. EST, the unreeling of the tether began under the watchful eyes of Hoffman and Guidoni. "The satellite is rock-solid," Hoffman reported. It was expected that TSS-1R would remain extended for 22 hours, before slowly 'creeping' back towards the Shuttle to dock back onto the mast at 1:43 p.m. on 26 February. An hour into the deployment, it eclipsed the limit of its predecessor. Gradually, the rate increased to 0.9 mph (1.6 km/h), then slowed briefly in order that the 40-degree deployment angle could be reduced to five degrees, placing the satellite almost directly 'above' the crew cabin. From 2,000 feet (610 metres), TSS-1R made several slow rotations in support of its science investigations and by 8 p.m.—four hours into the deployment—it had attained a distance of nine miles (15 kilometres).

Then things went badly wrong.

"THE TETHER IS BROKEN!"

The intent was that, as the tether neared its maximum extent, the deployment rate would have been gradually reduced. "It was within a kilometre of its final length, at which point we were going to put on the brakes and just let it sit there and start all the experiments," Hoffman said. "I was recording this huge arc in the tether through the camera, when I started to see little ripples." It reminded him of the jam on STS-46 and a horrible sense of *déjà vu* dawned. However, at 8:29 p.m. on 26 February, at a tantalisingly close-to-target distance of 12.2 miles (19.6 kilometres), it became clear that the tension was due to something else. The tether had not jammed, it had snapped.

The shocked astronauts recorded video footage of the incident and the breakage appeared to have taken place near the top of the mast. "The tether has broken at the boom!" Hoffman radioed urgently. "It is going away from us." In fact, tether and satellite were accelerating away from the Shuttle at a rate of about 420 miles (670 kilometres) during each 90-minute orbit. By the

Fig. 11.5: Franklin Chang-Díaz, Jeff Hoffman and Claude Nicollier eat a meal whilst halfway inside their coffin-like sleep stations.

morning of the 27th, it was trailing Columbia by 3,000 miles (4,830 kilometres). After winding the remaining tether back into the mechanism, the astronauts retracted the mast. The breakage close to the top of the mast, rather than further outwards, proved fortuitous. "If it breaks at the bottom, it will fly away from you and you're not in any danger," said Hoffman, "but if it breaks at the top you've got 20 kilometres of tether coming snapping back at you. We had practiced for that eventuality in the simulator. You've got to then cut the tether at the bottom and fly away from it" (Fig. 11.5).

Although nearly 24 hours of electrodynamic measurements had been lost, the $154 million reflight was far from a total failure. Before the break, its first performance run had successfully generated a current of 480 milliamps from the electrical charge that had collected on the satellite's surface. This was about 200 times greater than the levels obtained during STS-46. Other experiments in the payload bay continued to function in support of the satellite and tether until as late as 6 March. "We did get a lot of good data during the deploy," Hoffman told journalists. Currents measured during the deployment were at least three times higher than predicted in analytical models. Voltages as high as 3,500 volts were developed across the tether, achieving current levels of 480 milliamps. It was also possible for researchers to study the interaction of gas from the satellite's thrusters with the ionosphere. A first-ever direct observation of an ionised shockwave around the satellite—impossible to study or model in laboratories on Earth—was also accomplished. Moreover, as the satellite and its trailing tether sped through the ionosphere, it was possible to continue investigations in spite of the fact that it was no longer physically linked to the Shuttle.

It did not detract from the disappointment, however. "If you don't ever get your nose bloodied," said Chuck Shaw, paraphrasing Theodore Roosevelt's famous comment, "you're not in the game." CAPCOM Dave Wolf told the astronauts that it was too early to speculate on the cause of the tether breakage, but an investigative board was already getting established to explore the anomaly. The board was headed by Kenneth Szalai, director of NASA's Dryden Flight Research Center at Edwards Air Force Base in California.

On 27 February, as the satellite and tether flew above the Electronic Signal Test Lab at JSC, ground controllers transmitted commands to successfully reactivate the ROPE, TEMAG and RETE experiments. With a possible two additional days of data-gathering capability now re-established before the satellite's batteries were predicted to die, teams scrambled to assemble a last-minute research timetable to squeeze as much as possible from their dying payload.

Since the tether breakage, trajectory specialists predicted that Columbia would approach to within retrieval distance of TSS-1R on 29 February, and such a scenario was briefly contemplated, but discarded due to insufficient propellant margins aboard the Shuttle. Had it been approved, a retrieval would have consumed up to six days of crew time. But by this point, the satellite's batteries were failing. Very weak signals had been detected through the Merritt Island and Bermuda tracking stations and no further data was received after 1 March. Still, it endured for far longer than expected, prompting one manager to liken it to the Energiser bunny, for its capacity to keep going.

Clearly, a huge amount of data was gathered and Guidoni noted that the mission "demonstrated that tether dynamic applications work and we can generate electricity". But the mood remained sombre. Despite the measure of success gained on STS-75, the mood aboard Columbia remained sombre. "Every time I turn around and look through the window and I see this empty bay," said Cheli, "it's like a part of myself has left." Added Allen: "Scientists on the ground have lost a lot and we feel for them." And for Hoffman, who had seen TSS fail twice, albeit for different reasons, it felt like a punch to the stomach. Of course, as noted in scientific papers presented at an American Geophysical Union conference, TSS-1R's main scientific breakthrough was the discovery of tether currents three times higher than theoretically predicted. It was speculated that this might indicate some degree of ionisation, even when the satellite's cold gas thrusters were switched off. Overall, its current-collection and power-generation capabilities rose several times higher than predicted.

Columbia returned safely home on 9 March, wrapping almost 16 days in space, by which time Szalai's investigative board had begun digging into the failure. "Given the public investment in the tethered satellite, it is important that we find out what went wrong," explained Wil Trafton, NASA's associate administrator for the Office of Space Flight. "To do any less would be a disservice to the American and Italian people."

Published in June 1996, the board's 358-page report blamed "arcing and burning of the tether, leading to a tensile failure after a significant portion of the tether had burned away". The arcing itself was caused by either penetration from a "foreign object"—though not orbital debris or micrometeoroids—or a tether defect breached its insulating material. (Certainly, it was stressed that "the degree of vulnerability of the tether insulation to damage was not fully appreciated.") This had apparently triggered a local electrical discharge from the copper wire in the tether to a nearby electrical ground. "The board found that the arcing burned away most of the tether material at that location," *Flight International* noted, "leading to separation of the tether from tensile or pulling force." In his concluding remarks, Szalai highlighted that the problem was "not indicative of any fundamental problem in using electrodynamic tethers", adding that "constructing a tether that was strong, lightweight and electrically conducting took the project into technical and engineering areas where they had never been before".

For the astronauts, a short circuit had been at the forefront of their minds from the outset. "I was able to hook up a very powerful train of optics [and] telephoto lenses and take a close look at the broken end of the tether," said Hoffman. "I could see that it was brown and charred, so we knew before we ever came home that it almost certainly had been a short circuit that had melted the tether."

12

In the absence of *gravitas*

JOURNEYS TO THE ENDS OF THE EARTH

One summer evening in August 1492, three ships—the carrack *Santa María* and a pair of a smaller caravels, the *Pinta* and the *Santa Clara*—put to sea from Palos de la Frontera in southern Spain, to begin one of the most remarkable voyages in human history. Since the fall of the Mongol Empire and the collapse of Byzantine hegemony in the eastern Mediterranean, it had become increasingly difficult and dangerous for European traders to pursue overland silk and spice routes to India and China. Spain and Portugal had long vied with one another to search out seafaring routes to the 'East Indies'. In 1488, an eastward possibility had opened when the Portuguese explorer Bartolemeu Dias reached the Cape of Good Hope, but it was Christopher Columbus' expedition westwards in 1492 which garnered the most excitement. In the first of four epic voyages, Columbus sailed via the Canaries, in the expectation of reaching Japan, but after many weeks at sea dropped anchor in the Bahamas.

His expeditions aroused general European awareness of the Americas for the first time. Within a generation, Ferdinand Magellan would navigate the eastern coast of South America to reach a straight, which now bears his name, north of Tierra del Fuego, through which the first European passage into the Pacific Ocean was made. Those discoveries, five centuries ago, carried profound implications for the 'Old World' and introduced Europeans to a 'New World'. Fast forward to 1992 and a proposal by Hawaii Senator Spark Matsunaga to designate the year of the 500th anniversary of Columbus' pioneering expedition as the 'International Space Year'. Twenty-nine national space agencies and ten international organisations supported the theme of the year as a Mission to Planet Earth, with a focus on space exploration and sustainable technologies.

And for NASA, the year would be its most arduous since Challenger, rendered more difficult since the first Shuttle mission of 1992 had changed several times in terms of both crew and duration and was impacted by dreadful human tragedy. STS-42 carried the first International Microgravity Laboratory (IML-1), led by Mission Scientist Bob Snyder of MSFC, inaugurating a series of Spacelab flights dedicated to the life and microgravity sciences and featuring more than 200 scientists from over a dozen sovereign countries and six national or supranational space agencies: NASA, ESA, Germany's DARA, France's CNES, Japan's NASDA and the Canadian Space Agency (CSA). It was also several years overdue, with IML-1 having been requested in June 1981 and IML-2 in December 1983. In the Shuttle's March and June 1985 manifests, IML-1 was listed as a seven-day mission with a seven-member crew, scheduled to fly Columbia in May 1987 and destined for an orbit 186 miles (300 kilometres) high, inclined 57 degrees to the equator. At the time of the Challenger tragedy, the mission's hardware/software was classified as 70-percent complete, although no crewmembers had been assigned. IML-2 was requested for February 1989, though without a firm mission number. Interestingly, by the time the final pre-Challenger manifest was published in November 1985 IML-1's inclination had changed to 28.5 degrees. And at the time of the Challenger accident—less than 16 months prior to launch—it is notable that no crewmembers had been announced for IML-1, which would have given this mission one of the shortest training flows of any Spacelab mission.

Following the resumption of Shuttle flights, IML-1 remained attached to Columbia on STS-47 in April 1991, by now baselined for nine days. With plans afoot to add an Extended Duration Orbiter (EDO) 'kit' of cryogenic consumables to facilitate longer missions, IML-2 was pencilled-in for STS-63 in October 1992, but footnoted by NASA as expected "to extend beyond nine days". For the first time in the August 1988 manifest, an IML-3 made an appearance, requested for November 1994, with a notional IML-4— "for NASA planning purposes"—envisaged at some point after November 1996. Plans had sharpened yet further by January 1990, with IML-1 now aboard Columbia on STS-42 the following December, IML-2 in January 1993, IML-3 in June 1995 and IML-4 requested for April 1997. As the EDO concept evolved (see below), IML-2 was envisaged as a 13-day mission with IML-3 targeted for up to 16 days. However, by the time the first flight in the series took place, IML-3 and IML-4 had been definitively cancelled.

CHANGE AND TRAGEDY

The sensitive microgravity requirements of IML-1's experiments required STS-42 to operate in a 'pseudo gravity gradient' orientation, with the tail of the Shuttle facing Earth. This would allow positioning to be maintained primarily by natural forces, reducing the number of thruster firings. Many of these experiments resided inside the Spacelab long module and would be operated around the clock in two, 12-hour shifts. To facilitate planning for what was envisaged to be an early demonstration of space station research, IML-1's payload crew were announced in June 1989. Physician Norm Thagard and environmental engineer Mary Cleave were named as mission specialists and a cadre of four international scientists—physiologist and former Olympic high-jumper Ken Money and physician Roberta Bondar, both from Canada, were joined by U.S. physicist Roger Crouch and West German physicist Ulf Merbold—were selected for two payload specialist positions. Scheduled for launch in December 1990, IML-1 would last nine days, with the possibility of a tenth, dependent upon consumables usage.

Within months, however, the first changes unfolded. In January 1990, Cleave resigned her place on the crew "for personal reasons" and was replaced by Navy flight surgeon Manley 'Sonny' Carter. Years later, Cleave explained her rationale: between her two previous Shuttle flights, she had seen firsthand the effects of pollution upon the home planet and sought to relocate out of the astronaut corps to work on environmental programs. Also in January, Commander Ron Grabe, Pilot Steve Oswald and flight engineer Bill Readdy joined STS-42, as did Bondar and Merbold. Years later, Readdy recalled being summoned to the chief astronaut's office one morning. He and Oswald had been working on a long-duration orbiter study and assumed they had been called to brief management.

Suddenly, the director of flight crew operations pushed a press release in front of them. "You have any problem with that?" he grinned.

Readdy spotted Oswald's name first—"Well, congratulations, that's great!"—then paused at the sight of his own. Oswald laughed. "Well, buddy, you're going, too!" (Fig. 12.1).

One of Thagard's greatest challenges came relatively late training. The hydrogen leaks endured by Columbia in the summer of 1990 (see Chapter Seven) forced several of her downstream missions, including IML-1, to be delayed. By the time she took flight with ASTRO-1 in December 1990, she was only months away from a planned overhaul, in which she would be upgraded for longer EDO missions. The result was that IML-1 moved firstly

Fig. 12.1: IML-1 experiment containers are passed through Discovery's crew hatch for stowage in middeck lockers on the day before launch. After reaching orbit, the experiments were transferred into the Spacelab module.

onto Atlantis and then Discovery, with launch progressively slipping from December 1991 into January 1992. But the change in orbiter incurred a penalty. One of the reasons that Columbia was manifested virtually all long-duration Spacelab flights, lasting around ten days, was because she alone in the fleet carried five cryogenic tanks to supply her electricity-generating fuel cells. (At the time, Discovery and Atlantis carried four tanks each.) As shown in NASA's Shuttle manifests from December 1990 onwards, transferring STS-42 off Columbia to one of her sisters produced a duration of only seven days, cutting IML-1's science-gathering potential time in orbit by more than a third. According to Bondar, this change to the mission's duration made the crew's workload extremely frenetic. A silver lining was the orbital inclination, which had increased from 28.5 degrees to 57 degrees. "As a result of going from Columbia, which was the queen of the fleet, the oldest, but also the heaviest, to Discovery, we were able to go on an inclined orbit of 57 degrees instead," remembered Readdy. The result was that the Shuttle's flight path covered 57 degrees north to 57 degrees south on each orbit, enabling the crew to 'see' 1,500 miles (9,000 kilometres) in either direction, from parts of Alaska to tantalising glimpses of Antarctica. "From an Earth observation standpoint," he said, "it made the flight much, much more interesting."

Yet there was misfortune still to come. On the calm afternoon of 5 April 1991, Atlantic Southeast Airlines Flight 2311 approaching Glynco Jetport (today's Brunswick Golden Isles Airport) in Brunswick, Georgia, after an hour-long flight from Atlanta. As it neared the runway, in clear weather conditions, eyewitnesses noticed it was flying much lower than normal. The jet suddenly rolled sharply to the left and descended, nose-down, to crash into a patch of trees. All 20 passengers and three crew were killed. Amongst the dead were two small children, a Texas senator and IML-1's own Sonny Carter. Investigators later found an issue with the propeller control unit in the left-hand engine, a problem so sudden and catastrophic that the pilots did not even have time to declare an emergency.

Carter's death devastated the astronaut corps. Mike Mullane later wrote that it was "a gross violation of the natural order" for an astronaut to die not as a pilot, but as a passenger. Two weeks later, three-time Shuttle veteran Dave Hilmers, a Marine Corps engineer on the brink of retirement with plans to enter medical school, was named to replace Carter on IML-1. It was, said Don Puddy, JSC's director of flight crew operations, a decision he made with "regret". And when the STS-42 astronauts released their official crew patch later that year, it included a solitary gold star, honouring the memory "of our crewmate, colleague and friend".

In his memoir, *Man on a Mission*, Hilmers noted that he had been offered a fourth flight "as the payload commander for a mission that was a couple of years down the road", but turned it down, so intense was his desire to study medicine. Years later, Hilmers could not recall which mission it was, but judging from the timing of the offer in 1990, it was likely Spacelab-D2 or perhaps Starlab (see below). He was surprised when Chief Astronaut Dan Brandenstein offered him a seat on IML-1 and later learned that Grabe and Thagard "evidently made it known that they wanted me to take Sonny's place". Training was arduous, with barely nine months to prepare for a complex scientific mission, but Hilmers came right up to speed. "He just never missed a beat," remembered Readdy. "You couldn't throw too much information at him. The guy's just a sponge and able to absorb it all and then somehow figure out how to process it and spit it back to you."

STAYING AWAKE FOR SCIENCE

Weather concerns and a hydrogen pump anomaly with one of the fuel cells delayed the launch of STS-42 by almost an hour on 22 January 1992, but Discovery finally rose majestically into space at 9:52 a.m. EST. It was a relief for the Spacelab program, which had its heaviest booking of module-based flights in International Space Year. With three missions planned—IML-1, the first U.S. Microgravity Laboratory (USML-1) in June and the U.S./Japanese

Fig. 12.2: IML-1's tunnel adapter is lowered into Discovery's payload bay, ahead of connection to the Spacelab module at one end and the middeck airlock hatch at the other.

Spacelab-J (see Chapter Ten) in September—and only two modules available, STS-42 was committed to flying in January or February in order that its hardware could be turned around in time for the Japanese mission (Fig. 12.2).

Shortly after reaching orbit, STS-42's blue team of Grabe, Oswald, Thagard and Bondar set to work accessing IML-1 and transferring 11 lockers'-worth of research materials from the middeck to the module, which came alive about three hours into the mission. Meanwhile, the reds of Readdy, Hilmers and Merbold were scheduled for an abbreviated six-hour sleep period. "They tended to keep the pilot and the commander on the same shift," said Readdy, "so that meant that you had to have somebody else that was schooled in all the orbiter systems and piloting tasks on the other shift." To prepare for 24 hours of operations, the astronauts had shifted their sleep/wake cycles a couple of weeks prior to launch. "By the time we hit the launch pad, the crewmembers on my shift had already been up for eight or nine hours," Hilmers wrote. "I was exhausted, but while the plan was for my team to sleep soon after ascent, I would defy anybody to go through the launch of a Space Shuttle and then almost immediately try to go to sleep. It just didn't work like that!" That nap of six hours, he joked, was the hardest rest he had ever had.

The IML-1 experiments were devoted entirely to materials and life sciences in the microgravity environment and featured significant collaboration from U.S., European, Canadian and Japanese researchers. Of the life sciences complement, Europe's Biorack sought to understand the fundamental functions of organisms, including cell proliferation and differentiation, genetics, gravity sensing and membrane behaviour. To further this research, Biorack carried three incubators, a glovebox and a cooler/freezer to grow, handle and store hundreds of biological samples from investigators in Switzerland, the United States, Spain, Germany, France, the Netherlands, Scotland and Denmark. Embryonic mouse limb cells were studied to characterise the similarities observed between skeletal malformations in rodents and human children, with a focus on helping to clarify the processes by which bones heal in microgravity. Previous Soviet long-duration experience had already suggested that bone damage during extended space flights would be difficult to heal and a contributory factor in the breaking of weakened bones is the loss of calcium through prolonged microgravity exposure. Other experiments utilised the eggs of African clawed frogs, *Xenopus laevis*, and fruit flies, together with yeast, bacteria, lentil roots and plant shoots and the *Physarum polycephalum* slime mould, to understand the role of gravity in embryonic and cell development. The effect of radiation on soil samples and the eggs of stick insects was also closely studied and the German Biostack facility—housed in a rack under the Spacelab module's floor—analysed the influence of cosmic rays on bacteria and fungus spores, together with thale cress seeds and shrimp eggs.

Elsewhere was the Gravitational Plant Physiology Facility (GPPF), a space-based botanical laboratory, equipped with four centrifuge chambers, floodlights, three videotape recorders and plant compartments. Developed by ARC in Mountain View, California, the GPPF investigated the gravity-sensing mechanisms of oat seeds and reactions of wheat specimens to the effect of light stimulation. It supported a pair of plant experiments from the University of Philadelphia in Pennsylvania and University City Science Center in Philadelphia, the first of which investigated the oats' behaviour in altered gravitational fields, whilst the second explored the curving and 'straightening' of wheat plants when exposed to light in microgravity.

Bondar's presence on IML-1 was unsurprising, in view of the mission's substantial research from Canada. A group of space physiology experiments provided by CSA, McGill University in Montreal, the University of Calgary in Alberta, the London Ear Clinic in Ontario and the University of British Columbia at Vancouver focused on the adaptation of the human organism to the weightless environment, including the vestibular apparatus, the body's sense of position, energy expenditure, cardiovascular adaptation, eye-motion

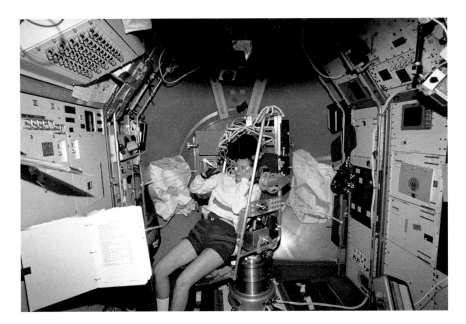

Fig. 12.3: Roberta Bondar prepares for a session in the Microgravity Vestibular Investigations (MVI) chair in the IML-1 module.

oscillations and back pain. The latter was devised in response to a typical incidence of spinal lengthening and back pain in over two-thirds of all astronauts (Fig. 12.3).

In support of these experiments, members of IML-1's payload crew utilised 'The Sled', in the centre aisle of the Spacelab module, to measure changes in the gravity-sensing part of the inner ear, the otolith. The sled moved gently backwards and forwards predictably, at a constant speed over the same distance, or at varying distances or at varying speeds with sudden stops and starts, stimulating the otolith. As crew members underwent the test, their response was monitored to gauge their ability to visually track moving objects and small electric impulses were applied to the subject's legs via an electrode.

The Microgravity Vestibular Investigations (MVI) chair, developed by 20 investigators spanning five nations, sought to provoke "interactions among the vestibular, visual and proprioceptive systems" to understand the integral changes in the adaptation process to microgravity. It required subjects to wear an accelerometer-instrumented helmet, whose independent visors fitted over each eye to furnish visual stimuli. The chair could be configured to position its subject 'upright', on their side or on their back and could be moved backwards and forward at constant speed ('sinusoidal') or over varying distances ('pseudorandom') or 'stepped' at various speeds, starting and stopping suddenly. During each test run, subjects' physiological responses were assessed in terms of the eyes' ability to track objects, perceive rotation during and after spinning and understand interactions between visual cues, vestibular responses and sensory perception.

"Part of the difficulty with this experiment was trying to keep your eyes open so that the experimenters could record the eye movements," remembered Hilmers. "The other problem was that I was asked to do it on Flight Day One, which is when astronauts are most prone to space motion sickness. Any head movement tends to make it worse." But worse yet for Hilmers was that it was very easy to fall asleep, with one's head secured in the darkened helmet. "Just imagine being tired and put into a dark place," wrote Hilmers. "Even with the rotations, it was easy to go to sleep, despite the efforts of the ground team, who could see my eyes from the cameras inside the helmet." From time to time, Hilmers could be heard breaking into song—"even the Marine Corps Hymn"—to keep himself from nodding off (Fig. 12.4).

Despite a tripped circuit breaker, which caused the chair to stop working a few seconds into its first run, the experiment produced pleasing results which "quantified the human vestibular function in the microgravity environment". Thagard found the chair easy to lift and reposition in different planes, even with the subject astronaut still seated in it. "It's sort of an operational hit," he said, "to have folks get out of that chair and reposition it." Curiously, Thagard

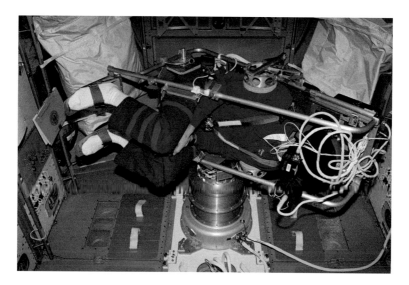

Fig. 12.4: Dave Hilmers, wearing an instrumented helmet, participates in an MVI investigation. Hilmers is positioned in the pitch position, which Norm Thagard noted still gave the chair's occupants a curious sense that they were sitting 'sideways'.

thought before the mission that in microgravity subjects would not notice the difference in the chair's orientation. But, in fact, when the chair was oriented in its 'pitch' position, even in the absence of gravity, Thagard still felt that he was sitting in a chair on his side. Other experiments required the crew to drink water, especially enriched with stable, non-radioactive isotopes of oxygen and hydrogen, to enable researchers to determine energy expenditure through post-flight urinalysis.

In the microgravity science arena, NASA experiments in vapour-driven protein crystal growth were undertaken, together with German Cryostat investigations which employed 'liquid diffusion' and offered researchers the flexibility of a temperature-controlled facility. Specifically involved in the experiments were Beta-galactosidase, a key enzyme found in the intestines of human and animal infants, which assists in the digestion of milk products. (This enzyme was the first protein ever crystallised in space in November 1983.) Also under study was the satellite tobacco mosaic virus.

NASA's Fluids Experiment System (FES) and Vapour Crystal Growth System (VCGS), previously flown on Spacelab-3 and comprehensively modified for IML-1, carried a range of investigations which grew crystals of triglycine sulphate and mercury iodide, as well as performing laser diagnostic recording and creating more than 300 three-dimensional structural holograms for post-mission analysis. Several of the mercury iodide experiments were supported by CNES and record-sized crystals were yielded.

The Japanese Organic Crystal Growth Facility (OCGF) sought to produce semiconducting crystals of tetrathiafulvalene (a compound with applications in molecular electronics) and nickel, whilst demonstrating the effectiveness of an epoxy cushioning material to damp accelerations otherwise known to disrupt the growth process, and the European Critical Point Facility (CPF) explored the behaviour of fluids when they reached the precise temperature-pressure stage at which the difference between vapour and liquid became indistinguishable. Samples of sulphur hexafluoride—used on Earth in the electronics industry—as part of a number of experiments were processed for as long as 60 hours during IML-1. Radiation and acceleration measurements were performed to determine the influence of Shuttle motions upon sensitive experiments (Fig. 12.5).

Despite the decision to reduce STS-42 from ten to seven days, an option was preserved to extend the mission by 24 hours if on-board consumables remained acceptable. As circumstances transpired, the crew's consumables usage remained below planned levels and an additional day was flown to complete the collection of scientific data. Early on 30 January, Hilmers led the deactivation of Spacelab and remembered that "coming down the tunnel for

12 In the absence of *gravitas* 299

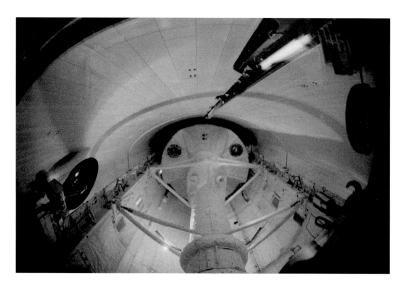

Fig. 12.5: Unusual view of the IML-1 module and tunnel in Discovery's payload bay, pictured during the closure of the payload bay doors on 30 January 1992.

the last time was kinda nostalgic, like saying goodbye to an old friend". The absence of half of their pressurised volume, following the closure of Spacelab's hatch, came as a surprise. "Suddenly, things for seven people become much more crowded," Hilmers said, "as we found out in deorbit prep." Readdy joked that traffic lights were needed in the tunnel, as it was virtually impossible for two crewmembers to traverse its length, passing each other at the same time. On one occasion, Bondar made to enter the tunnel to head back towards Discovery's middeck, only to meet Readdy approaching in the opposite direction. Quick as a flash, Canada's first female astronaut backflipped herself back into the lab. Traffic lights, it seemed, might not be such a bad idea, after all.

Touchdown eventually took place at Edwards Air Force Base at 8:07 a.m. PST on 30 January, after eight days. A California landing had always been manifested for IML-1, in part due to the heavyweight nature of the Shuttle—with the 23,200-pound (10,520-kilogram) Spacelab aboard, Discovery tipped the scales at 217,250 pounds (98,500 kilograms) on her return—and also to facilitate the immediate collection of physiological data from the payload crew. And if the payload crew was impressed by Spacelab's voluminous capabilities, so too were the pilots. "The systems integrated with the Space Shuttle, but they were different systems," Readdy said. "The software architecture was different. The displays were different. That took a while to learn, but it was a magnificent facility. Big as a school bus, shirt-sleeve environment, just a fabulous facility."

AS ONE MISSION CLOSES...

Sometime in September 1990, one of the strangest Spacelab missions vanished from NASA's Shuttle manifest like blip from a radar screen, to be replaced by another. Starlab, the only military-funded Spacelab to (almost) approach fruition, was always "an iffy assignment" for the payload specialist candidates tentatively assigned to it: astrophysicist Kenneth Bechis and Air Force engineers Denny Boesen, Maureen LaComb and Craig Puz. Starlab made its first definitive appearance in the August 1988 manifest, listed as a seven-day flight, targeted for launch in September 1990. Baselined as a Spacelab long module and a single pallet, it would orbit 200 miles (325 kilometres) high, inclined 33.4 degrees to the equator. Bechis and Boesen had been identified as prime payload specialists as early as July 1987. By January 1990, Shuttle delays had pushed Starlab firstly back to STS-49, scheduled for September 1991. And by March 1990, launch had slipped again to no earlier than January 1992. But before the end of 1990, the budgetary axe had fallen and this tantalising mission was gone for good, like so many Spacelabs before and after it.

Starlab's focus was for a seven-person crew to test systems for the Strategic Defense Initiative Organization (SDIO or 'Star Wars'), using laser equipment aboard the module and pallet to acquire, track and 'point' six Starbird rocket launches from Cape Canaveral in Florida and Peacock Point on Wake Island in Micronesia with red 'marker' and green 'illuminator' lasers. Originally conceived in 1985 as 'Air Force Program-513', it was envisaged for launch out of Vandenberg Air Force Base before the loss of Challenger, then rescoped for a Florida launch after the Californian site was mothballed and taken out of service. Inside the module would have been four experiment racks, an optical bench with mirrors, inertial sensors, phase modulators and diodes, the laser itself and the $85 million Wavefront Control Experiment (WCE), a deformable mirror capable of adapting at a rate of 4,000 times per second to compensate for atmospheric disturbances.

The optical path of the laser would run through a viewport in the aft end-cone of the module to a 2.6-foot-aperture (0.8-metre) telescope on the pallet, targeted by a steerable 'beam-walk mirror', fabricated by Ball Aerospace. Also affixed to the pallet was payload experiment package with a green illumination laser to track and 'mark' the four-stage Starbird rockets and a pair of deployable Space Test Objects (STOs). The latter, nicknamed 'starlets', were spherical capsules, each measuring 18.5 inches (47 centimetres) in diameter and weighing 150 pounds (68 kilograms). Their purpose was to align the marker and illuminating lasers before engaging the Starbird rockets and conducting other object-tracking exercises. All told, about 20 separate events (or 'engagements') were planned. But with delays to the Starbird launch infrastructure having already pushed the mission repeatedly back, and with other military projects—notably the 'Brilliant Pebbles' ballistic missile defence initiative—having assumed primacy in terms of SDIO funding, Starlab breathed its last in the late summer of 1990. Bechis, Boesen, LaComb and Puz resigned from the program shortly afterwards.

As this would-be Spacelab mission thus vanished, another moved in to take its place. A series of U.S. Microgravity Laboratory (USML) missions, up to four in total, had been planned for the middling years of the decade, devoted to materials science, combustion physics and fluid dynamics. According to the August 1988 Shuttle manifest, USML-1 was manifested aboard Shuttle Columbia on STS-55, scheduled for March 1992, with USML-2 on STS-73 in July 1993 and USML-3 and USML-4 listed "for NASA planning purposes" in November 1995 and November 1997, respectively. All four comprised a Spacelab long module and MPESS pallet and would have occupied 186-mile-high (300-kilometre) orbits, inclined 28.5 degrees to the equator. But since before the loss of Challenger, NASA had plans for 'extended' Shuttle flights, lasting up to 28 days, and the Extended Duration Orbiter (EDO) concept was born. As early as October 1979, MSFC issued requests for proposals for a 25-kilowatt power system to afford additional solar energy to keep Shuttles aloft for longer, with hopes that it might enter formalised development as early as 1982 and fly in 1984-1985. Other concepts included the Power Extension Package (PEP), an accordion-like solar 'wing', measuring 240 feet (73 metres) in length when fully deployed, converting raw solar power to regulated voltage and affording up to 15 kilowatts of electricity to enable missions beyond 20 days and possibly up to 45 days. The PEP hardware, weighing some 2,100 pounds (950 kilograms), would be mounted directly above the Spacelab module's tunnel adapter and unfurled and stowed by means of the RMS mechanical arm (Fig. 12.6).

However, as the Shuttle program dawned, budgetary realities caused many of these plans to be shelved. "We were very disappointed in [Spacelab], because when we first started doing our studies for Shuttle and the life sciences role…it was first advertised as 30-day missions," said John Stonesifer, chief of JSC's life sciences experiments program office in 1978-1982 and assistant for Shuttle support in the Space and Life Sciences Directorate in 1982-1987. "We went out with our first call for experiments, based on 30-day missions. Then, when they got farther along in the design of the Shuttle and it went from 28 days to maybe 21 days and then down to 14 days and down

Fig. 12.6: The cryogenic oxygen and hydrogen tanks of the Extended Duration Orbiter (EDO) pallet are clearly visible at the rear of Columbia's payload bay during USML-1 pre-launch closeout operations. The insulation-covered Spacelab module sits in the middle of the frame.

to ten days." Many of the principal investigators who had designed their experiments for month-long missions were unperturbed—"Give us three days if we can get it," was their can-do attitude—but the push within NASA to more than double Shuttle missions from a week to 16 days endured.

In June 1988, Shuttle prime contractor Rockwell International began developing the EDO hardware, most visibly manifested in a pallet of additional oxygen and hydrogen cryogens for emplacement at the rear of the payload bay. In January 1991, Columbia was selected for the $93.5 million EDO modifications and following her STS-40 mission the following June (see Chapter Thirteen) she was withdrawn from service for a year to receive the new hardware. At about the same time, Endeavour, built to replace the lost Challenger, came off the production line with EDO functionalities already in place and would fly her own long-duration Spacelab mission in March 1995 (see Chapter Seven). From a relatively early stage, therefore, it was envisaged that USML-1 would likely become the first EDO mission, baselined in both the August 1988 and January 1990 manifests for nine days in length, but footnoted with "plans to extend beyond nine days". The EDO modifications got underway within weeks of Columbia's return to Earth from STS-40; after several weeks of standard 'de-servicing' activities in Florida, she was flown to Rockwell International's factory in Palmdale, California. Over the next six months, she underwent more than 150 modifications, including the EDO pallet and a new Regenerative Carbon Dioxide Removal System (RCRS), the latter of which removed the crew's exhaled breath more efficiently than previously possible with lithium hydroxide canisters.

In the meantime, crew places for USML-1 began to fill. In August 1990, biochemist Larry DeLucas of the University of Alabama at Birmingham, chemical engineer Al Sacco of Worcester Polytechnic Institute in Massachusetts, mechanical engineer Joe Prahl of Case Western Reserve University in Cleveland, Ohio, and physicist Gene Trinh of JPL (who served as backup on Spacelab-3) were selected as candidates for a pair of payload specialist positions. A month later, veteran NASA astronaut Bonnie Dunbar was named as payload commander for USML-1, which had by now crystallised on the Shuttle manifest from STS-53 to STS-50 and was scheduled for mid-1992. By this point, USML-1 was anticipated to run for 13 days, longer than any previous Shuttle mission, with USML-2 and USML-3—by now scheduled for STS-79 in April 1994 and STS-102 in May 1996—both expected to fly for 16 days on Columbia. The final flight in the series, USML-4, had moved further to the right and was 'requested' for March 1998. All four missions would operate in orbits 160 miles (260 kilometres) high, inclined 28.5 degrees to the equator.

"WHAT? TOGETHER?"

"I think we've all been aware that living together and working together for 13 days is certainly a challenge," said Dick Richards, commander of STS-50, whose two previous missions had both lasted only four days. "We'll be sharing the same sleeping quarters. We have only one restroom on-board. We all have to co-operate to make life as good as possible." In the wake of his second Shuttle flight, Richards complained to Chief Astronaut Dan Brandenstein about the short nature of Shuttle flights. "I just raved to NASA about the fact that I can't believe we're going up there and spending four days," he said later. "So they said, okay, we've got this flight coming off called United States Microgravity Laboratory..." The rest was history. In December 1990, Richards was assigned to command USML-1, joined by pilot John Casper, mission specialist Carl Meade and flight engineer Ken Bowersox. Planning changed the following August, after the retirement of two veteran Shuttle commanders, the result of which saw Casper promoted to command another mission and Bowersox promoted to pilot USML-1. Physician Ellen Baker was assigned to the crew as STS-50's new flight engineer and in May 1991 DeLucas and Trinh were selected as prime payload specialists for a seven-member crew.

The relationships between NASA crew members and payload specialists has always been an interesting one. "You're a family when you're up there," said Dunbar, "so if there are any [interpersonal] problems to resolve, you need to resolve them on the ground." After one training session, she invited the four payload specialists—DeLucas, Sacco, Prahl and Trinh—out for dinner…and was surprised by the response.

"What? *Together*?" (Fig. 12.7).

It was a comical moment, perhaps, but it underlined in Dunbar's mind the importance of starting out as a team and bonding into something as close as possible to a family. "You have to get through to that so when you get on orbit, you know *exactly* what the other person is going to do and how they're going to react."

They would supervise 31 experiments in crystal growth, combustion science, fluid dynamics and biotechnology aboard the Spacelab module. More than one scientist considered USML-1 as a precursor to the work planned for a future space station. Around-the-clock science activities—carried out by a 'red' team of Dunbar and DeLucas, led by Bowersox, and a 'blue' team of Meade and Trinh, led by Baker, with Richards anchoring his schedule across

Fig. 12.7: Dr. Bonnie J. Dunbar, USML-1's Payload Commander, prepares to store material samples from the Crystal Growth Furnace (CGF). Dr. Dunbar chaired a committee which conducted a year-long study for NASA in 1987, resulting in a proposed mission called US-1. This later was renamed USML-1 and she was assigned as its first Payload Commander.

both shifts—would utilise the microgravity environment to produce new materials and protein crystals in support of new pharmaceuticals. Moreover, the USML-1 experiments were expected to help the United States to maintain a 'lead' in microgravity research and applications. Terrestrial spinoffs were expected to be the production of newer, faster semiconductors for the next generation of high-speed computers and the construction of more efficient chemical catalysts for converting petroleum into gasoline. Some of this work would be undertaken within a laboratory-style 'glovebox' and other research facilities lined the walls of the Spacelab module in refrigerator-sized racks.

Dunbar's role in USML-1 ran deep. Through her efforts, she had enabled NASA to gain funding for the mission. As chair of the agency's Microgravity Materials Science Assessment Task Force in 1987, she realised that much of the planning for Space Station Freedom centred on Spacelab missions with a life sciences bias, although there existed precious little emphasis on the development of furnaces for directional solidification or materials research. In fact, the only Spacelab missions under consideration to focus on the physical sciences were those which involved European and Japanese participation. Dunbar was keenly aware that such work had been pioneered aboard Skylab in the early 1970s and she felt that it was "a real loss of our investment and scientific discoveries of the future if we didn't build facilities for [the new] station". She worked closely with astronaut Sally Ride, who was leading a team to prepare a strategic roadmap for NASA's future, and one of the products of their efforts was USML. "We flew USML-1...with all new facilities that were destined for the International Space Station," Dunbar told the NASA oral historian. Those facilities included an entirely new variety of furnaces, developed by Teledyne Brown Engineering, Inc., along with a fluid physics module and a range of life sciences experiments and lower-body negative pressure apparatus. "What was thrilling for me was to be asked to be payload commander of that flight," she said. "In five years, I had an opportunity to see it go from a concept in a report to an actual flight of brand-new equipment and hardware."

As payload commander, Dunbar worked extensively on the mission timelines and would later derive satisfaction from the fact that 'handovers' between the two USML-1 shifts typically lasted no more than 15 minutes. It illustrated the excellence of their training. Much of that training was centred on the various institutions which developed the experiments: at the University of Alabama for protein crystal growth, at the University of Wisconsin for plant growth and at JPL for fluid physics. "You understand what the researcher wants to find, first," Dunbar explained, "because you're their hands, eyes and ears and if you don't understand what they're looking for, then when the

Fig. 12.8: Columbia roars to orbit on 25 June 1992.

unexpected comes up you also don't recognise that, either. You're part of the research team." By December 1991, six months before launch, the Spacelab crew were running fully integrated simulations; picking the busiest day of the flight, choreographing it down to the last minute and running through their tasks. "You train like you fly and you fly like you train," she said. "You don't have a chance to start over on orbit" (Fig. 12.8).

Five minutes later than intended, as bands of cloud and rain marched over the launch pad, Columbia roared aloft at 12:12 p.m. EDT on 25 June 1992. "When the engines lit, I didn't get too excited," recalled Bowersox, who was making his first flight, "but I did look over to the left to see what the gantry looked like and I noticed that it started *glowing* before we started moving, as we got the reflections of both the engines and the SRBs." Less than nine minutes later, STS-50 was safely in orbit and a couple of hours into the mission Dunbar and DeLucas set to work opening the tunnel and accessing the Spacelab module. Meanwhile, Baker's blue team bedded down for an abbreviated six-hour nap. Following a lengthy process of 16 steps to activate the lights, computers and experiments, Dunbar put in her first communications call to the POCC in Huntsville, Alabama.

"Huntsville, Spacelab," she began. "How do you read?"

Alternate payload specialist Al Sacco came straight back, with an up-tempo: "Spacelab, Huntsville has you loud and clear. How me?"

"Read you loud and clear, too."

"It's good to hear your voice from above," replied Sacco, "and we're ready for some science results!"

Neither Sacco nor Dunbar had long to wait.

CONSTANT SUPPLY OF CRYSTALS

With the arrival of the STS-50 crew in orbit on the afternoon of 25 June 1992, former astronaut Brewster Shaw, then serving as chair of NASA's Mission Management Team, quipped that "we're going to find out whether these kids are really friends or not". He had previously flown with Richards and need not have worried. Echoing the sentiment of the remainder of the crew, Bowersox later described his colleagues as "a joy to be around". The red team, plus Richards, who, although not attached to a specific shift, tended to align his work schedule with the reds, progressed smartly through the USML-1 activation process. Ten minutes ahead of schedule, at 3:42 p.m. EDT, Dunbar opened the hatch into the Spacelab, floated inside and turned on the lights.

Activation of the module was complete by 6:00 p.m. EDT and as Dunbar busied herself with preparing it for 13 days of research, Bowersox and DeLucas set to work unstowing and photographing a pair of Protein Crystal Growth (PCG) specimens on Columbia's middeck. It was hoped that data from these experiments would yield new knowledge about the proteins' molecular arrangement and assist in the production of more nutritious foods, highly resistant crops and better medicines with fewer adverse side effects. Moreover, the extended nature of USML-1 meant that slower-growing crystals could be produced. After their return to Earth, the three-dimensional structures of the crystals were to be carefully mapped. Crystals of proteins grown in terrestrial laboratories are large enough to study, but usually contain numerous flaws, caused by gravity-induced convection, buoyancy and sedimentation. Space-grown crystals, on the other hand, are of greater purity and have more highly-ordered structures which significantly enhances their X-ray analysis. On USML-1, protein crystal growth experiments were performed both within the middeck incubator and inside the glovebox of the Spacelab module.

Fig. 12.9: Larry DeLucas at work aboard USML-1.

The importance of this research to the pharmaceutical industry was illustrated by the presence of Larry DeLucas, already recognised as one of the United States' leading experts in crystallography. Consequently, USML-1 would be the first time that experiments of this nature had been performed with a human expert present to improve the production processes with a special controllable furnace. Previous experiments had been automated and the only astronaut interaction had been the mixing of the chemical solutions. "Someday," DeLucas said before the flight, "I feel confident we will find a drug that might help to prevent the complications of diabetes, maybe prevent hypotension, maybe find a cure for some types of cancer." It was considered possible that such research might eventually lead to treatments for emphysema and HIV-AIDS. Yet DeLucas was under no illusions; more than a decade of further work and a permanent space station would be essential in carrying such plans through to fruition. "My investigators don't need one crystal so we can make a breakthrough," he said. "We need a constant supply of crystals" (Fig. 12.9).

In addition to the protein crystal growth investigations, a Crystal Growth Furnace (CGF) was employed to solidify a wide range of materials—mainly semiconductors—which form the basis of electronic devices, including computers, timepieces and communications equipment. The furnace could process samples at temperatures of up to 1,260 degrees Celsius (2,300 degrees Fahrenheit) and enabled scientists to investigate the factors affecting crystal growth and the best methods for producing them. Post-mission analysis of the crystals from CGF revealed distinguishable differences from Earth-grown specimens and showed them to be of far higher quality. During USML-1, the furnace used 'directional solidification', whereby the solidification front proceeded in a specific direction along the sample, and 'vapour crystal growth', in which part of the sample was heated to make it sublime and the vapour was then allowed to flow towards and condense upon a substrate base in a cooler part of the sample ampoule. Specimens of cadmium telluride, mercury-zinc telluride, gallium arsenide and mercury-cadmium telluride, all of which have found applications in medical infrared sensors, night-vision goggles and telescopes, were produced.

The furnace itself could handle up to six samples at a time, each housed in its own quartz ampoule within a rotary changer. These samples were processed under computer control, although the experiments' principal investigators on the ground could uplink changes or adjustments whenever necessary. A flexible glovebox provided access to the interior of the furnace, but physical handling of the samples was only done if absolutely necessary, so as to minimise the risk of contamination to the Spacelab environment. Dunbar started the first CGF experiment, late on 25 June, when she initiated an 18-hour solidification run with six samples of mercury-cadmium telluride. Unfortunately, a circuit breaker tripped a few days later. Carl Meade followed a procedure and successfully revived it, but a subsequent mercury-zinc telluride crystal had to be stopped halfway through its growth cycle. Meade then inserted a cadmium-zinc telluride sample for 92 hours. Later in the mission, the CGF grew the longest gallium arsenide crystal—6.5 inches (16.5 centimetres)—ever produced. Moreover, the crystal contained a deliberate trace impurity, known as a 'dopant', to enable scientists to precisely engineer its electronic properties. On Earth, gravity-driven convection makes it very difficult to control the placement of dopants and if placed imprecisely the resultant crystal can yield inconsistent material properties. Another mercury-cadmium telluride sample was also removed from the CGF on 5 July, after six hours of growth. Its wafer-thinness enabled crystallographers to examine its structure with relatively little processing. By the end of the mission, the CGF had operated for 286 hours, produced seven crystals and achieved a peak temperature of 870 degrees Celsius (1,600 degrees Fahrenheit).

Elsewhere, crystals of 'zeolites'—complex arrangements of silica and alumina, which occur both naturally and synthetically—were grown both in Columbia's middeck lockers and in the Spacelab glovebox. Zeolites have 'open' crystalline structures which are selectively porous, enabling them to function as molecular sieves, capable of 'adsorbing' elements or compounds and forming an integral part of catalysts and ion-exchangers. On several occasions, Dunbar spoke directly to Sacco, principal investigator of the experiment, to discuss mixing procedures for test runs. "A person up there optimising the mixing of the crystal solution," said USML-1 Mission Scientist Don Frazier of MSFC, "is a milestone in research."

As the orbiter crew, Richards, Bowersox and Baker maintained Columbia in a 'gravity gradient' attitude, with her tail directed Earthward and her nose some 12 degrees off the direction of travel. This provided a stable platform for the USML-1 experiments and reduced the need for thruster firings. But as the payload commander, Dunbar had very strict rules. "Bonnie would have been happy if I never came in the lab," laughed Richards. "And, in fact, the only time I did go back in there, I came back in the lab maybe two times, and on my second trip I brought some coffee back there and I hadn't realised that Bonnie had made a rule that there'd be no coffee in the lab." As Richards entered the module, he accidentally spilled the coffee, producing a floating mass of globules in the pristine air. Dunbar glared at him as if to usher him out. Richards kept away from USML-1 after that. For his part, Bowersox enjoyed traversing the tunnel from the middeck to the Spacelab. "Cruising back and forth," he said later, "felt like Superman."

As the mission's commander, however, Richards had overall responsibility for the safety and success of the flight and as part of that mantle he insisted that crewmembers got to sleep at the end of their shifts. He was well aware of other Spacelab missions, in which astronauts stayed awake for 14 hours or more, labouring over troublesome experiments until their obsessive work ethics turned them almost into basket-cases. Richards' rule was simple: "I don't care if the lab is coming unglued. You'd better train the guy that's going to relieve you, because even if that's your experiment, you're going to bed. The other guy has got to be able to do it. You've got an hour to hand over and then you're going to bed." And 'bedtime' meant the coffin-like sleep stations, which Richards opined "gave you a little privacy for two weeks", although Trinh noted that it was often disorientating to find the door upon waking (Fig. 12.10).

But Richards was more than happy with USML-1's progress and the crew enthusiastically pursued their hectic around-the-clock schedule of research. In addition to the crystal growth experiments, a wide range of investigations examined the behaviour of fluids under differing influences, including the application of heat, in the hope that they might be used to produce high-technology glasses, ceramics, semiconductors, metals and alloys from ingredients mixed in a liquid state and cooled into solids. It was already known that fluid motions on Earth often introduces defects which restrict certain materials from achieving their full potential as lenses, computer chips, turbine blades

Fig. 12.10: A camera-toting Ellen Baker floats, seemingly inverted, out of the tunnel and into the brightly lit USML-1 module, a place which Bonnie Dunbar had decreed was a coffee-free (and pilot-free) zone.

and other products. In microgravity, the influence of this buoyancy-driven convection is drastically reduced and other, more subtle forces begin to prevail in fluid mechanics.

On 28 June, DeLucas began the Interface Configuration Experiment (ICE) inside USML-1's glovebox. This apparatus closely mirrored the kind of gloveboxes used in terrestrial research laboratories: it contained a pair of openings, through which a crew member could insert his or her hands, and a third port for the installation of experiment samples. Rugged gloves and finer surgical gloves were employed, dependent upon the degree of precision required. Most of the sample ampoules scheduled to be used with the glovebox had magnetic bases, enabling them to 'stick', and a large plastic window on the top of the unit allowed the astronaut to see the interior. Four still and video cameras provided coverage, which could be transmitted to Earth in real time. After mounting cameras for the ICE task, DeLucas filled an experiment vessel with an immersion fluid—hydrogenated terphenyl and an aliphatic hydrocarbon—to allow researchers to better determine the behaviour of fluids in differently-sized containers. On Earth, fluids behave in particular ways, but their flows in microgravity must be properly understood in order for better fuel tanks and containers for biological or human waste to be manufactured.

The Surface Tension Driven Convection Experiment (STDCE) and the Drop Physics Module (DPM) were fundamental facilities in the execution of this work. The former explored how fluids reacted in microgravity conditions when temperature differences existed along their interfaces, whilst the latter enabled 'containerless' studies of droplet behaviour in a weightless environment. During USML-1, the STDCE used a lightweight silicone oil, which was not susceptible to surface contamination, and more than 13 hours of video footage were recorded. Three runs of the experiment, each lasting around eight hours, were planned, but four were eventually conducted. For most of the mission, Trinh—"the wizard" of the DPM, said an admiring Richards—supervised the experiment. The only problem was an inability of the carbon dioxide laser to reach the required power level and the accidental introduction of bubbles into the sample chamber. Elsewhere, the DPM research saw water, glycerine and silicone oil droplets subjected to various external forces. Speakers mounted inside the experiment chamber allowed the droplets to be rotated, oscillated, merged, 'split' and suspended under acoustic pressure. It was hoped that such research could lead to new methods of 'encapsulating' living cells within membranes to protect them from harmful antibodies, potentially leading to revolutionary medicines. Drops were suspended, singly, for up to three hours at a time and at one stage, late in the mission, Dunbar was even able to 'bounce' droplets with the DPM speakers.

Flame behaviour was investigated, with the ignition of polyurethane foam cylinders, sealed in clear Lexan, and electrical wires were heated and burned within the glovebox. Paraffin and stearic acid candles were ignited to examine the physical appearance of flames in microgravity; they flared into spherical balls with glowing yellow cores. Typically, within ten seconds or so, presumably owing to the presence of soot, they turned blue and assumed hemispherical shapes, measuring around 0.6 inches (1.5 centimetres) across, then extinguished themselves within a minute or so. This data was consistent with short-duration studies aboard parabolic aircraft and drop-tower testing in the United States and Japan. Together with the USML-1 research, it led to new theories about the change in flame shape from spherical to hemispherical.

The final major group of USML-1 experiments focused on biotechnology, with the Generic Bioprocessing Apparatus (GBA) supporting more than a hundred studies of molecules and small organisms and 'Astroculture' demonstrating a prototype system for providing plant nutrition. Housed in a mid-deck locker, Astroculture was tended by Ken Bowersox, who darted downstairs from the flight deck during his shifts to ensure that its water and nutrients were satisfactory and check on its lighting and humidity controls.

As STS-50 neared its end, it was already being hailed as hugely successful. Early on 8 July, Carl Meade's blue team oversaw the deactivation of the Spacelab, but the progress of Tropical Storm Darby over the Baja peninsula

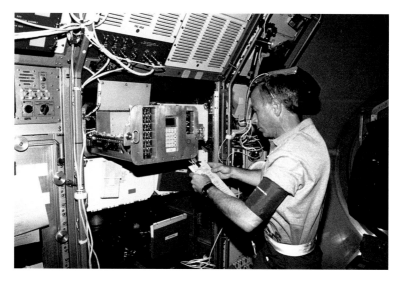

Fig. 12.11: Carl Meade reviews procedures before conducting an experiment inside the USML-1 module. This particular double rack housed the Generic Bioprocessing Apparatus (GBA), the Solid Surface Combustion Experiment (SSCE) and hardware for the EDO Medical Program.

threatened the option to land at Edwards in California. Mission Control waved off their landing attempt for 24 hours. Flight Director Jeff Bantle expressed preference to land in California, rather than KSC, since Columbia's heavier-than-normal weight of 225,900 pounds (102,500 kilograms)—including the 22,200-pound (10,000-kilogram) Spacelab—made an Edwards touchdown safer. When the rain in California did not clear by 9 July, Bantle opted to divert to Florida (Fig. 12.11).

With Spacelab closed up, the extra few hours aloft gave the astronauts some time to catch up on their notebooks, complete their flight recordings and gaze through the windows at the grandeur of Earth. For DeLucas, this pastime brought great pleasure. At one stage, he disappeared into the airlock to fetch some film for a camera, closed his eyes for a moment...and fell asleep. Dunbar went looking for him but could see nothing save a pair of feet sticking out from the midst of a floating mass of stowage bags. "If you were tired and you were floating," she later recalled, "you didn't have to lay down or sit down. You just sort of nod off once in a while."

The Shuttle streaked like a meteor over Houston, heading for Florida, and touched down on KSC's Runway 33 at 7:42 a.m. EDT on 9 July after almost two full weeks in space. Richards intentionally delayed the de-rotation of Columbia's nose gear to the runway on account of USML-1's excessive weight. Speaking to journalists afterwards, he apologised on behalf of blue team crewmates Baker, Meade and Trinh, none of whom had slept for 24 hours. "They just need to get home for a well-deserved rest, because it's a very tiring process, not just staying up over 24 hours, but getting used to gravity again," he explained.

In general, the crew had returned in good physical and psychological condition from their two weeks in orbit. Throughout the mission, they utilised equipment from the EDO Medical Project, which included the Lower Body Negative Pressure suit: an inflatable cylinder, four feet (1.2 metres) tall, which sealed around the waist and drew fluids into the legs to offset the upward fluid shift which tends to occur upon arrival in the microgravity environment. They also drank a litre of heavily-salted water before re-entry to help to keep body fluids in place until they could adapt to terrestrial gravity at a more leisurely pace. Richards, who did no exercise on his previous flights, took his physical wellbeing much more seriously this time. "During re-entry," he said, "you *have* to have a strong cardiovascular response, particularly during the final phase."

THE CURSE OF COMPETITION

The name 'Columbia' has long been associated with the people and culture of the United States, originating from the earliest arrival of European settlers, under the command of the Italian-born explorer Christopher Columbus, in the Americas in the autumn of 1492. Since then, the adjectives of 'pre-Columbian' and 'post-Columbian' have been routinely applied to the epochs before and after his arrival. Ships have been named in his honour, including two which profoundly altered humanity's fortunes in space. The accomplishments of both of these 'Columbias' were appropriately hailed by CAPCOM Mario Runco on the morning of 16 July 1994, as seven astronauts—the crew of STS-65 and IML-2—orbited Earth aboard Shuttle Columbia. By the time they landed, a week hence, they would have smashed the record for the longest Shuttle mission so far, almost 15 days strong.

Yet Runco's message paid tribute to *another* Columbia, too. "On this day, at this moment, 25 years ago," he began, "three of your predecessors began an epic journey that would change the way we viewed our world. Columbia's journey today, as her namesake did back then, is pushing the frontiers of knowledge and science for all mankind." Runco was, of course, referring to Apollo 11, whose command and service module (also named 'Columbia') had ferried them to the Moon in support of humanity's first piloted lunar landing. A quarter of a century later, STS-65 was midway through a two-week flight whose contribution to the space sciences was no less important.

That flight was to be Columbia's last, before she was scheduled for a year-long period of refurbishment and maintenance, and STS-65 again carried the EDO pallet of additional cryogens to support IML-2. The mission encompassed more than 80 experiments, provided by 200 scientists from 13 countries, including NASA, ESA, CNES, DARA, NASDA and CSA. Unlike IML-1, flown 30 months, the second flight was twice as long and carried twice as many facilities in its Spacelab long module. Several of these were controlled remotely by ground-based investigators, with U.S. and European research teams linked by intercontinental voice, video and data links. Such 'telescience' was seen as useful practice for future space station operations. In another indication of the importance of those experiments, the STS-65 crew was divided into two 12-hour shifts to run the laboratory around the clock. The red team comprised Commander Bob Cabana, Pilot Jim Halsell, Payload Commander Rick Hieb and Japanese Payload Specialist Chiaki Mukai, whilst the blue team consisted of Mission Specialists Carl Walz, Leroy Chiao and Don Thomas (Fig. 12.12).

The shift patterning caused some consternation, which Chiao attributed to decisions taken by Hieb. "We were not the most cohesive crew," he explained. "I really don't want to blame anyone, but maybe it was the payload commander set up the rivalry from the beginning—the red shift and the blue shift—because the mission specialists and the payload specialist were assigned first. We went through this training with this dynamic of a competition

Fig. 12.12: During the pre-flight Crew and Equipment Interface Test, payload commander Rick Hieb (outstretched hand) discusses the IML-2 payload with his crewmates. Facing the camera are Don Thomas and Chiaki Mukai.

between the two." Years later, Chiao was convinced that Hieb's competitiveness was good-natured, but it could be perceived that during training 'the reds' were choosing the 'good' shifts for themselves. In simulations, for example, it seemed to Chiao that the blue team always worked the undesirable night shifts in Houston.

Body clocks quickly became confused, as the astronauts travelled around the world, spending a couple of weeks in Houston, a couple of weeks in Europe, a couple of weeks back in Houston, a couple of weeks in Japan, and so on, working with experiment facilities and investigators. (It did not help that NASA Administrator Dan Goldin decided that his personnel could no longer fly in business class. "All those trips, except where we could beg our way up to the front, were done in economy class," Chiao lamented.) Still, working in shifts certainly engendered a measure of team bonding, particularly as there were four members of NASA's 1990 astronaut class (nicknamed 'the Hairballs') aboard…and *three* of them formed the blue shift. Fellow 'Hairball' Halsell was on the red shift. Every so often, Walz, Chiao and Thomas would offer him some advice. "You know, you *really* should be blue," they told him. "You *really* don't belong with those red guys!"

The inclusion of NASDA's Mukai, a physician, physiologist and surgeon who became the first female Japanese spacefarer, was illustrative of Japan's enormous contribution to IML-2. Backing her up was French physicist Jean-Jacques Favier. Having served as an alternate payload specialist for Spacelab-J (see Chapter Ten), Mukai was assigned in October 1992 to fly on IML-2. The announcement of crewmembers for this important life and microgravity sciences mission got underway a month earlier when aerospace engineer Hieb was chosen as payload commander. Eight weeks later came the naming of chemical engineer Chiao and materials scientist Thomas, as well as Mukai. Finally, in August 1993, the flight crew of Cabana, Halsell and Walz rounded out the seven-strong team.

As IML-2's payload crew, Hieb, Chiao, Thomas and Mukai supported a wide range of the facilities in the Spacelab module. Germany's Biostack sandwiched biological specimens between 'plates' of radiation detectors as part of a project to determine the impact of high-energy cosmic rays passing through the module's hull. The specimens would be examined after Columbia landed to identify the paths and entry points of heavy ions and assess physical changes or damage. The impact of such radiation, it was recognised, could prove detrimental to astronauts on future long-duration missions. Three sealed Biostack containers were filled with shrimp eggs and salad seeds to study their behaviour. Other radiation-measuring experiments were conducted using a Japanese monitor. More significantly, IML-2 marked the first time that such radiation data had been transmitted to Earth in real time. This immeasurably aided the research in ESA's Biorack, which explored the impact of microgravity and cosmic radiation upon genetically modified rapeseed roots, cress seedlings, fruit flies and human skin cells. Elsewhere, the Aquatic Animal Experiment Unit (AAEU) housed a complement of Japanese red-bellied newts, goldfish and Medaka fish. Obviously, the 'perishable' nature of these living specimens meant that loading them aboard the Spacelab module had to occur a few hours before launch. Fortunately, the processing of IML-2—and Columbia herself was exceptionally smooth and was even described as "fairly simple plug-in work" by NASA spokesman George Diller.

Chiao was sceptical that they would actually get to fly on their opening launch attempt on 8 July 1994. The night before, there had been reports of thunderstorms and lightning. Fuelling went ahead and the seven astronauts were strapped into their seats, certain that a scrub lay ahead. Then, unexpectedly, the mediocre weather began to clear and STS-65 rocketed safely into orbit at 12:43 p.m. EDT.

Minutes later, the sensation of three times a 'normal' gravitational load was replaced, in the blink of an eye, by the purity and unalloyed freedom of weightlessness. "For me, it felt like a forward tumble," Chiao remembered, "and then the full-headedness." But there was little time to think, only time to do. Helmets had to be removed and stowed, communications headgear had to be removed and stowed, cameras had to be unstowed. Those first moments in free-fall were puzzling and brought with them a little headache, a queer dizziness, but for Chiao at least, no disorientating nausea. Space sickness seemed not to affect him. Chiao was one of the lucky ones, for historically upwards of 25 percent of space travellers have reported an element of the debilitating malaise upon arrival in this strange new realm.

Halsell, also making his first flight, found the new environment unusual.

"This is Jim, trying to float for the first time in his life," the pilot joked as he rose from his seat, an hour after launch.

"Go for it, man," someone yelled.

"Feels really weird," Halsell replied.

Led by Hieb, the activation of IML-2 got briskly underway under the watchful eyes of the red team. "Smells like a lab," Hieb remarked as he opened the hatch from the Shuttle's middeck, firstly into the tunnel, then into the pressurised module itself. Meanwhile, Walz's blue team bedded down in their sleep stations for a six-hour nap. For Chiao, trying to sleep only hours after launching into space was like asking a child to go to bed on Christmas morning. He took a couple of Restoril, "which was the sleeping medication of choice, back then", but found himself floating, tightly zipped into his sleeping bag, for some period of time before the pills kicked in. Inside the coffin-like confines of the sleep station, he tried lying on his side, as he would on Earth, and was surprised when it worked and he dozed off.

WIDE-RANGING RESEARCH

"We're looking forward to a super two weeks up here," Cabana exulted, as within three hours of launch Chiaki Mukai activated the Advanced Protein Crystallisation Facility (APCF) in a pair of middeck lockers. This device employed vapour-diffusion, liquid-to-liquid diffusion and dialysis to grow high-quality protein crystals and during the course of STS-65 more than 7,000 video images of the slowly growing crystals would be obtained. In the meantime, Chiao spent part of his first blue shift initiating Biorack. Mukai had already transferred many of the perishable specimens from middeck storage lockers into the facility, allowing Chiao to kick off a five-day experiment to examine calcium loss from human bones. Similar work had been undertaken on IML-1 and had suggested that bones typically did not lose a sizeable amount of calcium, as long as they were exposed to periods of 'compression', often through vigorous exercise. However, more data was needed to counteract the effect of microgravity on the human skeletal structure. Biorack housed 19 experiments, including a Norwegian investigation into the growth of genetically modified rapeseed roots and cress seeds and a study of premature aging in fruit flies, possibly induced by their difficulty in coping with weightlessness. Indeed, at the start of the mission, Hieb reported that the flies were "buzzing around with excellent vitality", but that after ten days or so their bull-at-a-gate response had slowed substantially and tailed off and they were acting more like their ground-based 'control' counterparts. Other Biorack work included observations of cultures of human skin fibroblasts—cell-producing connective tissues—and bacterial cells.

Germany's Slow Rotating Centrifuge Microscope (NIZEMI) focused upon the response of living and non-living specimens to various levels of gravitational influence. The centrifuge applied loads ranging from a thousandth of terrestrial gravity to around 1.5G on slime mould, Moon jellyfish (*Aurelia aurita*), cress roots, lymphocytes, green algae, immune system T- and B-cells and a transparent material known as 'succinonitrile-acetone'. All of the researchers involved with the NIZEMI research praised its "great success", which mirrored the comments from other investigators. Canada again flew its back-pain experiment (previously carried aboard IML-1), which sought to explore the possibility that problems caused by the lengthening of the human

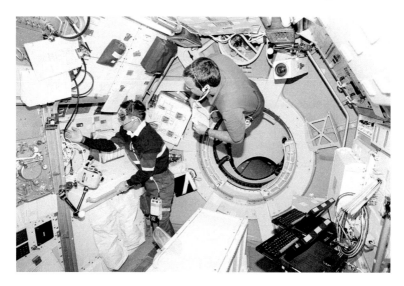

Fig. 12.13: STS-65 Commander Bob Cabana, wearing protective goggles, opens a science drawer aboard IML-2, as payload commander Rick Hieb looks on.

spine in weightlessness might be induced by changes in the function of the spinal cord or spinal nerve 'roots'. IML-1 found that astronauts 'grew' slightly during missions and experienced a 'flattening' of their normal spinal contours. During IML-2, the spacing between discs in the astronauts' spinal vertebrae was measured and, each day, the entire crew filled in questionnaires to describe occurrences of back pain or symptoms of spinal column dysfunction. They also took stereo photographs of themselves in differing postures, subjected themselves to electrical impulses to stimulate their sensory nerves, squeezed hand grips as electrodes gathered data on their blood pressures and heart rates and would surrender to MRI scans after landing (Fig. 12.13).

Japanese scientists were particularly impressed with the work in the AAEU, which offered the opportunity to examine the spawning, fertilisation, embryonic stage, vestibular function and general behaviour of live fish and small amphibians. IML-2 carried almost 150 pre-fertilised newt eggs, as well as four adult newts, in cassette-sized containers. Unfortunately, two of the newts died, but the others seemed to develop at approximately the same rate as their ground-based 'controls'. Other AAEU passengers included goldfish and Medaka fish, both of which were shown to respond well to light stimulation within their tank.

Although life sciences consumed a sizeable portion of IML-2 crew time, a significant thrust of the mission's research surrounded materials science, processing and fluid physics. One of these was ESA's Critical Point Facility (CPF), which employed five high-precision thermostats to investigate the behaviour of substances at the peculiar stage in which they are technically neither a liquid or a gas. (More accurately, the material's properties fluctuate to and fro from a liqueous to a gaseous state, such that its 'bulk' state is indistinguishable.) During the mission, the CPF was used to measure the wave motion of heat within sulphur hexafluoride and at one stage a wire was charged inside the test cell to 500 volts in order to simulate the pressures induced by terrestrial gravity. "The effect," said investigator Richard Farrell, "was like turning the gravity on and off."

Elsewhere in the Spacelab module was the Electromagnetic Containerless Processing Facility, known by its German acronym of 'TEMPUS', which provided a levitation melting device for processing metals. Containerless processing was also considered hugely advantageous, because Earth-based processes involving liquids can cause them to be affected by the properties of their holding vessel; in microgravity, on the other hand, positioning and control can be effected with greater precision. This, in turn, reduced motions within the sample liquid and was less intrusive upon the physical phenomena under study. Moreover, containerless processing was known to eliminate contamination of the sample caused by the material in the container's walls. For IML-2, TEMPUS carried 22 spherical specimens, housed on a storage disk. This rotated until the desired sample was positioned over a mechanism that transferred it to the processing area, where it could be placed into either a vacuum or an ultra-pure helium-argon 'atmosphere'. Most of the experiments were run by computer command, with the members of the payload crew keeping a close eye on their progress. At one stage, Hieb watched over a zirconium-cobalt alloy as the TEMPUS ground team sent telescience commands to levitate, and then melt, the small metal sphere inside the processing chamber. Such materials were expected to aid the production of near-perfect metallic 'glasses' with unique mechanical and physical properties. Other samples included alloys of niobium-nickel, aluminium-copper-cobalt and nickel-tin and on one occasion all thruster firings were temporarily suspended to provide the experiments with an ultra-pure acceleration environment.

European involvement in IML-2 was represented also by RAMSES—a French acronym translatable to 'Applied Research on Separation Methods Using Space Electrophoresis'—which separated and collected ultra-pure components of biological samples, including haemoglobin and bovine serum albumin, according to their electrical charges. Meanwhile, the Bubble, Drop and Particle Unit (BDPU) explored bubble growth, evaporation, condensation and temperature-induced thermocapillary flows in microgravity. Video and still cameras mounted in close proximity to the device enabled scientists

12 In the absence of *gravitas* 321

Fig. 12.14: Leroy Chiao (upper) and Don Thomas work at respective experiments aboard IML-2 during STS-65's blue-shift activities.

to carefully scrutinise the behaviour of fluids under various conditions, including the injection of vapour bubbles into a test cell filled with an alcohol-water solution. Liquid refrigerants were also boiled and the coalescence of large bubbles, apparently unaffected by gravity-induced buoyancy, were captured by Spacelab's video cameras (Fig. 12.14).

With a Japanese crew member aboard, it seemed inevitable that the involvement of her parent nation in IML-2 was strong. In addition to the aquarium, a Large Isothermal Furnace (LIF) was used to heat and rapidly cool materials—including ceramic-metallic composites and semiconductor alloys—in various temperature ranges to identify relationships between their structures, processing and physical properties. Usually, the furnace followed a pre-programmed heating and cooling cycle, reaching a maximum temperature of around 1,600 degrees Celsius (2,900 degrees Fahrenheit).

Working around the clock, the payload crew closely monitored the IML-2 experiments, whilst Cabana, Halsell and Walz kept watch over Columbia's systems, which performed with near-perfection. Their return home was originally scheduled for 22 July, after a mission of 13 days and 17 hours, which would have fallen slightly short of the 14-day record set by the STS-58 crew the previous November (see Chapter Thirteen). However, both landing opportunities at KSC on the 22nd were waved-off, due to unacceptable cloud cover and although weather conditions were fine at Edwards in California mission managers opted to wait for Florida to improve. Seventeen hours later, at 6:38 a.m. EDT on 23 July, Columbia settled gracefully onto Runway 33 at KSC, wrapping up a spectacular voyage of almost 15 days. The enormous success of IML-2 received official acknowledgement in a review meeting at the European Space Operations Centre (ESOC) in Darmstadt, Germany, in early November 1994. The consensus from all of the scientists and investigators was hugely positive. "This is the important part of the mission, where the scientists get their samples," said IML-2 Mission Scientist Bob Snyder. "In some cases, this is going to take many months, up to possibly several years, in the cases where this huge amount of data has to be analysed."

GLIMMERS OF THE FUTURE

If IML-2 marked the end of the line for the International Microgravity Laboratory, hope sprang eternal for a longer lifespan for the U.S. Microgravity Laboratory. When USML-1 ended in July 1992, three more missions were baselined—USML-2 aboard Columbia on STS-74 in September 1995, followed by USML-3 in September 1997 and USML-4 'requested' for September 2000—and all would benefit from the longevity afforded by the EDO hardware, all three targeting 16-day missions. Sadly, those hopes evaporated after 1993, when USML-3 and USML-4 were formally deleted from the Shuttle manifest, leaving USML-2 aboard Columbia in late 1995 as the final dedicated mission. But if USML-1 occurred after a lengthy overhaul for Columbia, so too did USML-2, producing a mission and vehicle which were among the most up-to-date in the Shuttle fleet.

After STS-65, Columbia was scheduled to begin her modifications at KSC early the following year, which would have enabled her to fly STS-67 with ASTRO-2 in December 1994 (see Chapter Seven) and be ready for USML-2 in September 1995. But following procedural changes and a March 1994 decision to use Rockwell International staff in Palmdale, California, to perform future upgrades, rather than the KSC workforce, it became increasingly clear that the time needed to remove the ASTRO-2 hardware, fly Columbia out to California, conduct six months of work and prepare her for a September 1995

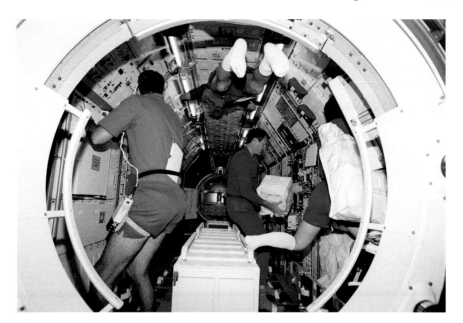

Fig. 12.15: Spacelab proved a highly versatile research facility, and a roomy one, as this IML-2 view of the module's interior reveals.

mission would not be practical. ASTRO-2 was therefore shifted onto Columbia's sister Endeavour and flew successfully in March 1995. This provided the workforce with a broader window to accomplish the work on Columbia and in October 1994 she was transported to California. There, the queen of the Shuttle fleet underwent 66 modifications to enhance her performance, meet new mission requirements and reduce turnaround times. Among them, Columbia received new wiring to allow future Spacelab crews to monitor downlinked data on laptop computers. Returned to Florida in April 1995, she was put directly back into processing for USML-2. But the mission, originally targeted for late September, found itself delayed no fewer than six times, due to sensor malfunctions, high winds and thunderstorms, a glitch with a master events controllers, a failed computer and suspected damage to oxidiser ducting in the main engines. By the time the much-troubled STS-73 eventually took flight at 8:53 a.m. EST on its seventh attempt on 20 October—"Catapulting scientific knowledge through microgravity research," exulted the launch announcer as Columbia leapt from the pad—the USML-2 mission already seemed snakebitten (Fig. 12.15).

Assembly of the STS-73 crew got underway in March 1994, when physicist Kathy Thornton was named as payload commander and Air Force polymer scientist Catherine 'Cady' Coleman as a mission specialist. Announced the following June as prime payload specialists were Al Sacco of Worcester Polytechnic Institute (who served previously in an alternate role on USML-1) and atmospheric scientist Fred Leslie of MSFC. Aside from their scientific credentials, both men had other impressive details on their resumes: Sacco a keen scuba diver and instructor, Leslie was a parachutist with several records in group freefalls. Leslie had been a co-investigator for the Geophysical Fluid Flow Experiment (GFFE), flown on Spacelab-3 in April 1985, and was also intimately involved in the similarly-named Geophysical Fluid Flow Cell (GFFC) investigation aboard USML-2. "We're primarily researchers and scientists," Leslie said later, "and the disadvantage is that often, you only fly once." But in his case, one advantage was that he actually received an assignment more quickly than 'professional' NASA astronauts: from entering the USML-2 training pool in June 1994 to launch in October 1995, Leslie went from ground to space extremely fast as a payload specialist. Sacco, on the other hand, led several of the mission's zeolite crystal growth studies, in addition to his past USML-1 expertise. Backing the pair up were physicist Glynn Holt of JPL and materials engineer David Matthiesen of Case Western Reserve University. Finally, in November 1994 Ken Bowersox, who served as USML-1 pilot, was named as the mission's commander, with first-timers Kent Rominger as pilot and Mike Lopez-Alegria as flight engineer.

Emerging into a blaze of media flashbulbs on the morning of 20 October, Bowersox insisted that his crew wore their USML-2 baseball caps back-to-front; Sacco later explained that they wanted to demonstrate that science was not for geeks. Ninety minutes after liftoff, Columbia's payload bay doors were opened, the port-side door positioned uniquely at 62-degrees-open. As had been the case on several previous Spacelab flights, the Shuttle would operate in a gravity gradient attitude, with her tail directed towards Earth, to ensure a stable microgravity environment for the USML-2 experiments. In furtherance of this aim, the partially open door helped to minimise the risk of micrometeorites or other debris impacting the delicate radiator panels.

Elsewhere, the science began in earnest, with Thornton—clad in her self-proclaimed "lucky socks", emblazoned with the Stars and Stripes—and Sacco leading the effort to open the tunnel, access the Spacelab module and activate the experiments. On an earlier launch attempt, Thornton wore a pair of St. Christopher medals, which did not seem to bring them the luck they needed, but on their seventh try, she laughed, "the socks did it". As Bowersox, Rominger, Thornton and Sacco on the 'red' team laboured to get USML-2 up and running, their blue counterparts, Lopez-Alegria, Coleman and Leslie, bedded down in the middeck sleep stations for a six-hour nap. That, recalled Coleman after the flight, "was probably the hardest thing we had to do the whole mission". In view of the fact that STS-73 had five first-timers, with only Thornton and Bowersox having flown before, episodes of space sickness were anticipated and the mission's first day was not overloaded with tasks to accommodate crewmembers who might be feeling under the weather. But there were no instances of sickness, allowing the reds ample opportunity to get USML-2 up and running.

During USML-2, Leslie had ample opportunity to work with GFFC, a reflight of the geophysical fluid flow experiment from Spacelab-3, which revealed several types of convection difficult to study on Earth, as well as enabling researchers to observe their structures, instabilities and turbulence. For its second mission, it modelled conditions in Earth's mantle—a region of predominantly magnesium-rich silicate rock—as well as plasma flows on the surface of the Sun. After activation, Principal Investigator John Hart and his team at the University of Colorado at Boulder controlled the experiment remotely from the ground. Time-lapse movies were collected from a series of still images, produced every 45 seconds, as fluid swirled between GFFC's two rotating hemispheres. Despite minor problems relaying temperature parameters, Leslie expressed happiness with its performance, calling it "a planet in a test tube". Later in the mission, efforts were made to simulate activity in Jupiter's atmosphere, which was of particular interest as it radiates significantly more heat than it receives from the Sun, contracting and liberating gravitational potential energy as heat. During each experiment run, voltage, rotation-rate and temperature parameters were adjusted to create unstable and turbulent flows to better explore ocean and atmospheric dynamics (Fig. 12.16).

Elsewhere in the Spacelab module were several other experiments modified or improved in the wake of their first flight. ESA's Glovebox had been outfitted with a larger working area and better lighting and supported seven separate investigations. Experiments on USML-1 in June 1992 had yielded protein crystals of much higher quality than had been previously achieved and by 21 October Sacco had already gotten his first zeolites up and running. Zeolites are used on Earth for the purification of fluids in life-support systems, as well as in the petroleum refinery process and in waste-management and biological fields. They act as molecular sieves to separate specific molecules from solutions and may someday enable gasoline, oil and other petroleum products to be refined less expensively. In addition to operating experiments in the Spacelab module, Sacco tended his own Zeolite Crystal Growth (ZCG) furnace in Columbia's middeck, which processed 38 sample containers to create large, near-perfect crystals. Results from USML-1 had shown that zeolite crystals whose nucleation and growth were carefully controlled from the outset produced greater levels of purity than Earth-grown ones.

Fig. 12.16: Kathy Thornton was convinced that her lucky U.S. socks aided in USML-2's success, after six scrubbed launch attempts.

Other experiments stored in middeck lockers included the Diffusion-Controlled Crystallisation Apparatus for Microgravity (DCAM), which pioneered a method of autonomously running long-duration protein crystal growth experiments aboard the ISS. More than 1,500 protein samples were processed during the USML-2 mission, despite a requirement to lower the Shuttle's cabin temperature when one of the thermoelectric coolers slightly overheated. Another middeck experiment was Astroculture, which sought to grow and provide nutrients for potato plants. It grew ten small tubers to evaluate the extent to which microgravity affected starch accumulation. Starch is an important energy-storage compound, but evidence from previous missions had suggested that its accumulation was more restricted in space. This did not appear to be the case, at least in the first few days of the mission, but by 1 November video footage shows that their leaves were beginning to wilt. However, Astroculture successfully demonstrated its ability to provide proper nutrients, light, water and humidity levels and after USML-2 its hardware was made available commercially for sale or lease.

As with most Spacelab missions having a microgravity research emphasis, the module was outfitted with devices to measure the influence of Shuttle accelerations on very sensitive experiments. As Thornton powered up USML-2 after arriving in orbit, Sacco activated two accelerometers, one of whose sensor heads had been deliberately placed next to Case Western's STDCE experiment. Previously aboard USML-1, this allowed scientists to view in great detail the behaviour of fluid flows in microgravity, where temperature differences existed along their interfaces. Such 'thermocapillary' flows occur in many industrial processes on Earth and when they manifest themselves during melting or resolidification they can create defects in crystals, metals, alloys and ceramics. Gravity-driven convection overshadows the flows in ground-based tests, making them difficult to measure. High above Earth, on the other hand, it became possible to explore them in greater depth. Fluid physicists expected this knowledge to provide clearer insights into bubble and drop migration, which in turn could aid the development of better fuel-management and life-support apparatus. STDCE used silicone oil as the test fluid and a laser diode and several cameras monitored its transition from a 'steady', two-dimensional flow into an 'oscillatory', three-dimensional one. Smaller, ground-based tests had highlighted periodic variations in the fluid's motion and temperature.

By comparing conditions for the onset of this oscillation in microgravity and on Earth, it was hoped to identify its cause. New optics provided precise images of oil surface shapes and flow patterns. Early in the mission, Leslie reported the onset of oscillations during the very first STDCE run, as ground controllers watched three different views simultaneously downlinked by Spacelab's new television system. By 23 October, the astronauts drew down the volume of silicone oil to create a concave surface. As a laser gradually heated the surface, investigators identified the transitional point at which oscillations began to occur in each run, prompting Principal Investigator Simon Ostrach to speculate that the onset was affected by heat sources, temperature distributions on the surface and the dimensions of the container. In subsequent STDCE runs, Coleman lowered and raised oil levels to create deeply concave, then convex, surfaces. The 'size' of the laser beam on the oil surface was also adjusted to determine its effect on the direction and nature of fluid flows, as well as lifting its temperature to introduce oscillations. This led to several instances of new phenomena. During one test, erratic flows with no apparent organisation or pattern appeared as Thornton increased oil temperature beyond the point at which flows began to oscillate. Later, Sacco handed control to the ground, who ran the experiment via telescience.

Sacco also demonstrated 'surface tension' in space by squeezing orange juice from a tube. It promptly formed a ball, thereby highlighting how surface tension makes fluids assume spherical shapes when they no longer have to contend with terrestrial gravity. The six-channel, simultaneous video footage from STDCE was only possible thanks to the new High-Packed Digital Television Technical Demonstration (HI-PAC). Prior to USML-2, video from Spacelab could only be downlinked on a single channel, but as part of testing the new system the STS-73 crew was able to acquire live images of their colleagues sitting at their consoles in Mission Control.

Other fluid physics investigations were carried out in the Drop Physics Module (DPM), another facility carried over from the USML-1 mission, for which alternate payload specialist Holt was a co-investigator. It sought to explore future ways of conducting 'containerless' processing and encapsulation of living cells for pharmaceutical research. One investigation was the Drop Dynamics Experiment, provided by Spacelab-3's Taylor Wang, which gathered high-quality data in support of such applications. During USML-2, the experiment investigated the breaking-apart of distorted drops under a fluid of varying viscosity, as well as attempting to encapsulate and create 'compound' drops. It was hoped that such tests might lead to methods to insert living cells for the treatment of hormonal disorders into polymer shells to protect them from immunological attack and provide timed releases, perhaps in case of diabetes. A second DPM experiment examined the influence of 'surfactants'—substances which alter a fluid's properties by aiding or inhibiting the way it mixes with other substances—on the behaviour of single and multiple drops. Early in the mission, investigators watched video of the first deployment as Thornton released a small glob of water and used precisely controlled sound waves from a pair of loudspeakers to manipulate its movements. Later, Leslie succeeded in encapsulating an air bubble inside a floating water droplet, whilst Sacco manipulated drops treated with an organic surfactant known as bovine serum albumin, deploying two drops and bringing them perfectly together until they coalesced into a single blob; and Coleman stretched a 'bridge' of liquid between the DPM's two injector tips and centred a drop of water inside a glob of silicone oil, whilst the interfaces of both were moving in opposite directions (Fig. 12.17).

Elsewhere, the Crystal Growth Furnace (CGF)—for which David Matthiesen was a research lead and served as principal investigator for one of its experiments, a study of dopant segregation behaviour during the growth of gallium arsenide crystals—proved capable of achieving very high operating temperatures, in excess of 1,000 degrees Celsius (1,800 degrees Fahrenheit) and processing large semiconducting crystals, together with metals and alloys. During USML-2, it was employed to grow samples of cadmium-zinc-telluride, mercury-cadmium-telluride, gallium arsenide and mercury-zinc-telluride, having achieved great success in producing surprisingly defect-free specimens on USML-1 three years earlier. For the second mission, the primary CGF sample chamber was outfitted with a spring-loaded piston, which moved to reduce the volume of the cylinder as the material contracted whilst cooling. This helped to eliminate air voids in the crystal and ensured that it maintained

Fig. 12.17: Al Sacco is pictured at work aboard USML-2.

an 'even' contact with the container walls along its entire surface. Cadmium-zinc-telluride is routinely used as a base on which infrared-sensing mercury-cadmium-telluride crystals can be grown. USML-1 had produced cadmium-zinc-telluride crystals 1,000 times purer than those grown on Earth. The materials flown on the second mission carried potential terrestrial applications in high-speed, low-power digital circuits, optoelectronic-integrated circuits and solid-state lasers. All samples were processed satisfactorily, together with a fifth specimen, of germanium with a trace of gallium, to demonstrate the influence of changes in the Shuttle's attitude. In total, eight crystals were grown, one of which saw the CGF operating at its highest temperature (1,250 degrees Celsius or 2,300 degrees Fahrenheit) and another which took just 90 minutes to complete. In total, in the middeck and the Spacelab module, more than 1,500 protein samples were processed during USML-2. These included feline calcivirus, akin to a virus responsible for causing digestive problems in humans, which proved to be so well-formed that it elicited applause from team members on the ground. Sacco also set up initial conditions for the growth of a collagen-binding domain protein, which is important in the study of arthritis and joint disease, and Coleman activated a protein known as duck delta crystallin, which is similar to the protein responsible for causing a rare, but deadly disease in humans.

Combustion science was another focus of STS-73's Glovebox research and, in recognition of his efforts, Fred Leslie received a collective 'high-five' for his work on 2 November, according to team members of the Fibre-Support

Droplet Combustion experiment. He successfully placed droplets of fuel onto a thin fibre, using needles in the experiment module, then igniting them with a hot wire. Despite early difficulties with clogged drop accumulators, he performed several 'burns', varying the quantities of methanol and methanol-water fuel on each run.

As the mission's second week in space wound down, the astronauts were asked to place the Shuttle into several different attitudes to increase temperatures across her belly. Flight controllers were concerned that her tyre pressures were lower than acceptable for landing. Thermal conditioning of the belly brought them up to within the required minimum landing pressure limit. Early on 5 November 1995, a few hours before Columbia's 6:45 a.m. EST landing at KSC, the final experiments aboard the Spacelab module were shut down and the payload was deactivated and the hatch closed.

"Huntsville, Spacelab," radioed Thornton.

"Go ahead, Kathy," came the reply.

"We're gonna head to the middeck and get dinner and get to bed, 'cos we don't have much time left up here," Thornton said, wistfully. "But I wanted to tell you and all the friends in Huntsville and Marshall how much we've enjoyed working with you, from the time we started training a year and a half ago. We've appreciated your determination and your humour and particularly your patience with us. We're happy to have been part of the team that's made all this possible. We'll see you on Earth."

"We'll see you guys soon," came the response. "It's been a great couple of weeks."

TWO-FACED MISSION

At 9:49 a.m. EDT on 20 June 1996, Columbia rocketed into orbit on a mission which would virtually mirror the research to be conducted aboard a future space station. In the moments before liftoff, her seven astronauts received a fitting send-off from launch controllers: to "have more fun than a barrel of monkeys". Designated the Life and Microgravity Spacelab (LMS), their 16-day flight—according to Mission Manager Marc Boudreaux of MSFC—had "the key ingredients to take us into the next era of space exploration." One of those ingredients was the relatively short period of time it took from the initial blueprints for the mission until its realisation: just 18 months. Historically, Spacelab missions were intricately complex affairs, involving hundreds of scientists and engineers from numerous institutions spread across the globe, and took up to four years to prepare from conception to launch. In

12 In the absence of *gravitas* 331

Fig. 12.18: Unusual view of the USML-2 module, with Columbia's port-side payload bay door positioned at 62-degrees-open to facilitate micrometeoroid protection.

the wake of IML-2 in July 1994, it was recognised that there would be a conspicuous lack of life and microgravity research flights until the space station became operational at the turn of the new millennium (Fig. 12.18).

With this in mind, in December 1994, NASA announced its intention to stage a one-off, 16-day mission in mid-1996. Partway through the following May, seven astronauts were assigned to the flight, which became known as STS-78: payload commander Susan Helms, mission specialists Rick Linnehan, a veterinarian, and Chuck Brady, a Navy flight surgeon, and a quartet of French, Canadian, Spanish and Italian candidates to fill a pair of payload specialist slots. Of those four, Canadian physician Bob Thirsk and French physicist and metallurgist Jean-Jacques Favier were named to serve as prime payload specialists, with Italian Air Force physician Luca Urbani and Spanish aeronautical engineer Pedro Duque as alternates. In addition to her role as payload commander, Helms also served as the flight engineer for ascent and re-entry. In October 1995, Spacelab-D2 veteran Tom Henricks was named as STS-78's commander, with Kevin Kregel as pilot.

The mission was intended to be 'international' in its flavour, providing yet another ingredient in preparing for the ISS, and many of its payloads were carried over from IML-2 and SLS-2. Principal investigators for the experiments were based at four institutions in Europe and four in the United States and a broad use of telescience would be employed to effect control from the ground. However, unlike many earlier missions, the LMS crew would operate a single-shift system. The benefit of this approach was that life science experiments, which, by their very nature, required a significant amount of crew time and video coverage, but only minimal power and energy, could be executed when all seven astronauts were available as test subjects and operators. The automated and remotely controlled materials science investigations, on the other hand, could then be run via telescience as the crew slept. Full telecommand and video facilities would still be available, but at the same time the absence of movement from the astronauts would not disturb the highly sensitive microgravity investigations. Developed at a cost of $138 million, LMS comprised 22 major experiments, supported by NASA, ESA, CSA, CNES and ASI, together with adjunct research teams in ten other nations. Specifically, the life science experiments explored the responses of living organisms to the weightless environment, with emphasis on musculoskeletal physiology. The second side of the LMS coin—microgravity research—probed the subtle influences at work whilst processing a variety of materials and examining the behaviour of fluids.

Columbia's 174-mile-high (280-kilometre) orbit, inclined 39 degrees to the equator, enabled the crew to maintain their terrestrial sleep/wake circadian rhythms. Ninety minutes after launch, the payload bay doors were open and the tunnel and Spacelab module were activated. "Today was the busiest first shift of activities we've ever had for Spacelab," explained LMS Mission Scientist Patton Downey of MSFC on the evening of 20 June. "Virtually every experiment on board either had its equipment activated or checked out." Henricks and Kregel oriented the Shuttle into a 'gravity gradient' attitude, with her tail pointing Earthward, in support of the sensitive microgravity experiments.

Yet there was precious little time for Earthgazing, as LMS took priority. One of its most important payloads was ESA's Advanced Protein Crystallisation Facility (APCF), which carried 11 experiments to study three different growth methods. This device employed vapour-diffusion, liquid-to-liquid diffusion and dialysis and, in total, more than 5,000 video images would be acquired during STS-78 of the development of key protein crystals. The APCF had flown aboard IML-2 and was activated six hours into the mission. The crew

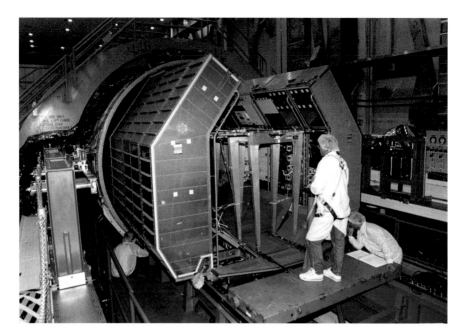

Fig. 12.19: Research racks for the Life and Microgravity Spacelab (LMS) are rolled inside the pressurised module's cylindrical shell, ahead of electrical, mechanical and fluid connections.

typically provided daily reports of the status of its displays, as the facility had no space-to-ground telemetry capability of its own (Fig. 12.19).

Also provided by ESA was the Advanced Gradient Heating Furnace (AGHF), which successfully processed 13 samples for one semiconductor and five metallurgical experiments. The furnace ran near-continuously throughout the mission, performing better than in ground-based tests. Its objective was to solidify alloys and crystals in a number of investigations designed to understand the conditions in which the structures of freezing materials changed in the solidification process. Investigators hoped that AGHF research would increase scientists' knowledge of the physical processes involved in solidification and lead to enhancements in ground-based materials research. The first actual experiment run in the furnace got underway when Helms inserted a sample cartridge to examine the transition in solidifying metal mixtures from order, column-like grains to unordered, round ones. Processing of AGHF samples was sequential and required the exchange of experiment cartridges and the activation and deactivation by a crew member. On 22 June, a sample of pure aluminium, reinforced with zirconia particles, was placed into the furnace as part of an investigation into the physics of liquid metals containing ceramic particles as they solidified. Other experiments included a polycrystalline sample used to gather information on how to combine liquid metal alloy components into precise, well-ordered solid structures. It was anticipated that knowledge from melting and resolidifying such compounds could help manufacturers make higher-quality metal alloys and semiconductors. Later, an aluminium-copper mixture was solidified as part of a French experiment. Astronauts Brady and Helms also ran an experiment which sought to control the internal structures of aluminium and indium alloys during solidification, which was expected to have terrestrial benefits in producing new materials for engineering, chemical and electronics applications.

The third European microgravity facility was the Bubble, Drop and Particle Unit (BDPU), which was used on STS-78 to observe and record the behaviour of fluids under differing temperature levels and concentrations. Early in the mission, it was utilised for one fluid physics investigation to explore the processes controlling evaporation and condensation. In its specially designed test cell within the BDPU, a small heater emitted an electrical charge into liquid freon, supersaturated with gas, which produced a single bubble through boiling. Although it required some minor in-flight maintenance by Kregel and Favier after suffering a blown fuse, the device functioned well and processed all nine of its test containers. One experiment examined surface tension and interactions of gases and liquids in microgravity. Various sizes of air bubbles were injected into a water and alcohol solution with temperature gradients ranging from 'hot' to 'cold'. Another study observed the behaviour of inert gas bubbles within silicone oil. From such experiments, it was hoped that new insights might be gained into controlling defects in many aspects of materials processing, perhaps leading to the production of stronger and more resilient metals, alloys, ceramics and glasses. A second glitch with the BDPU led to a remarkable repair on 28 June, the procedure for which was uplinked to the crew in the form of a video from Mission Control. Their efforts proved successful when they activated the Electrohydrodynamics of Liquid Bridges experiment, which focused on changes occurring in 'bridges' of fluid suspended between a pair of electrodes. Fluids under study included castor oil, olive oil, eugenol and silicone oil. Another, two-part, investigation was provided by the University of Naples and looked at the interaction of moving, pre-formed bubbles and the melting and solidifying 'edge' of a solid, whilst its second segment examined the ways in which droplets were captured by, or pushed away from, a moving solidification front. The experiment called for the injection of water droplets of differing diameters into a liquid alloy, in order to study their behaviour during the application of heat. Despite difficulties with the injector, which refused to retract from its deployed position, ground-based engineers devised a solution and the experiment ran successfully.

Elsewhere in the LMS module, the crew also concentrated on the second complement of activities: the life science investigations. These were further categorised under five disciplines: human physiology, musculoskeletal, metabolic, neuroscience and space biology. Two hours after reaching orbit, Favier and Thirsk kicked off the human physiology research by donning electrodes and sensors to monitor their eye, head and torso movements for the Canadian Torso Rotation Experiment (TRE). It had been known for three decades that many astronauts suffer motion sickness in space—particularly during their first few days aloft—and the aim of the Canadian study was to identify and ultimately avoid movements which contributed to this sense of illness.

Brady and Linnehan joined their colleagues in the torso rotation study and participated in initial tests to evaluate their muscle strength and control. It was already known that muscle fibres became smaller, or 'atrophied', in microgravity, which resulted in a steady loss of muscular mass. Many of these observations tend to be short-lived, generally vanishing when astronauts return to Earth, but the potential impact of longer, six-month tours aboard the ISS on human muscles remained unknown. Using an ESA-built device known as the Torque Velocity Dynamometer, the astronauts were able to take precise measurements and calculate their muscle performance and function, including strength, amounts of force produced and resultant fatigue. Blood samples were taken throughout the mission to enable ground-based physicians to better understand metabolic and biochemical changes in their bodies in space. Additionally, the Astronaut Lung Function Experiment was used to gauge the influence of microgravity exposure on lung performance and respiratory muscles during rest and periods of heavy exercise. On 22 June, the human physiology studies entered the first of two specialised three-day 'blocks' to probe changes in sleep and performance patterns. It marked the first-ever comprehensive study of sleep, circadian rhythms and task performance in space. As astronauts circle Earth every 90 minutes—experiencing 16 'sunrises' and 'sunsets' in each 24-hour period—their 'normal' timing cues are significantly affected. Investigators hoped that such work might be beneficial to workers on Earth, as well as sufferers of jetlag. Typically, during the sleep experiments, all four science crew members wore electrode-laden skullcaps to observe their brainwaves, eye movements and muscular activity; "a real fashion statement," Brady quipped. A second block of time for the experiment began on 2 July when they filled in questionnaires at the start and end of their shifts and again wore the skullcaps whilst asleep. Following Columbia's landing, data from

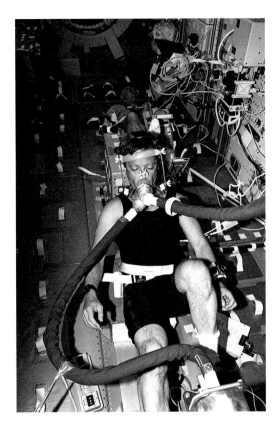

Fig. 12.20: Canada's Bob Thirsk participates in one of LMS' many life sciences investigations, as Chuck Brady (background) conducts another. Note the multitude of foot restraints positioned across the Spacelab module's 'floor', to assist in securing crewmembers at each research facility.

both blocks was compared to create a picture of how the astronauts' sleep patterns had changed during the mission (Fig. 12.20).

Extending this physiological research into the musculoskeletal arena were a range of experiments to explore the underlying causes of muscular and bone loss, which featured the first-ever collection of muscle tissue biopsy samples before and after the mission. Almost immediately after Columbia touched down at KSC on 7 July, the science crew underwent MRI scans and biopsies for comparison with pre-flight samples. It was expected that this might lead to improved countermeasures to reduce in-flight muscular atrophy. The astronauts routinely used a bicycle ergometer in the Shuttle's middeck, which was fitted with a large, weighted flywheel, surrounded by a braking band to resist imparting their pedalling to the hull, enabling the crew to exercise without disturbing the sensitive microgravity experiments. The Torque Velocity

Dynamometer was used to measure calf-muscle performance during these exercise periods. Additionally, Linnehan and Favier wore electrodes on their legs, which applied precise electrical stimuli to cause involuntary muscular contractions. Data from these experiments was expected to yield new insights into why muscles lost mass in space. The dynamometer was also employed for musculoskeletal tests to measure the astronauts' arm and hand-grip strength. On 24 June, they strapped their arms into the machine, curling and extending them as it provided resistance. Overall, the device performed near-flawlessly, with the exception of a few mechanical set-up problems and software glitches and operated for a total of 85 hours during STS-78.

In light-hearted reference to all the electrodes worn and blood and tissue samples taken from, Linnehan, Brady, Favier and Thirsk jokingly called themselves "the rat crew". Even their food and drink intakes were carefully monitored, as part of ongoing investigations of metabolic changes and calcium loss during space missions. They typically took non-radioactive calcium isotopes at each meal, from ten days before launch until a week into the mission, and by tracking its relationship to food-and-drink intake, scientists were able to distinguish calcium intake and excretion and determine the total amount used by their bones. Their efforts, pain and discomfort did not go unnoticed, particularly by an admiring Henricks. "I'm one of those non-scientist pilots who believes that doing medical experiments in orbit is part of an astronaut's job," he said later. "When so few humans go to space, we'd better get as much information as we can out of every person who goes. In my opinion, people who don't want to participate in those experiments picked the wrong profession." For STS-78's payload crew, he had the utmost respect. "The most painful thing they had to endure was using an electric charge to contract their calf muscles to their maximum contraction. That's why I'm disappointed when I see astronauts who don't want to participate in minor experiments that cause some discomfort. I saw those guys on STS-78 being so brave and doing things that were extremely uncomfortable."

Many of the LMS research facilities were cross-disciplinary and Canada's TRE was also employed for some of the neuroscience studies, by providing for the first time an opportunity to bridge the gap between space motion sickness and the causes of disorientation and nausea on Earth. It was hoped that this could help scientists to learn more about the problems associated with postural disorders and vertigo leading to falls and broken bones. The device precisely tracked the positions of the astronauts' eyes, heads and upper bodies as they went about their everyday activities. The most important observations came soon after entering orbit and they reported symptoms associated with adaptation to the microgravity environment. Voluntarily fixing the head to the torso with a neck brace has acquired the name 'torso rotation' because the subject has to turn their entire body in order to move their head. On Earth, this gradually leads to motion sickness in an example of deliberate 'egocentric' motor strategy, during which the subject concentrates on a body frame of reference, rather than an external world reference. Similar motor strategies are often adopted by astronauts, thereby exacerbating the onset of space sickness symptoms. "The findings will make a contribution to a further understanding, countermeasures and rehabilitative programmes for not only astronauts, but also for people in hospitals on Earth," Thirsk explained in a space-to-ground news conference on 30 June. "With this information, we can figure better ways to keep people in space healthier and fight off muscle and bone degeneration and also use the information on Earth." Another device used as part of the neuroscience research was a piece of headgear for the Canal and Otolith Integration Studies, which explored the impact of microgravity exposure on the vestibular system of the inner ear and resulting changes in eye-hand-head co-ordination. Throughout the experiment, astronauts wore high-tech modified ski goggles, which carefully tracked their eye and head motions as they watched a series of illuminated targets. Typically, they remained either in a fixed position on the bicycle ergometer or 'free-floating' in the Spacelab module and the targets displayed themselves across the inner surface of the goggles.

Seven astronauts were not the Shuttle's only living passengers on STS-78. Also hitching a ride were embryos of the hardy Medaka fish, provided by Columbia University College of Physicians and Surgeons in New York, which were flown as part of experiments into gravity's role in the development of animals. At intervals, an on-board video microscope provided television viewers with pictures of the growth of the transparent embryos. It was recognised that understanding the impact of microgravity on vertebrate development would become increasingly important as long-duration space station missions got underway. A total of 36 embryos of the Medaka—which is known to be particularly tolerant of reduce temperatures—developed during the mission. Judging from the video images downlinked from Columbia, the specimens in orbit appeared to develop at a slower rate than the ground controlled specimens. Other living creatures included 20 loblolly pine seedlings in a special plant-growth facility. When trees growing on Earth 'bend', then right themselves, they form so-called 'reaction wood', which is structurally inferior. During LMS, biologists carefully examined the cellular structures of the pine seedlings as part of efforts to devise ways to prevent reaction wood formation on Earth, which would prove enormously beneficial to the paper and lumber industries. After several days of growth, Favier and Helms harvested the seedlings, applied a chemical fixative, photographed them and stored them for landing.

Columbia's marathon mission was expected to get within hours of the 16.5-day duration of Endeavour's STS-67 flight (see Chapter Seven). On 29 June, it became official that STS-78 would snare the new record, when NASA told the crew that they would remain aloft until 7 July, producing a 17-day mission. The announcement was accompanied by background music from the movie *Mission: Impossible*, to which Henricks responded that his crew was "willing, able and eagerly anticipating" the extension to their flight. As the days wore on, the LMS science received nothing but praise from scientists on the ground. "We have 41 principal investigators involved and all but very few have 100 percent, if not 200 percent, of the data they hoped to collect," Patton Downey explained. By the time Columbia alighted at KSC at 7:37 a.m. EDT on 7 July, just two hours shy of 17 full days, STS-78 ended as the longest Spacelab mission in history and the second longest Shuttle flight of all time. It was, to be fair, a magnificent accomplishment, but even 17 days was a far cry from the month-long missions originally envisaged for Spacelab two decades before.

BROKEN ANKLE, BROKEN PLANS, BROKEN RECORDS

Early in April 1997, Columbia stood ready for a complex Spacelab mission, one of whose astronauts had begun that year fearful that he might not fly at all. But by the end of July, Don Thomas and the whole STS-83 crew had flown not once, but twice. Their payload—the Microgravity Science Laboratory (MSL)-1—was expected to be second-to-last voyage of the Spacelab module and one of the last Shuttle flights devoted to materials processing, combustion science and fluid physics research, with future experiments to be performed aboard the space station. In Many ways, MSL-1 was the final roll of the dice for physical scientists and its importance was highlighted by an experienced crew. In January 1996, payload commander Janice Voss, mission specialist Don Thomas, newly returned from IML-2, and payload specialists Roger Crouch, who served as an alternate crewman on IML-1, and Greg Linteris, a mechanical and aerospace engineer from the National Institute of Standards and Technology in Gaithersburg, Maryland, were named to support more than 25 experiments on a planned 16-day mission. Backing up Crouch and Linteris was aerospace engineer Paul Ronney of the

Fig. 12.21: The Microgravity Science Laboratory (MSL)-1 module is readied for integration into Columbia's payload bay for STS-83.

University of Southern California. In May 1996, Jim Halsell, Susan Still and Mike Gernhardt were named as STS-83's commander, pilot and flight engineer, targeting launch early the following year (Fig. 12.21).

But the mission's first wave of misfortune hit home on 29 January 1997, when Thomas slipped down some stairs and broke his ankle, following a training session at JSC. Unlike the Russians, NASA no longer assigned backup crews, but on this occasion (and barely nine weeks before launch) it was considered necessary to quickly train another astronaut to stand in for Thomas, if necessary. By mid-February, it was official: USML-2's Cady Coleman would serve as his backup, shadowing the final weeks of training. "I'm in a period of pretty heavy physical therapy, spending about five or six hours a day walking in swimming pools, walking with the cast and without the cast, getting my mobility strength back," Thomas said in mid-March. "We've got three weeks until launch and there's no doubt in my mind or the doctor's mind that I'll be ready in time."

Liftoff was routinely postponed by 24 hours past the originally scheduled 3 April and Columbia (with Thomas thankfully aboard) rose into space at 2:20 p.m. EDT on the 4th. Within a few hours, STS-83's blue team of Voss and Crouch, led by Gernhardt, opened the hatch into the Spacelab module and started activating its systems and experiments. The primary thrust of MSL-1 was threefold: crystal growth, combustion physics and the development of techniques to produce stronger and more resilient metals and alloys.

Additionally, STS-83 would evaluate some of the hardware, facilities and procedures that NASA intended to employ aboard the ISS. It had already become clear that physical processes ordinarily masked by gravity on Earth were virtually eliminated in space, thus making it possible for scientists to conduct hitherto impossible experiments.

An important element of this work was the ability to grow larger and purer crystals of proteins ranging from insulin to HIV-Reverse-Transcriptase and determining their three-dimensional structural blueprints. By unlocking the crystals' structural details in this manner, biochemists hoped to better understand how they fitted into the overall biology of the human body. It was anticipated that MSL-1's growth facilities would process upwards of 1,500 protein crystal samples, ultimately to address the 'social cost' of diseases such as cancer, diabetes, Alzheimer's and AIDS, which were estimated in 1997 to cost the United States around $900 billion per year.

Elsewhere, experiments focused on the differences of the combustion process in microgravity and its significance was highlighted by the presence of combustion expert Linteris on the STS-83 crew. At the time of his selection to join the MSL-1 crew, Linteris was working on advanced fire suppressants and the inhibition mechanisms of chemical inhibitors at the National Institute of Standards and Technology in Gaithersburg, Maryland. It was anticipated that developing a better understanding of the peculiarities of different types of fuel, and the fires they produce, could lead to increased efficiency and reduced emissions within internal combustion engines. In the United States, in 1997, the annual expenditure on crude oil was estimated by the American Petroleum Institute as close to $200 billion and the process of combustion was a major player in converting the chemical energy in fuel into useful thermal and mechanical energy, but was also a major contributor to air pollution, through the emission of nitrogen oxides, carbon monoxide, unburned hydrocarbons and particulates. "Minimised emissions," said Fred Dryer of Princeton University, "and best miles-per-gallon require us to carefully control and tailor the combustion process." Prior to MSL-1, this control and tailoring could only be done with sophisticated computers but experiments aboard STS-83 provided an opportunity to analyse theoretical predictions and develop new models. For example, theories held that small fuel droplets should go through three separate 'regimes' during combustion. One of these, known as a 'quasi-steady state', had been frequently studied on the ground: the square of the droplet/flame diameters decreased with time in a linear fashion and eventually extinguished itself when it became too small to support itself. The MSL-1 experiments not only provided additional data on that regime in far greater depth than was possible on Earth, but also investigated the other two regimes.

On Earth, a mere one-percent increase in fuel efficiency translated into saving of 100 million barrels per year or $5.5 million per day.

One of the most important facilities was the Combustion Module (CM)-1, developed by NASA's LaRC in Cleveland, Ohio, which supported the Laminar Soot Processes (LSP) and the Structure of Flame Balls at Low-Lewis Number (SOFBALL) experiments. Despite minor troubles with a cable configuration, LSP was activated by Linteris on 5 April. The experiment sought to gather data on flame shapes, together with the quantities, temperatures and types of soot produced under varying conditions. It supported ongoing efforts to develop methods of controlling fires on Earth and limit the number of deaths caused by carbon monoxide poisoning, associated with soot. Within a day of powering-up the experiment, Linteris was rewarded with his first glimpse of the concentration and structure of soot from a fire burning in the microgravity environment. Another major study in this area was the Droplet Combustion Apparatus (DCA), which occupied much of Voss' time during one of her early blue shifts. It housed a variety of experiments to investigate burning drops of different fuels and monitor conditions at the instant of their extinction. Inside the DCA, which filled one rack in the Spacelab module, was the Droplet Combustion Experiment (DCE) to explore the fundamental combustion aspects of isolated fuel drops under varying pressures and oxygen concentrations. In most practical combustion devices, liquid fuels are mixed with oxidisers and burned in the form of sprays. An essential prerequisite for an understanding of such 'spray combustion' and its application to the design of efficient and clean combustion systems is knowledge of the laws governing droplet combustion. In the absence of buoyancy-induced convection currents, a droplet ignited in microgravity burns with spherical symmetry and yields a simple, one-dimensional system capable of being very precisely modelled. Previous experiments on the ground had produced data for drops no larger than 3 mm in diameter, but the microgravity environment available in the MSL-1 module provided scientists with an opportunity to better investigate the complicated interactions of physical and chemical processes during droplet combustion (Fig. 12.22).

"Everything's going great!" exulted MSL-1 Mission Manager Teresa Vanhooser of MSFC as these and other experiments got underway. But STS-83's fortunes were about to drastically alter for the worse.

Shortly after reaching orbit, the pilots and Mission Control had been monitoring erratic behaviour from one of Columbia's three fuel cells. Mounted beneath the payload bay floor, these cells used a reaction of oxygen and hydrogen to generate electricity in order to support the Shuttle's systems, run the MSL-1 experiments and yield drinking water for the crew. One cell was

12   In the absence of *gravitas*   343

Fig. 12.22: Don Thomas, pictured during IML-2 in July 1994, would almost miss out on flying at all in 1997, but that year he went on to log two more missions. He remains one of only a handful of astronauts to have flown three times with the Spacelab pressurised module.

technically sufficient to support orbital and landing operations, but flight rules dictated that all three had to be fully functional for a mission to continue. Each cell had three 'stacks', made up of two banks of 16 cells apiece. In one of Fuel Cell No. 2's stacks, the difference in output voltage between the two banks had shown signs of increasing. It had been noted before launch but had been cleared for flight. Late on 5 April, Halsell and Still adjusted Columbia's electrical system configuration to reduce demands on the ailing cell. This allowed mission controllers to stabilise it for ongoing analysis. Overnight, the rate of change in the cell slowed from five millivolts per hour to around two millivolts but continued to exhibit a slight upward trend. Early on the 6th, Halsell and Still manually purged the cell at Mission Control's request. However, as the situation showed no sign of improving, it was decided later that afternoon to bring Columbia home at the earliest possible opportunity.

Fuel Cell No. 2 was accordingly shut down, followed by several other pieces of non-critical equipment to provide additional power to support the MSL-1 experiments for as long as possible. The lights in the Spacelab module were turned off and the astronauts found themselves running experiments by torchlight. In Thomas' words, working in the darkness with a torch clamped between his teeth became "standard operating procedure" and made for a "most interesting work experience". Nevertheless, the entire crew reacted with "shock and disbelief", according to Thomas, at the decision to cut short the flight. The fuel cell problem devastated a carefully choreographed 16 days of research, as scientists scrambled to reprioritise their schedules to make the most of the one or two more days available before coming back to Earth. Already, efforts were underway to lobby NASA to stage a reflight of MSL-1, later in 1997.

Certainly, in spite of the troubles, the research work was proceeding superbly. Crouch—a jolly bear of a man, described by Halsell as "one of the heaviest Southern accents I've ever had the opportunity to fly with"—worked with the SOFBALL investigation, which was design to explore the conditions under which a 'stable' flame ball can exist and if heat loss is responsible for its stability whilst burning. During the first run, a mixture of hydrogen, oxygen and carbon dioxide burned in the facility for the entire 500-second limit. This was particularly significant, because they represented the lowest-temperature and weakest flames ever burned, with the most diluted mixtures, according to alternate payload specialist and principal investigator Paul Ronney. "These mixtures will not burn in Earth's gravity," he said. "We have known that burning weaker mixtures increases efficiency, but not much is known about the burning limits of these mixtures." As well as offering insights into the combustion process, it was anticipated that SOFBALL results would help to enhance theoretical models. By the time of MSL-1, anomalous 'flame balls'—essentially stable, stationary, spherically symmetric flames in combustible gas mixtures—were receiving significant attention. By examining their behaviour in the microgravity environment, it was hoped that more efficient combustion engines, emitting fewer atmospheric pollutants, could be produced. During typical SOFBALL runs, a chamber was filled with a weakly combustible gas (hydrogen and oxygen, highly diluted with an inert gas) and ignited. The resultant flames and their motions were then imaged and recorded by video cameras, radiometers, thermocouples and pressure transducers. Unfortunately, the shortened nature of STS-83 meant that only two of 17 planned SOFBALL experiments could be completed.

Elsewhere in the Spacelab module, other experiments focused on materials processing, as part of efforts to develop techniques for manufacturing stronger and more resilient metals, alloys, ceramics and glasses. Two facilities used to support this work—Japan's Large Isothermal Furnace (LIF) and Germany's Electromagnetic Containerless Processing Facility (TEMPUS)—had previously flown aboard IML-2. The former heated and rapidly cooled materials, including ceramic-metallic composites and semiconducting alloys, in various temperature ranges to identify relationships between their structures and physical properties. Usually, the furnace followed a pre-programmed cycle, reaching a maximum temperature of about 1,600 degrees Celsius (2,900 degrees Fahrenheit), and was activated by Thomas on 5 April. Among its early investigations were a study of the diffusion of impurities in molten salts and the Liquid Phase Sintering (LPS) experiment, which tested theories of how liquefied materials form a mixture without reaching the melting point of the new alloy combination. Other LIF activities diffused molten semiconductors as part of efforts to explore how uniformly their constituents mixed during the cooling process. Diffusion studies have many terrestrial applications, from very small movements in plasma to massive depletions in Earth's ozone layer.

Meanwhile, TEMPUS provided a levitation melting device for processing metals. It was activated by Crouch, late on 4 April, a few hours after Columbia entered orbit. Containerless processing was also considered hugely advantageous, because Earth-based processes involving liquids can cause them to be affected by the properties of their holding vessel; in microgravity, on the other hand, positioning and control can be effected with greater precision. This, in turn, reduced motions within the sample liquid and was less intrusive upon the physical phenomena under study. Moreover, containerless processing was known to eliminate contamination of the sample caused by the material in the container's walls. TEMPUS was used to study the 'undercooling' and rapid solidification of metals and alloys, which typically occurs when a solid is melted into a liquid, then cooled below its normal freezing point without solidifying. The phase change is delayed and the material enters a 'metastable' state.

Following the decision to bring STS-83 home early, TEMPUS Assistant Investigator William Hoffmeister oversaw the effort to reprioritise several of its experiment runs to acquire as much data as possible. The crew was able to activate, observe and complete one run by melting a zirconium metal sample and levitating it, as part of efforts to examine the relationship between internal flows in liquids and the amount of undercooling that a sample can tolerate before it solidifies. TEMPUS provided the scope for physically manipulating samples, controlling rotations and oscillations and even 'squeezing' them through the application of an electromagnetic field. The experiments involving zirconium, in particular, were expected to determine the behaviour of this strong, ductile, refractory metal, which has found terrestrial applications in nuclear reactors and chemical processing equipment. With the impending landing of Columbia, the TEMPUS team had good knowledge of how to reprioritise their schedule to make the best of unexpected events (Fig. 12.23).

Other areas of research aboard MSL-1 included plant growth, which focused on the Plant Generic Bioprocessing Apparatus (PGBA), provided by

Fig. 12.23: The MSL-1 facility, pictured during its short maiden voyage in April 1997.

BioServe, one of NASA's commercial space centres at the University of Colorado at Boulder. It offered a highly controlled environment with lighting, temperature and gas-exchange functions, together with a nutrient pack to supply water and other nutrients to nine different plant species. One of these, selected through a project with a Brazilian research group, was a member of the black pepper family. Also making its second Shuttle flight was the Middeck Glovebox, which supported experiments in fluid physics, materials processing and combustion science.

In spite of the disappointment at the early return to Earth, informal plans were already afoot whilst Columbia was in orbit to refly the mission, later in 1997. "There were rumour already flying," noted Thomas on his website, OhioAstronaut.com, "that after fixing the problem NASA would be re-flying our crew in a few months to complete our Spacelab science mission. That definitely helped ease the sting of coming home early." His thoughts were echoed by Crouch. "In some ways, this could make for a more meaningful flight in the long run," said Crouch, "but, certainly, *this* one was a bummer!" Ironically, ISS delays had pushed the first Shuttle construction mission into the summer of 1998 and made the reflight possible.

So it was that when Halsell departed Columbia after completing a picture-perfect 1:33 p.m. EDT touchdown at KSC on 8 April, he was approached by KSC Director Roy Bridges with a handshake and the words "We're going to try to give you an oil change and send you back". With a duration of three days, 23 hours, 13 minutes and 38 seconds, STS-83 established itself as the shortest operational Spacelab flight. Not surprisingly, the seven astronauts were in favour of a reflight. Just three days later, on the 11th, Space Shuttle Program Manager Tommy Holloway released a statement, authorising planning for a reflight in early July. "The Shuttle manifest for the remainder of the year, while tight, appears able to accommodate a reflight of Columbia and its MSL payload," he explained, "with reasonable impact to Shuttle launch dates for the rest of calendar year 1997." Internally designated 'STS-83R' (for 'Reflight') in the early days, the mission was eventually assigned the number 'STS-94' and its target launch on the first day of July was formally announced by NASA on 25 April. "We're ready to go fly," said Halsell as these plans came together. "If it were up to me, I'd like to give the guys a week or two off to let them decompress from this flight and then we'll come back and start ramping up again for the next flight."

NASA normally spent around $500 million per mission, although a substantial proportion of that figure was devoted to hardware testing, processing, training, planning and simulations which did not require repetition. Holloway quoted about $60 million for STS-94 and stressed that flying Columbia within three months offered "a very good test of a capability we should have in place for the station, to bring an element of the station back, for whatever reason, and turn it around in as reasonable time as practical". True to the predictions, the MSL-1 reflight *was* cheaper: $55 million for the actual processing of Columbia, plus $8.6 million for expenses associated with the turnaround of the Spacelab payload itself. "Our approach has been to treat this flight as a launch delay," said Lead Flight Director Rob Kelso. "The crew is exactly the same, the flight directors are all the same and the flight control team is almost identical. It's a mirror-image flight in many respects." Even the embroidered patch, worn by Halsell's crew, was exactly the same, albeit for a different-coloured border: red for STS-83, blue for STS-94.

Naturally, the MSL-1 investigators were ecstatic at the chance to fly again so soon. The Spacelab module remained in Columbia's payload bay, although the tunnel was removed to allow technicians better access to its interior. Ordinarily, between flights, the modules were transferred to the Operations and Checkout Building, but during the short MSL-1 turnaround technicians successfully completed many critical tasks, including replenishing fluids for the myriad combustion science experiments. Normally, the Shuttle processing team supervised an orbiter for 85 days, but the reflight required a stay time of just 56 days in the OPF. To accommodate it and ensure that necessary work was completed, they deferred certain structural inspections until her next mission.

For Halsell, flying twice in quick succession was "a marvellous, once-in-a-career opportunity". As forecasters continued to assess the Florida weather in the final days of June, the prospects of achieving an on-time liftoff seemed grim, with thunderstorms anticipated on 1 July and only marginal scope for improvement on the 2nd and 3rd. As part of efforts to avoid the thunderstorms, on 30 June NASA opted to bring the launch time *forward* by 47 minutes, opening the 'window' at 12:50 p.m. EDT, rather than 1:37 p.m. This removed one end-of-mission daytime landing opportunity at Edwards in California, but also enabled two more opportunities at KSC. After a 12-minute delay, Columbia rose perfectly into the Florida sky at 1:02 p.m (Fig. 12.24).

Fig. 12.24: Technicians in the Operations and Checkout Building conduct servicing of Columbia's airlock and beneath her payload bay floor in May 1997, as preparations ramp up to refly MSL-1 the following July.

"It looks like Columbia's performing like a champ," radioed Capcom Dom Gorie from Mission Control, shortly after orbital insertion.

"Roger, Houston, we copy," replied Still, "and thanks to the whole ascent team for getting us to a safe orbit."

Within four hours of launch, Voss and Crouch entered the Spacelab module to begin what they hoped would be 16 days of around-the-clock research. "It's good to be back," Voss reported, as Crouch backflipped behind her in the voluminous lab. Among the most important unfinished business from the April flight was the completion of 144 scheduled experiments in the Combustion Module. In the event, more than 200 were completed during STS-94. Among the payloads getting a reflight was SOFBALL, whose off-the-shelf gas chromatograph performed flawlessly, successfully verifying the composition of a variety of pre-mixed gases prior to combustion and determining the remaining reaction and other combustion products. Unfortunately, in light of the constraints imposed upon the chromatograph by the Combustion Module, it was not possible to measure the gases to the required accuracy. Still, SOFBALL achieved the weakest flames ever burned, mixing a variety of gases which were too small to be flammable on Earth, but which burned for more than eight minutes in CM-1. During the inaugural SOFBALL run on 8 July, the mixture was so weak that it only produced what one researcher described as "flame kernels". A later test, using a richer mixture of hydrogen, oxygen and sulphur hexafluoride, proved more successful, burning for 500 seconds and allowing investigators to understand how different concentrations of fuel and oxidiser affected the flame balls' stability and existence. Many of the flames were so weak that they equally barely a single watt of energy, compared to about 50 watts normally produced by a single candle on a birthday cake.

Working on SOFBALL, Thomas received the distinction of having a newly-discovered combustion 'effect' named in his honour. One surprising finding from the experiment involved what happened when two small fuel droplets burn in close proximity: they initially moved away, then approached each other in a phenomenon dubbed the 'Thomas Twin Effect'. Elsewhere, the Laminar Soot Processes experiment used ethylene gas as part of research into more environmentally-friendly fuel-burning engines. Already, STS-83 had offered an important insight into how soot particles formed in microgravity and the shortened nature of the mission enabled teams to enhance LSP for STS-94. One experiment run on 3 July involved a propane-fuelled study of soot, producing a "beautiful and steady flame", according to Linteris. As well as using different fuels—propane, then ethylene—the experiment also burned them under differing atmospheric pressures, to determine the amount of soot

Fig. 12.25: Susan Still and Don Thomas move a piece of experiment inside the MSL-1 module during the STS-94 reflight of the laboratory mission.

produced. Differing fuel types often made substantial differences. Natural gas tends to make little soot, whilst propane produces somewhat higher quantities and ethylene (which is typically used in diesel engines) generates even more. In total, on STS-94, no fewer than 17 separate experiment runs were conducted with LSP (Fig. 12.25).

Related research focused on the DCE, which involved several phases of observations of the burning characteristics of heptane fuel drops under a range of atmospheric pressures. These phases were 'normal' terrestrial sea-level pressure, together with half and one-quarter atmospheric pressure. The astronauts had a tougher time igniting the droplets at the lower pressures, but when they *did* ignite they burned stronger and more vigorously than anticipated in pre-flight models. Despite three brief malfunctions by DCE's computer, the experiment performed exceptionally well and by the time that it was shut down on the evening of 14 July heptane droplets had been successfully burned.

Elsewhere, payloads were moved from Columbia's middeck to the Spacelab module and back, thanks to a new storage rack, dubbed 'EXPRESS', being tested on MSL-1 as a means of getting experiments and equipment quickly to the ISS. The idea behind the new facility was to enable researchers to essentially plug their hardware, electrical power and video and data connections into the EXPRESS rack, giving them an easier and more generic interface and the chance to get experiments into orbit less than a year after conception, rather than 3.5 years ordinarily required for Spacelab missions. One possible advantage was that graduate students seeking to complete master's or PhD degrees in a few years could get their research projects into space in a reasonable time. For MSL-1, the rack supposed the PGBA plant-growth facility and the Physics of Hard Spheres (PHaSE), both of which were transferred from middeck lockers to EXPRESS when Columbia entered orbit. Unfortunately, plans to observe plants growing on a live video feed to the University of Colorado hit a snag when the lens clouded up. Several plants, including a species of sage, native to south-east Asia, were being flown as part of research into future anti-malaria and chemotherapy treatments. Meanwhile, PHaSE studied the transitions involved in the formation of colloidal crystals—collections of fine particles, suspended in liquids, such as milk or ink—in microgravity, as part of research into new 'designer' materials to produce future semiconductors, optics, ceramics or composites.

In materials processing, LIF and TEMPUS took centre-stage. Minor teething troubles hit LIF early in the mission, when it was found that the facility was using up helium from its cooling purge faster than predicted, although this did not impact its ability to collect data. The LPS study subjected tungsten, nickel, iron and copper to intense 1,500 degrees Celsius (2,700 degrees Fahrenheit) heating to create solid-liquid mixtures. On Earth, the process of 'sintering' is employed form very hard, very dense solids, which can be used to make cutting tools, car transmission gears and radiation shields. With four times longer in orbit on STS-94 than on STS-83, the scientific teams associated with both LIF and TEMPUS were gathering so much data that they expected to spend at least a year analysing it all. During the mission, TEMPUS successfully yielded the first measurements of specific heat and thermal expansion of glass-forming metallic alloys and, in so doing, obtained the highest temperature and the largest undercooling ever achieved in a space-based furnace. So revolutionary was the facility that NASA managers guaranteed TEMPUS a 'free' second flight on MSL-1, in exchange for U.S. participation in using it. In total, 20 investigations were performed and almost a quarter of a million commands were transmitted during STS-94. The shortened flight in April had actually proven beneficial for TEMPUS investigators, since they were able to examine the characteristics of the facility and their samples and make adjustments and improvements where needed. On the whole, it performed superbly, with the exception of a problem with one of its video cameras, although a few experiments were terminated earlier than intended, when their samples inadvertently came into contact with the side of the container (Fig. 12.26).

The peculiar sense of *déjà vu* was not lost on Columbia's crew. "This is the first chance to refly a payload and a crew altogether in the same group so

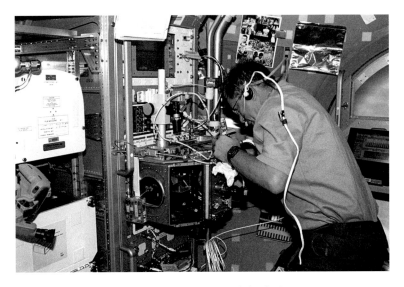

Fig. 12.26: Roger Crouch works in the MSL-1 module during STS-94.

quickly," said Voss. "It's been much easier to get back into the swing of things and all the experiments are going great and we all feel extremely comfortable and well-prepared because we've done this so recently." However, she added that the communications between the red and blue shifts were difficult at times. "There's a lot of issues, like where everything is stowed, and people always find their favourite places. On a single-shift flight, where we all sleep at the same time, it's a little bit easier to negotiate, because everyone's awake and you can ask them where they put something, but on a dual-shift flight you have to be very careful to work together as a team across those few hours when you're both awake."

In total, during the reflight, 206 fires were set, more than 700 protein crystals were grown and a variety of fluid physics and combustion science experiments were performed. A record 35,000 telecommands were relayed to the Spacelab research facilities and, although Columbia carried sufficient consumables to support a mission extension to 17 days, a request to lengthen the flight was not submitted. Early on 17 July, Halsell and Still performed the de-orbit burn to return to Earth. Columbia settled onto Runway 33 at 5:46:36 am EDT, wrapping up a mission just shy of 16 full days. Summing up the mission, Halsell had nothing but praise for Columbia, which had performed flawlessly. "Days have gone by," he told journalists, "without having to do an 'error log reset', which is our way of saying there have just been no problems, whatsoever. Our flavour for this flight is that it has done what it set out to do."

13

The human factor

A PROBLEM WITH MONKEYS

Few missions waited as long from inception to orbital insertion as the first Spacelab Life Sciences (SLS-1). Payload specialist candidates for this dedicated medical research flight—cardiologist Drew Gaffney of the University of Texas Health Science Center in Dallas, biochemist Millie Hughes-Fulford of the University of California Medical Center in San Francisco, physiologist and veterinarian Bob Phillips of Colorado State University and NASA biophysicist Bill Williams—were announced in January 1984, targeting launch aboard Columbia on STS-61D in January 1986, under the mission designator of 'Spacelab-4'. But for physician Rhea Seddon, one of the first women selected into NASA's astronaut corps in January 1978, the wait had been even longer. She entered the space agency, keen to be useful in the life sciences arena, but by early 1984 had been assigned to a relatively 'vanilla' Shuttle mission to deploy a pair of communications satellites. With Spacelab-4 also on the horizon and piquing her interest, Seddon doubted that she could train for both missions in such rapid succession. One evening at an astronaut office happy hour, she got chatting to George Abbey, JSC's director of flight crew operations. She told him that she had long been interested in Spacelab-4 but was already on another flight.

Abbey frowned and thought for a moment. "Well," he said, "Spacelab flights take a long time. That might not be a problem."

Indeed it was not.

Early in February 1984, Seddon and fellow mission specialists Jim Bagian, also a physician, together with Air Force engineer John Fabian, were assigned to STS-61D. Fabian would serve as the mission's flight engineer. "I was the designated chief-in-charge, because the other two were medical doctors and there were going to be in the back doing the medical experiments," he said. "Somebody needed to co-ordinate these activities and training and also to make sure that decision made early on, before the commander was named, didn't start off in the wrong direction. That was kind of my role." Finally, in January 1985 astronaut Vance Brand was named as STS-61D's commander, with Dave Griggs as pilot, but their places on the crew did not last. Spacelab-4 ended up meeting with significant delay, in part due to plans to fly four squirrel monkeys and 48 rats as research subjects on the mission. As detailed in Chapter Five, questions about the risk of herpes in the monkeys and problems associated with the cages of the Research Animal Holding Facility (RAHF) caused great consternation at NASA and the mission found itself pushed back. Years later, Fabian recalled attending many meetings with the life sciences community to gain assurances that the astronauts could not catch the monkeys' herpes and that the cages would not release any contaminants into the Spacelab atmosphere. "They had not performed very well on their first flight, so they kept being *on* the flight, *off* the flight, *on* the flight, *off* the flight; there was that uncertainty," said Seddon. Fabian remembered one particularly waspish remark from NASA Administrator James Fletcher—"Your monkeys are *never* going to fly again with my astronauts," he reportedly thundered to the life sciences community—and the cages proved a major sticking-point in the question of whether Spacelab-4 would fly or not. It was, remembered Bonnie Dalton, acting chief of ARC's Life Sciences Division, very much a 'make-or-break' mission. "If we had any failures with that research animal holding facility," she said, "it meant we probably wouldn't be able to fly animals in Spacelab anymore." As the redesign of the cages progressed, Seddon wrote in *Go for Orbit*, the rats were tentatively added back onto the mission as a hardware engineering test.

By the time Seddon returned from her first Shuttle flight in April 1985, therefore, it was "kind of a scramble" to get ready for a mission which had by now been renamed from Spacelab-4 to SLS-1. "It turned out to be what we called *over-subscribed*," she said, with too many experiments added and an excessive demand upon the Shuttle crew's time. By NASA's June 1985 manifest, SLS-1 was listed as "under review", with SLS-2 scheduled for February 1987 and SLS-3 and SLS-4 'requested' for July 1988 and September 1989. Neither Seddon nor Bagian had expertise in planning mission timelines, but by November 1985 SLS-1 had moved to no sooner than March 1987, which afforded them some breathing space. "But there was a lot of work to be done

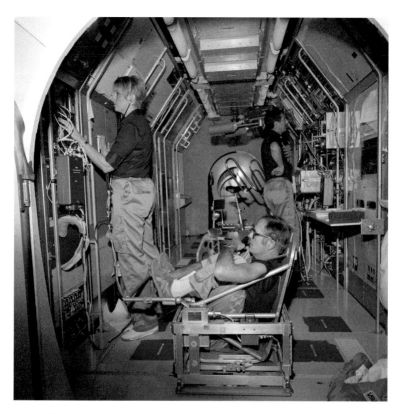

Fig. 13.1: Millie Hughes-Fulford, Bob Phillips (seated) and Drew Gaffney participate in SLS-1 training in a mockup of the pressurised Spacelab module.

as we realised we needed to descope," said Seddon. "What they ended up doing was dividing it into two missions" (Fig. 13.1).

On the day of Challenger's destruction, 28 January 1986, Seddon, Bagian and the payload specialists were at a meeting with a support contractor at an offsite facility. Late that morning, they stepped outside to watch the launch on television and saw the disaster unfold. The group was silent for a moment. Then Hughes-Fulford spoke. This month, she reminded them, was the month they should have launched on STS-61D. "This," she said, pointedly, "should have been *our* flight."

The loss of Challenger, perversely, lengthened the breathing time to get ready for SLS-1 still further. In papers published by the Rogers Commission hearings, as of January 1986 SLS-1 was identified as 80-percent complete and SLS-2 was 60-percent complete. And by the August 1988 manifest SLS-1 found itself assigned to STS-40 in June 1990, with SLS-2 on STS-59 in July 1992 and SLS-3 and SLS-4 requested for October 1993 and June 1995, the latter unfunded, but listed "for NASA planning purposes". With the Extended Duration Orbiter (EDO), it was noted that there were "plans to extend beyond nine days" on SLS-2 and at least the first pair of missions were targeted for orbits 148 miles (239 kilometres) high, inclined 42 degrees to the equator. This inclination was later reduced to 39 degrees, remembered Seddon's husband, fellow astronaut Robert 'Hoot' Gibson, "so that you wouldn't have the circadian shift…where you'd wind up getting up at the equivalent of one in the morning on re-entry day because they would shift your sleep schedule". On the SLS missions, the investigators desired a 'normal' daytime launch and 'normal' daytime landing, with no interference into the astronauts' sleep patterns and no sleep stations. "We wanted to keep our sleep and wake cycles the same, because changing circadian rhythms can change your physiology," Seddon explained. "So we want to be able to go to bed and get up sort of at the same time every day. And when we looked at how much equipment we had to have for the post-flight testing, we really only had one set of equipment, so that said we needed to plan to go to the landing site that was the most predictable." That landing site happened to be Edwards in California. "NASA had to bite the bullet on that, because it cost extra money to get the orbiter back to Florida," added Seddon, "but it's more chancy to fly into the Cape."

By January 1990, SLS-1 had moved to August 1990, with SLS-2 now on STS-60 in September 1992, SLS-3 on STS-88 in February 1995 and SLS-4 and a 'new' SLS-5 requested for June 1996 and June 1998, respectively. Both SLS-2 and SLS-3 were identified as EDO missions, the first scheduled for 13 days, the second for 16 days, "dependent on prior long-duration flight experience". The final mission in the series, the notional SLS-5, was unfunded and added "for NASA planning purposes".

With several higher-priority missions for NASA and the Department of Defense sitting ahead of Spacelab on the flight manifest, it is unsurprising that SLS-1 slid further backwards in the post-Challenger timeframe. Bill Williams had resigned from the payload specialist pool in February 1985 to spend more time with his family and John Fabian left NASA in the summer of 1986, leaving only Gaffney, Hughes-Fulford and Phillips with Seddon and Bagian. In April 1985, Gaffney and Phillips were selected as SLS-1's prime payload specialists, with Hughes-Fulford as an alternate, provisionally pointed at the prime seat on SLS-2. As the mission met with added delay, Brand and Griggs were themselves removed from STS-61D in September 1985 and reassigned to other missions. "The science all got shoved downstream and it kept being later and later," recalled an irritated Seddon. "We kept training more and more and it got pretty frustrating." Every month or two, she and Bagian and the payload specialists got together to do some training or look at SLS-1 checklists or timelines, until eventually, in the spring of 1989, the time came for the crew to reassemble.

"THREE WOMEN OR THREE DOCTORS?"

When Seddon and Bagian were assigned to STS-40 in February 1989, their launch aboard Columbia was planned for June of the following year, scheduled initially as an eight-day mission, but later extended to nine. "It was time for us to get back together and nail everything down," Seddon said of those first few months after assignment to STS-40, "the timeline, the equipment, what we could do, what we could sign up to do." During this early period, Phillips was grounded by a medical issue, but elected to remain as a non-flying backup crewman, supporting SLS-1 from MSFC on the ground. Not only was he no longer qualified according to NASA's medical rules, but Phillips' health made him a poor subject for several of the experiments. "He really wasn't a backup, but he trained with us; he knew everything we knew," said Seddon. "We just thought the world of Bob, so we were pleased that he was going to stay on with us." When the crew designed their mission patch, it included seven stars—one for each of them—which were carefully shaped into the letter 'P', to recognise Phillips' long contribution to the flight.

Hughes-Fulford's addition to the crew meant that STS-40 would earn a new record as the first spaceflight to include as many as three women. In April 1989, NASA assigned Bryan O'Connor as the mission's commander, John Blaha as pilot and astronomer Tammy Jernigan as flight engineer. One journalist asked O'Connor if he was afraid of having three women on his mission. "No," replied the straight-arrow Marine Corps colonel. "I'm more worried about having three *doctors* on the flight!"

That sly humour, though, masked O'Connor's real concern about the compatibility of his payload specialists. "Sometimes, I've thought that Millie and Drew were like oil and water and it was a pleasant surprise for me when, seeing how they operated—or *didn't* operate—together in the office or after the training's done or whatever, to where they would take *all* that baggage, old concerns with one another's performance, disagreements about the science, and we'd get into the simulator and there was *none* of that," O'Connor said. "These two trained like two professionals." Yet when the pair left the simulator, O'Connor likened their relationship to the 19th-century feud between the Hatfields and the McCoys. In her NASA oral history, Seddon agreed there was "some stress, just some frictions" in the crew relationship. "Each thought the other had taken inappropriate actions that jeopardised the success of the mission," Seddon later wrote. But O'Connor wondered if some of their "underlying issues" affect work in space, during a critical time. He put the whole crew through a personality assessment, which revealed a broad range of personalities which required a correspondingly broad understanding. "You've got to be a little more forgiving of certain things," O'Connor reflected, "and be more sensitive to other things to communicate properly and to operate as a crew" (Fig. 13.2).

The personality assessment enabled them to smooth the road to a much better working relationship. "I think everybody just assumed that everybody would get along," remembered Seddon. "Sometimes that's hard to do, especially when you've been training together in close quarters for a long time. Everybody has their little quirks." It underlined not so much a need for the astronauts to attempt to alter their personalities, but to adjust their approaches to how they 'led' in certain areas and 'followed' more harmoniously in others. By the time STS-40 closed in on its launch, O'Connor was satisfied that his crew was an oiled machine, with "no bad baggage". He was not trying to get

Fig. 13.2: The SLS-1 module, minus its connecting tunnel, is pre-positioned in Columbia's cavernous payload bay.

Gaffney and Hughes-Fulford to be best buddies—"after the mission was over, they probably never spoke again"—but to complete their best work in orbit. It worked and the payload specialists' performance was nothing short of outstanding, with a full plate of intense medical investigations completed.

Joy and sorrow can be close bedfellows and, certainly, in the case of STS-40, happiness and sadness pervaded their first few weeks together. In June 1989, Dave Griggs was killed in an aircraft accident and John Blaha was drafted in to replace him as pilot on another Shuttle mission, which took him out of the running for STS-40. Sid Gutierrez was assigned as O'Connor's new pilot. The pair had worked together in the aftermath of the Challenger disaster, jointly leading an action centre at NASA Headquarters to plan the agency's response to the Rogers Commission recommendations. Seddon remembered Gutierrez's humour and this veteran Air Force fighter pilot derived his own set of down-to-Earth nicknames for several of SLS-1's clumsily-titled experiments: the mission's body-mass measurement device, in Gutierrez-speak, became known as "weighing yourself", whilst a pulmonary function test was renamed "blow in the pipe" and the Baroreflex collar (see Chapter Six) became "suck on the neck".

Although the payload crew of Bagian, Seddon, Gaffney and Hughes-Fulford were responsible for SLS-1's research, the flight crew of O'Connor, Gutierrez and Jernigan showed an early willingness to participate in blood draws and other medical tests. "Tammy had already signed up for everything," O'Connor recalled. "She was a scientist herself and certainly was interested and very engaged in the science training." For the pilots, it was more problematic. Much of their training revolved around flying the Shuttle, which meant that their physical and mental health was paramount. As a result, they objected only to experiments which might pose a risk to their ability to fly the orbiter, such as those focused on the behaviour of the eyes or the vestibular system.

ANIMALIA, GALORE

When the crew began training in the late summer of 1989, they confidently expected to fly the following June, but delays pushed their mission to August and, with the hydrogen leaks experienced by Columbia that year (see Chapter Seven), eventually into late May 1991. Problems with leaky transducers, an inertial measurement unit and a balky computer delayed STS-40 until 5 June. The crew arose early that morning and began suiting up for launch; as part of one of the medical experiments, a central venous catheter—a thin plastic tube—was inserted into Gaffney's arm at the crook of the elbow which extended to his superior vena cava, the large vein which carries blood to the heart from the upper body. The experiment sought to monitor Gaffney's cardiovascular changes, fluid shifts and blood-pressure behaviour during ascent and was connected to a sensor box in his space suit's pocket. Weather posed something of an obstacle, before the Shuttle finally roared aloft at 9:24 a.m. EDT. "Thanks for the great ride," O'Connor told Mission Control after arriving in orbit.

Normally, the opening hours of a Shuttle mission were a flurry of activity and STS-40 was no exception, particularly since—unlike most Spacelab flights—this would operate on a single-shift system, rather than having two teams working around the clock. Having said this, the astronauts averaged 14-hour working days, their circadian rhythms closely monitored to provide for a more uniform set of biomedical data points (Fig. 13.3).

SLS-1 consisted of 18 experiments to investigate the fundamental problems affecting the biology of humans, animals and fish in the microgravity environment of space. Ten of these investigations required the astronauts as direct subjects, whilst seven utilised a complement of 28 rats and one used almost 2,500 jellyfish. During the mission, the crew explored how the heart, blood vessels, lungs, kidneys and hormone-secreting glands responded to the new conditions, studied the causes and effects of space sickness in greater depth and examined minute changes to muscle and bone structure whilst in

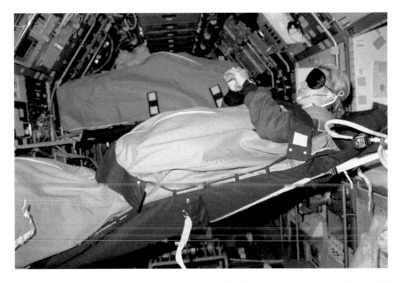

Fig. 13.3: SLS-1 was one of the few single-shift Spacelab missions, dictated by the needs of its life sciences payload. Here, Rhea Seddon has secured her sleeping bag and donned an eye-patch to get some rest inside the module at the end of a long workday.

orbit. Scientists were particularly interested in their physiological response in the first few hours of flight. In support of this objective, Bagian, Seddon, Gaffney and Hughes-Fulford underwent an extensive battery of demanding tests before launch, and after landing, with a full 24 hours of 'head-down' bed rest to evaluate whether this ground-based analogue for space flight produced similar cardiovascular responses. Additionally, vestibular tests assessed their sensitivity to linear acceleration. The mission also had a pronounced international flavour, with researchers from France, Russia, Germany and Canada participating in a biospecimen-sharing project. This kind of co-operation would feature with increasing prominence and regularity in subsequent Spacelab flights.

SLS-1, in its original guise as Spacelab-4, got underway in April 1978, when NASA issued an announcement of opportunity for medical and biological experiments to be flown on future Shuttle missions. Testing of the equipment began on Spacelab-3 in April 1985 and highlighted a number of operational flaws, notably contamination, leaks and odour problems from the RAHFs (see Chapter Five). Late in 1985, the *Newsletter of the American Society for Gravitational and Space Biology* revealed that these problems needed resolution before the dedicated medical research flight on Spacelab-4, then planned for early 1987. Before Challenger, NASA hoped to fly at least one RAHF on Spacelab-4, housing as many as 24 rats, and transfer them, in space, to a General Purpose Workstation (GPWS). When it flew on SLS-1, the redesigned hardware performed admirably; its single-pass auxiliary fan—"our vacuum cleaner," joked Bonnie Dalton—could indeed capture crumbs, flecks of the rats' hair and faeces, swooshing them down to the back of the unit and emitting no noticeable odours. Moreover, at one stage in the flight, when the science crew moved rats from the RAHF to the GPWS, it marked the first time that any rodents had floated freely in space, outside their cages. "He didn't want to get stuck out in the middle of nowhere with nothing to hold on to," Seddon recalled of one rat's experience outside the cage. "He would just hold onto your hand and then once we got him turned loose, we didn't want to throw him; we just wanted to get him off our hand. He floated around until he could grab onto the cage." The rat was very docile, she continued, and did not try to bite her.

The 28 rats (*Rattus norvegicus*) were part of a larger group of 74, of which 45 were kept on the ground as 'control' specimens and another had been dropped from the flight, just hours before launch, when a clogged water line failed in his cage. ("I imagine he's having a happy life now," NASA Test Director Mike Leinbach joked at the time.) The rats weighed an average of 0.6 pounds (275 grams) and were about nine weeks old; the 20 (soon to become 19) flight specimens were loaded into the RAHF a few days before launch. The device required an entire Spacelab module rack and its cages provided the rats with food, water and waste-management facilities, as well as controlling temperature and humidity, temperature and ventilation. The rats' movements were monitored and recorded by infrared light sources and sensors mounted inside each cage. During the mission, the RAHF was extensively tended by the science crew, who evaluated its capabilities. Generally, it proved favourable and maintained high levels of particulate containment.

Its environment was so good, in fact, that it was even possible for several of the rats to be moved to the GPWS in a portable transfer unit. "To our surprise, when Drew got the cage into the workbench and opened the door, the rat hung onto Drew's hand for dear life," remembered Seddon. "He did a twirl, then floated back into his cage." (On SLS-2, scientists would oversee the first dissections of rats in space for studies of their vestibular mechanisms.) Nine other rats were carried in a pair of Animal Enclosure Modules (AEMs) in Columbia's middeck. Like the RAHF, these provided food, water, waste-management, air and lighting, but did not enable the astronauts to actually handle the rats. These nine rats flew in the middeck because they were loaded aboard so late in the countdown—only 15 hours before launch—that technicians could not gain easy access to the Spacelab module. At first, in orbit, the rats clung to the sides of their cages, but upon realising that they would not fall, they became more relaxed and floated freely around.

One fundamental thrust of the SLS-1 rodent research were studies of the effect of microgravity on their muscles and bones, particularly through measurements of changes in circulating levels of calcium-metabolising hormones and the uptake and release of calcium in their bodies. These changes posed a considerable concern, because they appeared to be similar to those observed in human patients with osteoporosis and a clear understanding of the mechanisms behind them was seen as vital in the planning and execution of future long-duration space missions. Bone growth in the rats' legs, spines and jaws was observed and the loss of calcium and phosphorus was measured, revealing decreased skeletal growth and reduced leg-bone breaking strength and spinal mass. Other experiments explored decreases in the strength and endurance of muscles. In general, the rats returned from STS-40 much more lethargic than they had before launch, with reduced muscle tone, and were found to use their tails much less frequently as an aid for balance. Additionally, their red and white blood cell quantities decreased, although they were in far better overall shape than anticipated.

FIRST MEDICAL RESEARCH FLIGHT

The astronauts participated in the calcium studies, too, by carefully monitoring their daily intakes of food, fluids and medication and weighing themselves frequently. In particular, the science crew took blood samples to better determine the role of calcium-regulatory hormones on the observed changes in their calcium balance. For Bagian, Seddon, Gaffney and Hughes-Fulford, to say that SLS-1 *was* their 'home' for most of the nine-day mission would not be an exaggeration, for as well as working in the Spacelab module they also *slept* there, finding it much more comfortable than the cramped middeck. "They all thought it was a great place to sleep," O'Connor told Mission Control on one occasion. "It was nice and dark and quiet back there," agreed Seddon. "We were doing single shifts, so the lab was essentially buttoned up for the night. It was dark and we could cool it down, so we just hung our hammocks back there." Every so often, the quiet would be disturbed—the module was situated near the end of the payload bay, so the science crew could hear the *boom boom boom* of the orbiter's manoeuvring thrusters going off, as well as the mice in their cages and the refrigerators switching on and off—but Seddon considered it a far more peaceful place to relax than the flight deck or middeck (Fig. 13.4).

Moreover, all seven astronauts devoted everything to the science, typically beginning their duties ahead of time, working through meals and continuing

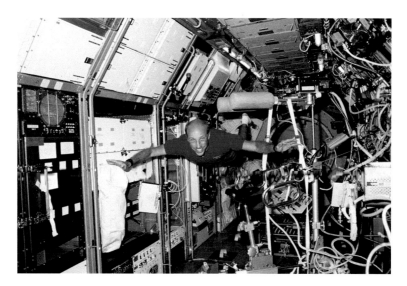

Fig. 13.4: Jim Bagian appears to 'fly' through the Spacelab module during a moment of levity on the busy mission.

well past bedtime. The result was a substantial increase in the overall scientific yield from SLS-1, whose effects would be amply illustrated in the analysis of results. In some quarters, the work in the Spacelab module was described as 'slavish', owing to its unending intensity, but the reality was rationalised by Seddon. "Jim and I had each flown a flight," she told the NASA oral historian, "but the things that happened on those flights were pretty much stand-alone chunks of time. And suddenly, on *this* one, I don't think we completely understood how things could go awry if one piece of your equipment didn't work and they had to re-plan the day." In fact, so delicate was the need to keep to the timeline that it had even been thrown into some disarray by the launch delay of 85 minutes or so on 5 June, which meant the astronauts had to follow an 'off-nominal timeline' for Day One of the mission.

As a consequence, they were frequently kept up late, troubleshooting or working an experiment which had earlier experienced difficulties. Every so often, the voice of O'Connor would be heard over the module's intercom, calling them back up to the flight deck for lunch. The unanticipated busyness of the mission perturbed Seddon. "Not enough sleep, getting up early, staying up late, having to fiddle with balky gas analysers, freezers that always needed defrosting and broken connectors" conspired to make SLS-1 one of the most challenging flights she ever flew.

One particularly significant experiment focused on the heart, lungs and blood vessels and was slightly delayed on 5 June when the Gas Analyser Mass Spectrometer (GAMS) developed difficulties and needed additional time to stabilise itself for operations. Detailed observations of the cause of the light-headedness reported by many spacefarers were made and one experiment investigated the theory that it might arise when the 'normal' reflex system responsible for regulating blood pressure behaves differently after adapting to a weightless state. Crew members wore a special neck cuff, a little like a whip-lash collar, which recorded blood pressure levels. Other experiments evaluated how quickly the astronauts adapted to the new environment, through prolonged expiration and 'rebreathing'—inhaling previously exhaled gases—whilst at rest or pedalling a stationary bicycle ergometer. This produced surprising new data on the amount of blood being pumped out by the heart, together with oxygen usage and carbon dioxide production rates. Still other tasks involved differing gaseous mixtures and examined the influence of terrestrial gravity, and its absence, on lung function. Measurements of blood pressure in the 'great veins', close to the heart, were conducted using the catheter in Gaffney's arm. This experiment was part of a cardiovascular adaptation study, developed by Gunnar Blomqvist of the University of Texas Southwestern Medical Center in Dallas.

The catheter study had an interesting history in itself. "Drew was one of the scientists who proposed the measurement as part of a study," wrote Seddon, "before he knew he would be the subject on it." In pre-Challenger days, still known at that time as Spacelab-4, concerns had been raised about the safety aspect of having a catheter threaded into Gaffney's arm towards his heart, but it became far riskier in the months and years after the accident, when the Shuttle program did away with the practice of sending astronauts into space in lightweight flight suits and introduced a new partial-pressure suit, made by the David Clark Company. This pumpkin-orange ensemble was bulky and cumbersome and, although it provided hyperbaric and cold-water immersion protection for its wearer, it caused a headache for the catheter experiment. "Once you put it under a pressure suit," Seddon recalled, "where you can't get to the catheter that's threaded up at the crook of the elbow, you had to go through *another* round of safety assessments."

In an emergency, it might slip further into his arm and directly into his heart, potentially irritating the heart's muscle and inducing a rhythmic disturbance. Or its connection could work loose, allowing air into the line and impeding blood flow through the heart valves. Most dire was the remote risk of a blood clot or infection at the insertion site. It placed Gaffney in the difficult situation of potentially having to bail out of the Shuttle, in an emergency situation, with the catheter embedded in his arm. "How do you sit in your chair during ascent," asked O'Connor, rhetorically, "with a catheter to your heart, hooked up to some electronics that goes into the orbiter's recording systems and *then* be able to bail out or egress from the cabin on the launch pad if you have to go very quickly?" It was a difficult question to answer and one which consumed many months of effort and training and planning and modifications to the flight hardware.

Notably, disconnects had to be configured to ensure that Gaffney would not bleed to death if the catheter pulled loose in an emergency. The catheter had been inserted on 4 June, a full day before launch, and was removed by Bagian about four hours after reaching orbit. Its data indicated the degree of body fluid redistribution and the rate at which this redistribution occurred. It revealed, unsurprisingly, that his blood pressure rose whilst on the launch pad, rose sharply and peaked during ascent and steadily returned to normal after a few minutes in space. This appeared to refute earlier suggestions that a rise in blood pressure resulted from fluid shifting into the upper body as a result of weightlessness (Fig. 13.5).

A substantial number of blood-related experiments were performed to investigate the mechanisms responsible for decreasing numbers of circulating red blood cells (known as 'erythrocytes') in space and a subsequent reduction

Fig. 13.5: Millie Hughes-Fulford prepares to spin-up Jim Bagian on a rotating chair for a physiological study.

in the oxygen-carrying capability of the blood. Samples were taken from each crew member before, during and after the flight and their relative volume to plasma was measured to check the rate of production and destruction of blood under 'normal' terrestrial and microgravity conditions. Although the SLS-1 data did not provide conclusive answers, it did indicate that a drop in red blood cell production *was* a contributory factor. Televised downlinks from the Spacelab module revealed what appeared to be a modern torture chamber, with all of the blood work in progress. One time, Gaffney calmly offered his arm to Hughes-Fulford for a needle and submitted for *another*, wielded by Bagian, an hour later. The blood work was frustrating at times, Seddon reported, as samples took longer to acquire in microgravity "and veins are not co-operating", but progress was otherwise exceptionally smooth.

Another set of experiments focused on the brain, central nervous system, eyes and inner ear and included studies from a joint U.S./Canadian project to investigate the impact of space sickness on the performance of the astronauts. On STS-40, members of the science crew placed their heads inside a rotating dome, which induced a sense of half-rotation in the direction opposite to that of the dome's own rotation. The astronaut subject then used a joystick to indicate his or her perception of self-motion.

Awareness of position in space is important, particularly during re-entry and landing, when astronauts need to be able to reach levers and switches. The general results of the SLS-1 motion experiments pointed to a loss of sense of orientation and limb position in the absence of visual cues. On several occasions, the members of the science crew were also blindfolded and asked to describe the positions of their limbs in reference to their torsos and point towards familiar structures inside the Spacelab module. Other work investigated changes in the inner ear, which has long been known to be highly sensitive to gravity and responsible for inducing disorientation in space.

The measurements of the nervous system involved the first flight of jellyfish on the Shuttle; 2,478 Moon jellyfish (*Aurelia aurita*), to be precise, which were housed in a pair of containers—"Small flasks of the tiny jellies," remembered Seddon, "which were barely big enough to see"—in one of Columbia's middeck lockers. They were chosen because they are one of the simplest organisms known to possess a nervous system, employing structures known as 'rhopalia' to maintain a correct orientation in water, very similar to mammalian otoliths. The main aim of the jellyfish experiment was to determine their reproductive abilities and the impact of the new environment on their gravity-sensing organs and swimming behaviour. "Jellyfish tend to swim in circles and arcs," remembered Tammy Jernigan, "whereas on the ground they tend to swim to the top of the flask and float down." The tiny jellies were videotaped throughout the flight and finally 'fixed' on 12 June to preserve them for their return to Earth. Overall, the jellyfish polyps, which developed into sexually reproductive ephyrae in space, proved 'normal' in most respects, despite hormonal changes and abnormalities in their swimming behaviour after landing.

However, the problem-free appearance of the flight was deceptive. Scientifically, SLS-1 was proving to be a grand success, but a potentially catastrophic problem with Columbia's thermal insulation cropped up shortly after arrival in orbit. Within minutes of opening the payload bay doors on 5 June, a camera revealed that a number of thermal blankets on the aft bulkhead had become detached. Moreover, part of the payload bay door seal strip was displaced. The danger, of course, was that the seal and insulation could hamper the successful closure of the doors prior to re-entry. Lead Flight Director Randy Stone assured the press that there was no cause for alarm. "We don't believe this to be any issue with respect to safety or mission duration," he said. "The latches on these doors are very strong and we believe that even if the seal *was* in the way, we could collapse the seal and close the doors safely with no problem." By 8 June, Mission Control told the crew that they did not believe a contingency spacewalk by Bagian and Jernigan to manually collapse the seal was needed and were surprised by a rare note of disagreement from Bryan O'Connor. He was worried that the seal might snarl on a mechanical 'fork' which assisted in the closure of the doors. After several discussions, and an uplinked explanation of the procedures, he acquiesced that "you've answered our big questions".

Despite their busy workload, none of the astronauts tired of the astonishing view of Earth through the windows. On her final evening in orbit, Hughes-Fulford floated alone in the darkened Spacelab module, working out to the haunting echo of Irish singer Enya from her Walkman. At length, tired and hot, she stopped and paused by the window to look out. "The Earth was in darkness and I saw great lightning storms below," she recalled. "The storm cells were almost a hundred miles across! When the lightning struck, the *entire* cloud canopy would light up underneath, like giant hot air balloons." Startled, she looked around for a camera to capture this amazing scene. There were none within reach and Hughes-Fulford knew that by the time she had returned to the flight deck to get one, the view would be gone…to be replaced by another. On that last night of SLS-1, she put her nose to the window and planted a memory of what she saw in her brain, where it would remain for life (Fig. 13.6).

At length, on 13 June, the orbiter crew set to work readying Columbia's systems for the impending return to Earth. The moment of truth for the payload bay doors came early the following morning and it was decided to orient

Fig. 13.6: Columbia touches down at Edwards Air Force Base in California on 14 June 1991, a landing location specifically selected to afford the SLS-1 payload crew immediate access to post-flight research facilities and personnel.

the troublesome seal towards the Sun to 'thermally condition' it. Bagian entered the Spacelab module for the final time, partly to store a few bags of blood samples and also to videotape the closure of the doors. The port-side door closed and latched without incident, followed by the starboard door, and Jernigan confirmed that they were both properly sealed shut. An hour later, at 7:39 a.m. PDT, O'Connor and Gutierrez guided their heavyweight ship to a safe landing at Edwards' concrete Runway 22. "We couldn't land on the more commodious lakebed runway," recalled Seddon, "since our tyres would dig into the surface too much and perhaps damage the landing gear."

More than 7,500 people had gathered at the California site to witness the touchdown but would not actually see the astronauts disembark from Columbia. On this occasion, a specially designed airport-style 'people mover'—the Crew Transport Vehicle (CTV)—had been commissioned to whisk them away for medical checks. "It's like a big trailer house on high that can be jacked up," explained former space suit technician Jean Alexander. "They [the astronauts] come inside and take the suits off and kind of get their land legs back and the doctors check them out. If there's any medical experiments that have to be done after landing, there's gurneys in there." The rats, too, were quickly taken away, but to a somewhat different fate, for 15 of them (and their ground-based 'controls') were euthanised and dissected a few days later for analysis of their inner ear mechanisms.

For Hughes-Fulford, the landing after nine days in weightlessness produced peculiar sensations. The position of the Shuttle's nose meant that the entire cabin was angled slightly downwards and she found it hard to get out of her seat and stand up. After removing her suit in the CTV, she finally had the chance to walk down the steps to see her husband and daughter. "Because my equilibrium was totally gone," she said later, "I was holding onto the rail, trying to move as quickly as possible and not walk like a little old lady." When her daughter arrived, Hughes-Fulford gave her a hug and whispered in her ear: *"Help hold me up!"*

CONTROVERSIAL DISSECTIONS

"John, we're going to fly you one of these days," Launch Director Bob Sieck called over the communications loop on 15 October 1993. The disappointment of another scrubbed launch attempt was evident in his voice. "Just hang in there."

"Nice try," replied STS-58 Commander John Blaha from Columbia's flight deck, as he and his six crewmates prepared to disembark from the orbiter after two and a half hours on their backs in bulky, uncomfortable pressure suits, harnesses and parachute equipment. It was the second time that they had been through this routine in trying to get into space for what was to be NASA's longest Shuttle flight to date, lasting a little over two weeks. For now, however, Columbia was living up to her unenviable reputation: an immovable bear, difficult to get off the ground, but once in orbit imbued with the beauty and grace of a swan. The delays in getting her previous mission off the ground (see Chapter Six) had already pushed Columbia's SLS-2 mission from late August 1993 into mid-September, before eventually settling on a launch at 10:53 a.m. EDT on 18 October (Fig. 13.7).

For Rhea Seddon, veteran of SLS-1 and payload commander of SLS-2, one of the worst aspects of a Shuttle mission were the bulky, cumbersome pressure suits that post-Challenger crews were obliged to wear for ascent and re-entry. Her concerns were amplified in May 1993, when she broke four metatarsal bones in her left foot, whilst practicing a fully-suited training evacuation from the orbiter. It was fortuitous that the injury was a minor one and that the

Fig. 13.7: Columbia launches on SLS-2 on 18 October 1993.

training was of the refresher nature and therefore could be quickly caught up on after her recuperation. Seddon was named as payload commander of SLS-2 in October 1991, with a projected launch two years later in July 1993. Her expertise from the SLS-1 mission was a crucial factor in the assignment... and it was an assignment that Seddon had actively sought. "There was some controversy about my being on the next flight," she told the NASA oral historian, "because I was...on the SLS-1 flight and they wanted four other subjects...They wanted to get eight subjects altogether and if *I* flew, they were only going to get seven subjects, because I was a repeat. They already *had* data on me. They weighed the pros and cons of that, but I had been following SLS-2 as long as I'd been following SLS-1 and continued to follow it after the first flight." There were other physician-astronauts in the office, but several were assigned to other missions or did not want to do SLS-2. Seddon had also established good working relationships with the SLS investigators and had the benefit of having flown recently. "There were just so many things that came out of SLS-1," she said, "that they wanted to capture and I think they knew that if I wasn't on SLS-2, I would probably be busy with another flight and not be able to help them as much as they liked."

In December 1991, astronaut Shannon Lucid, a biochemist, and physician Dave Wolf were assigned as mission specialists. Seddon was relieved that a decision had been taken by NASA to assign a payload commander for SLS-2. Her previous mission, SLS-1, had worked well thanks to the good working relationship she developed with crewmate Bagian. However, having no one in overall authority made payload decisions "a little awkward". For Seddon, it meant that she could attend meetings, represent her mission's payload and make "reasonable decisions on our behalf". There were other issues, too. "We'd had a problem on SLS-1, with people unvolunteering for some experiments after assessing all the risks," she said. "This was within their rights but being limited to a few subjects for each experiment sure threw a monkey-wrench into some of the investigators' plans." Seddon recalled being asked by the chief astronaut to pre-brief potential SLS-2 crewmembers on the experiments they might be required to participate in. "Although the experimenters and some NASA managers felt it an astronaut's duty to participate in any assigned research," she said, "federal law stands against making that a job responsibility."

Also announced to the SLS-2 training complement in October 1991 were three candidates for a single payload specialist position on the flight: physician Jay Buckey of the University of Texas Southwestern Medical Center, electrical engineer Larry Young, director of the Man-Vehicle Laboratory at MIT, and veterinarian Marty Fettman of Colorado State University. (Of note,

Young came highly recommended by Bob Phillips, whilst Buckey developed much of the equipment used in SLS-1's heart and blood studies.) "Some felt it cruel that we made them compete, but choosing a crewmember based on how they'd worked in their own labs and how they performed in the short, one-hour interview for the job was hard to imagine," wrote Seddon. "Much of what was important to a space mission was teamwork and an ability to perform all parts of the payload science well."

In October 1992, Fettman was announced as the primary payload specialist, making him the first professional veterinarian ever to travel into orbit, and with a very specific purpose. "NASA's series of SLS missions play a central role in our program of space biomedical research," explained Lennard Fisk, NASA's Associate Administrator for the Office of Space Science and Applications. "The experiments that Dr. Fettman and his fellow SLS-2 crew members conduct will give us valuable information on how living and working in space affects the human body." Several months after the announcement of the SLS-2 science crew, in August 1992 the three-man 'orbiter' team was named. John Blaha would command STS-58, having already flown three times and trained briefly for a spot on the SLS-1 mission. Joining him as pilot was Rick Searfoss, with Bill McArthur as flight engineer.

Due to the life sciences slant of the SLS-2 flight, the crew timeline was planned as a single-shift operation and Blaha was clear that although he was responsible for the safety and success of the mission, it would be Seddon who led the research in the Spacelab module. Unlike some Shuttle commanders, who viewed their role as little more than a truck driver, Blaha saw the mission's goal as obtaining good science from SLS-2 and he willingly offered himself, Searfoss and McArthur as subjects for non-invasive experiments. "In other words," said Seddon, "they wouldn't do anything that would make them sick or weak, because they might have to fly us home at any point in time." Years later, Blaha remembered that SLS-2 gained 30-40 percent of additional data by using himself, Searfoss and McArthur.

Before launch, McArthur was keen to do any experiment asked of him, but he was cautioned both by the chief astronaut and by Seddon. *Don't volunteer for everything or anything*, they told the affable Army aviator. *Don't sell your soul to the devil quite yet.* "I avoided some of the more intrusive medical experiments," admitted McArthur. However, he did participate in a 'glucometry' investigation, which required him to prick each of his fingers at least once to measure his blood glucose levels. "Once or twice, I made a mistake," he said later. "I would go one side of the tip of the finger and if I made a mistake, I'd have to then go to the other side of the tip of the finger. Eventually, my fingertips were all tender."

The inclusion of veterinarian Fettman on the crew had been on the cards since before the SLS-1 mission, since it would involve extensive physiological examinations with 48 male rats (*Rattus norvegicus*), caged in a pair of RAHF cages. It would also controversially feature the first-ever in-flight decapitation and dissection of six rats. "While the crew would rather have given the rats injections to put them to sleep," said Seddon, "the chemicals interfered with some of the changes caused by weightlessness." This made the use of a small guillotine the most humane option. As the payload commander, and a surgeon by training, Rhea Seddon assigned herself and Fettman to oversee the dissections. Not surprisingly, this had drawn much public criticism, but according to Fettman and NASA Associate Administrator for Life Sciences Harry Holloway it was an essential tool in measuring ongoing changes in the rats' body tissues during flight. "This is really a unique opportunity to collect biological specimens," said Fettman before launch. "We believe these tissues will provide some answers to questions that potentially will change our interpretation of past observations." It was rationalised that examinations of rats brought home from SLS-1 had been unable to conclusively differentiate between the effects of microgravity exposure and the effects of their readaptation to terrestrial conditions. The SLS-2 dissections would enable researchers to more precisely trace tissue changes in the rats (Fig. 13.8).

Still, Holloway called for an unscheduled pre-launch assessment of the plans, led by Deputy Surgeon-General Robert Whitney of the Department of

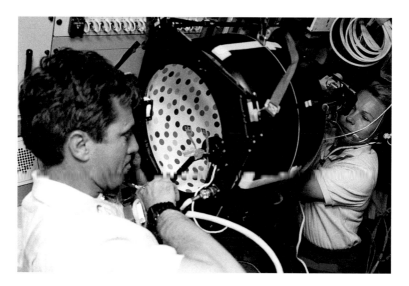

Fig. 13.8: John Blaha and Rhea Seddon work with the rotating dome experiment aboard SLS-2.

Health and Human Services. Holloway denied claims that the assessment was forced upon him by the White House or NASA Administrator Dan Goldin. Whitney's investigation described NASA's animal-care provisions as "superb" and commended the agency's use of the fewest number of rats as necessary to satisfy the needs of more than a hundred investigators. And the presence of Fettman on SLS-2 added another layer of sensitivity to the animal rights protesters in that he was able to assess procedures to ensure that they could be properly achieved in a weightless environment.

Shortly after entering orbit on 18 October, Lucid and Fettman began taking the first blood samples and Wolf took their blood pressures to acquire data on early adaptive processes to the microgravity environment. The findings correlated SLS-1 data by revealing a slightly lower central venous pressure than had been predicted in ground-based studies, coupled with a larger volume in the heart's left ventricle than would be expected with the lower pressure. This shed new light upon the basic physiology of the human heart in space. Immediately after the Spacelab module had been opened for business, Seddon took ultrasound measurements of Lucid's heart with a new echocardiograph imaging device. Both Lucid and Fettman had—like Drew Gaffney on SLS-1—ridden into orbit with catheters threaded into their arms, which ran to the tips of their hearts. Lucid's catheter was removed late on 18 October, followed by that of Fettman the following day. Data dropouts from the echocardiograph led to the crew resorting to the portable American Flight Echocardiograph (AFE), which Wolf helped to design in his pre-astronaut days.

Within hours of activation, the SLS-2 payload was already shaping up to be a tremendous success. NASA had invested $175 million in the payload, which featured 14 major experiments, of which eight focused on the crew and six on the rats. Body tissues from the latter were to be preserved for distribution to U.S., French, Russian and Japanese medical scientists after the mission as part of an extensive biospecimen-sharing project. In fact, Russia had long been courted as a partner in space sciences research and in 1993 was being approached to play a leading role in the space station effort. Indeed, in August 1991, when President George H.W. Bush and Soviet Premier Mikhail Gorbachev met in Moscow to talk about potential co-operation in human space exploration, it was suggested that a physician-cosmonaut might fly SLS-2, in exchange for a NASA astronaut making a long-duration visit to Mir. "The missions will increase knowledge about life sciences and data will be shared by both countries," explained *Flight International* on 14 August. "They are also seen by some observers as the first step towards joint flights to Mars." Although there were no Russians on SLS-2, cosmonauts were included in two other flights, firstly in 1994.

Seddon and the members of the SLS-2 science team took frequent blood draws from the rats' tails during the early stages of the mission and performed additional radioisotope and hormone or placebo injections to measure plasma volumes and track protein metabolism. This was part of a study into how red blood cell masses changed in weightlessness. Ultimately, the six unlucky rats destined to meet their maker in orbit were decapitated by Fettman and Seddon on 30 October, using a modified laboratory dispatcher. Pre-flight studies had already concluded that it was best to decapitate, rather than anaesthetise, the rats, because the chemicals would have degraded their neural tissues and impaired subsequent observations. "Things went pretty well," Fettman recounted, after the six-hour procedure ended. "We're happy to accomplish this. It was a big day for us" (Fig. 13.9).

Despite the science, the mood aboard the Shuttle was sombre. At one stage, only Seddon and Fettman were at work inside the Spacelab module. Blaha poked his head over Fettman's shoulder once or twice to check on their

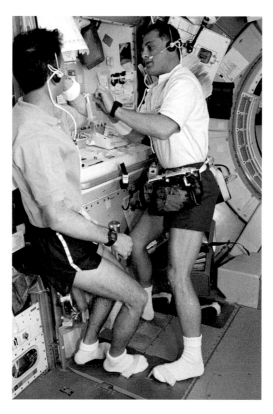

Fig. 13.9: Dave Wolf draws blood from Marty Fettman, the first professional veterinarian to fly in space.

progress and was quickly gone. After landing, the rat tissues were used as part of a series of neurovestibular and musculoskeletal investigations to explore changes in their gravity-sensing organs and the effect of microgravity upon their limb muscles and bones. "We were able to collect bone marrow and spleen and thymus," recalled Fettman, "which are the blood-forming organs, in order to study the progenitor cells for those blood cells in their organ of origin."

It was originally intended that the crew would extract a dozen or so organs from each rat and then dispose of the carcass, but NASA issued a research announcement for interested scientific parties to use the other body parts. "It became known as the Parts Program," Seddon remembered, and the astronauts found that it was no more difficult to remove eyeballs and lungs and insert them into little bags of fixative to preserve them. "Some had to be frozen," she said, "and some of them had to be refrigerated and some of them just needed to be put in the fixatives." Significantly, the inner-ear mechanisms had to be placed into fixative within two minutes of dissection, and with the inner ear buried deep within the skull, this required immense skill from Seddon and Fettman. Limb muscles, too, had to be attached to muscle clamps and fixed within ten minutes. "It was just the choreography that was incredible," Seddon added, "and Marty was just terrific at this stuff."

UNDERSTANDING THE HUMAN ORGANISM

In addition to the rat research, a series of joint U.S./Canadian experiments, originally carried aboard Spacelab-1 and SLS-1, were reflown to explore motion sickness and human vestibular changes. For SLS-2, the hardware included a rotating chair mounted in the centre aisle, which examined changes in the astronauts' reflexive eye motions. Seddon was the first to use the chair on 21 October, as part of studies of the vestibulo-ocular reflex in the eye, which enables us to see whilst we are in motion. Other experiments featured a rotating dome, placed over the astronauts' heads, whose interior face was coated with a pattern of dots which seemed to 'rotate' in an opposite direction. The subject used a joystick to indicate their perceived direction and velocity. Dave Wolf also donned a special skull-cap, the Acceleration Recording Unit, which was fitted with motion sensors and was used to record the time and severity of space sickness symptoms. Investigators hoped that if the science crew wore the cap throughout their working day it might enable them to correlate instances of sickness with periods of provocative head movement.

Studies of muscular atrophy included the ingestion of amino acids, labelled with non-radioactive isotopes of nitrogen, enabling the astronauts to track protein metabolism. Urine, saliva and blood samples were routinely acquired to determine the rates of protein synthesis and catabolism. Other experiments upon the rats looked at the performance of their hind limbs in microgravity, which showed an almost 40 percent reduction of muscle fibres at the end of the 14-day mission. The rats tended to rely more heavily on their forelimbs for bipedal locomotion and used their hind limbs only as grasping aids. After their return to Earth, they exhibited slow motions and an abnormally low body posture, all of which pointed clearly to a weakened muscular state, fatigue and co-ordination difficulties. Moreover, muscle protein 'turnover' in rats is much more rapid than in humans and two weeks of weightless exposure for them was roughly equivalent to two *months* for a human.

Other experiments focused on cardiovascular and regulatory systems. Data from SLS-1 highlighted increases in heart rate, size and output, which researchers attributed to the initial increase in central blood volume caused by fluid shifts within the body. Three SLS-2 studies assessed the functional capabilities of the system by monitoring the astronauts' cardiac outputs, heart rates, arterial and venous blood pressures, blood volume and the amount and distribution of blood and gases in the lungs. Cardiovascular 'deconditioning' had long been recognised as a problem after the return to Earth. Astronauts complained of light-headedness, an increased heart rate and decreased pulse pressure. Echocardiograph data, together with the catheters, an exercise bicycle and GAMS supported much of this research (Fig. 13.10).

As with SLS-1, STS-58 followed a single-shift system, although the whole crew typically put in 14-hour working days. In view of the long flight, and in line with EDO protocols, each astronaut received several periods of free time to relax. "The crew members are an important part of these investigations," stressed SLS-2 Mission Scientist Howard Schneider. "We want to assure ourselves that we continue to study the physiological effects of space flight and *not* the physiological effects of fatigue! If the crew is up there and is overly stressed, we don't get good science."

Early on 1 November, the final SLS-2 experiments were concluded and then Dave Wolf supervised the deactivation of the Spacelab module in time for the first landing opportunity at Edwards. Several hours later, Blaha landed Columbia at Edwards to conclude a mission of 14 days, the longest Shuttle flight at that time. For the astronauts, and particularly the science crew, it was the start of a week of post-mission medical experiments. NASA put them up in a resort called 'Silver Saddles' and despite the discomfort of frequent blood draws—"We began to look like drug addicts," joked Seddon, "because they

Fig. 13.10: Payload commander Rhea Seddon prepares to 'spin-up' Marty Fettman on the rotating chair.

kept drawing blood from us"—it was a pleasant time, being able to relax in the evenings and eat with their families.

THE TRAUMA OF NEUROLAB

Long before he piloted the final flight of the Spacelab module in April 1998, Scott Altman's claim to fame was that he provided a stand-in for Tom Cruise during the making of the movie *Top Gun*. In 1986, attached to Fighter Squadron 51 at Naval Air Station Miramar in California, flying the F-14A Tomcat, Altman performed several aerial manoeuvres for Cruise's character, Pete 'Maverick' Mitchell. "*Top Gun* was a real thrill," he remembered, years later. "The word was going around town that Hollywood was coming to Miramar and they were going to do a movie and we were all excited. My squadron had just gotten back from a 7.5-month cruise, so our airplanes were at home, we were available, we weren't too highly tasked. It turned out they picked my squadron to fly the F-14s. Then the skipper got together and tried to pick four guys that he thought were mature enough to handle the capability that they were being given in working with the movie. The flying was incredible. Most Navy pilots don't get to buzz the tower, like in the movie. If you did, you could just peel your wings off and throw 'em at the door, because you probably wouldn't be flying anymore! But since it was Hollywood, they wanted the scene."

As was normal practice for Spacelab flights, the payload crew—"four brilliant docs," according to STS-90 Commander Rick Searfoss—was named in advance of the flight crew, with LMS veteran Rick Linnehan announced as payload commander in August 1996, together with Canadian physician Dave Williams as a mission specialist. Four candidates for two payload specialist positions had also been identified the previous April: SLS-2 alternate crewman Jay Buckey, then serving as an associate professor of medicine at Dartmouth Medical School in Hanover, New Hampshire, Jim Pawelczyk, a physiologist at Pennsylvania State University, IML-2 veteran Chiaki Mukai and NASA veterinarian Alex Dunlap. Finally, in April 1997, a year before launch, Buckey and Pawelczyk were selected as prime payload specialists. Rounding out the seven-strong crew were SLS-2 veteran Rick Searfoss as the mission's commander, Scott Altman as pilot and Kay Hire as flight engineer. "Two Ricks on one Shuttle," Linnehan quipped. "It'll be fun."

Their mission, originally identified as SLS-4 but renamed 'Neurolab', was among the most complex ever attempted, focusing on the nervous system and investigating the effects of microgravity on blood pressure, balance, movement and sleep regulation. Original plans called for an SLS-3, SLS-4 and a notional (though unfunded) SLS-5. Of these, SLS-3 was requested as early as August 1982 and intended as a collaboration with France's CNES in February 1996, but met its demise due to budgetary cuts in January 1994. Its research focuses would have emphasised the human musculoskeletal system and performance. That left SLS-4, always intended for investigations in the neurosciences, for which an announcement of opportunity had been issued by NASA to the life sciences community in July 1993. At that time, its launch was projected for the last quarter of 1997 or the first quarter of 1998.

Nor was the mission exclusively focused on its human subjects, for the crew also transported over 2,000 animals into orbit and, controversially, euthanised and dissected several of them to perform complex surgical procedures. However, as a veterinarian, Linnehan insisted that all were treated respectfully. "I can guarantee the animals are well-fed, well-housed and well cared for," he said. "It's my duty to check every day to make sure everything looks good as far as their food, water and general health. I have absolute authority, on-orbit, if I need to, to stop an experiment if an animal becomes sick" (Fig. 13.11).

The $99 million Neurolab came about following President George H.W. Bush's declaration of the 1990s as the 'Decade of the Brain', in recognition of advancements made in those years into neurological science. The 26

Fig. 13.11: Rhea Seddon, payload commander of SLS-2, works with a camera at the rear end of the Spacelab module. She participated to a significant degree in developing cardiovascular experiments for Neurolab.

experiments inside the Spacelab module were broadly grouped into eight sets: four of which used Columbia's crew as test subjects, whilst the others focused on mice, snails, crickets and fish. The mission attracted participation from the United States' National Institutes of Health, NASA, ESA, CSA, DARA, CNES and NASDA. Key questions still required answers, in spite of 37 years of sending humans into space, including the mystery of how astronauts adapted so quickly to the weightless environment, even though all of our basic movements were learned in the presence of gravity. Moreover, physiologists were intrigued to learn how gravity-sensitive organs, such as the inner ear, cardiovascular system and muscles, coped in space and how astronauts' sleep patterns were affected as the Sun 'rose' and 'set' 16 times in each 24 hour period. It was as part of these studies that STS-90's animal passengers entered the equation. How would their inner-ear mechanisms—so vital for balance and motion—change in microgravity...and what of animals actually *born* in space? Would their gravity-sensitive organs develop differently to their Earth-born counterparts? The ultimate question was: *Must* gravity necessarily be present at the point in an animal's life when basic locomotion skills, such as walking or climbing, were being learned? The STS-90 crew would also participate in this research, through studies before, during and after the flight.

Preparing Columbia for launch was as complicated as the experiments themselves. After having been transferred to the pad in March 1998, tracking an opening launch attempt on 16 April, many of her mammalian and aquatic passengers had to wait until the final hours of the countdown before they could be loaded into their cages inside the Spacelab module. This proved difficult since the Shuttle was oriented vertically. Working from the middeck, two technicians were lowered, one at a time, in sling-like seats, down the tunnel into Neurolab. One technician sat in the tunnel, whilst the other entered the module to await the cages and aquariums, which arrived in separate slings. The delicate procedure was completed without incident, although telemetry from a couple of fish subjects—whose heads had been 'wired' with tiny transmitters—proved spotty. When one fish 'hid' in the corner of its tank, for example, the antenna used to gather telemetry data occasionally refused to function properly. After a day's delay to replace a faulty network signal processor, Columbia roared uphill at 2:19 p.m. EDT on 17 April.

The astronauts and their 2,000 mammalian and aquatic crewmates were almost joined by another in the final hours before launch, when a wayward bat unwittingly latched onto the ET. "We did take his body temperature with an infrared camera," said Launch Director Dave King. "He was 68 degrees [Fahrenheit, or 20 degrees Celsius] and the tank surface was 62 degrees [Fahrenheit, or 16 degrees Celsius], so we've decided he was just trying to cool off. Some have said he may have heard the crickets in Neurolab, but it was his choice whether to hang around when we started the engines or not!" Hopefully, for the bat's sake, he took flight long before Columbia's three main engines thundered to life, for his sensitive ears would not have survived the acoustic shock. It was the first time that crickets had flown into space. However, there no possibility of any chirping from Neurolab, for they were not yet old enough have the wings needed to produce sound. As well as humans, rodents and crickets, STS-90 also carried several hundred snails and fish, whose gravity-sensing organs are either similar to our own or very simple and easy to analyse. It was hoped that the results of such experiments could lead to advances in the development of electrodes as connections to the nervous systems of people with deafness caused by hair-cell damage. As with previous life sciences flights, Neurolab functioned on a single-shift system, with all seven astronauts waking up at the same time, to preserve their circadian rhythms and support the labour-intensive experiments.

After a busy first day activating Neurolab, the crew was awakened on 18 April by CAPCOM Chris Hadfield, with the invitation to "get those neurons into action". Overall, for the first part of the mission, Columbia performed without incident, but on 24 April a valve malfunctioned in the Regenerative Carbon Dioxide Removal System (RCRS), which triggered an alarm and raised the very real possibility of a premature return to Earth. "It was a noise I did not recognise," admitted Hire, upon hearing a strange sound at about the same time that the valve failed. Already running two hours behind schedule, the crew initially suspected one of two electronic controllers, but switching to a backup achieved little success. As a precautionary measure, Mission Control asked them to load two lithium hydroxide canisters, thereby providing ample time to evaluate and recover from the problem. Searfoss and Altman opened the RCRS, removed a hose clamp and used aluminium tape to bypass a faulty check valve, which was apparently allowing cabin pressure to leak into the system and throw off its electronic controllers. The fix worked and the risk of a shortened mission was thankfully averted.

Elsewhere, the Neurolab research was progressing well, including the dissection of several mice in the GPWS by Williams and Buckey on 18 April. They euthanised four adult mice to recover tissue samples from their brains and inner ears, which were needed by members of Neurolab's neutral plasticity team to investigate how nerve cells 'rewire' themselves under the stress of a new and unfamiliar environment. It was hoped that such work could lead to new insights into neurological disorder, such as stroke, Parkinson's disease and balance problems. Pawelczyk added that the surgical work had to be performed quickly to avoid the onset of degradation in certain nerve fibres. Although it was not the first time that animal dissections had been performed aboard the Shuttle—the SLS-2 crew had done similar experiments in October 1993—many scientists considered the previous work to have been messy and time-consuming. Neurolab demonstrated the ability of trained medical specialists to conduct intricate surgery in a weightless environment. Later, embryos were taken from euthanised pregnant mice as part of studies into the role of gravity in their early development. The results of such studies were expected to prove crucial in determining whether or not humans or animals could someday be born in space, thus allowing colonies to be established and long-duration voyages to the stars, which would require several generations of space travellers. Before dissecting them, the astronauts injected the pregnant mice with cell 'markers' to label the brain cells in their embryos, track the

Fig. 13.12: Neurolab, the final flight of the pressurised Spacelab module, is readied for launch.

development and migration of the cells and compare the results with data gathered from Earth-born mice (Fig. 13.12).

Typically, the pea-sized foetuses ranged between ten and 14 days old and, after removal from their mothers, were preserved whole until the end of the mission. Other mice, with 'hyperdrive' units fitted to their heads, connected to the hippocampus area of their brains, were placed into the GPWS on two maze-like 'racetracks' as part of the Escher Staircase Behaviour Testing experiment. The hippocampus helps to develop spatial maps to help the mice to navigate from place to place and the investigation sought to explain the disorientation frequently experienced by Alzheimer's sufferers. The hyperdrive units did not create unpleasant side effects for the mice, because their brains had no pain receptors, and they had been taught to use the racetracks before launch. They received rewards of sweetened condensed milk after completing a certain number of correct turns. In space, however, the rules changed and investigators monitored their neural activity as the mice struggled to recognise 'home-base' after making fewer turns than they had been taught. Several of the mice exercised their 'space-legs' for the first time on 22 April, using their forelimbs to scoot around a small jungle gym, but hardly using their hindlimbs. As they moved, the astronauts videotaped the mice and marked their joints with ink to allow each motion to be meticulously analysed after landing.

"It's amazing to watch these animals behave in orbit," said Linnehan on 25 April. "They act just like we do. They learn very fast how to get around their cages and get to food and water. Just last night, we were checking on some of the younger rats, watching them eat. It was kind of akin to the way we eat. We float around and hold our food to feed ourselves. One of the rats was holding onto a piece of food with his front paws, munching on it leisurely, letting it go and going over to drink some water, coming back and grabbing the food again." Later in the mission, the first survivable surgical procedure was conducted on six of the mice. Buckey and Williams anesthetised and injected them with dye markers in a muscle development study. Although the procedure was relatively straightforward, it paved the way for more complex work aboard the ISS.

However, problems were brewing. Electrical problems were noted with transmitters implanted in four oyster toadfish and on 27 April NASA revealed that dozens of the neonatal mice had died unexpectedly. Others were in such bad shape that they had been euthanised by the astronauts, whilst a few were resuscitated and nursed back to health with bottle-feeding of nutritional and other supplements; at one stage, Linnehan thanked his crewmates for helping to hand-feed and care for the ailing mice and checking on them each evening before bedtime. At first, Linnehan suspected that the deaths may have been caused by the design of the rodents' cages; in the free-floating environment, perhaps they found it harder to move around and get to their mothers, eventually succumbing to dehydration and depression. Back on Earth, the immense death toll—more than 50 percent of the neonatals—drew condemnation from animal rights groups, including People for the Ethical Treatment of Animals (PETA), which charged NASA with abuse and pressed Congress to ban experiments aboard the Shuttle.

The disaster could not have occurred at a worse time for the neonatals, who were particularly vulnerable and just learning to move and search for solid food in their new environment. The poor state of some of them obliged Linnehan to cancel several experiments, telling Mission Control pointedly that "there is no meaningful data to be gained with these animals at this point". In total, almost two-thirds of a total complement of 96 mice were lost.

"It was an unforseeable event and it's regrettable that it happened," said Linnehan. "However, we still got back most of our primary science." It later became apparent that the ill-fated neonatals were victims of a higher-than-normal rate of maternal neglect, from mothers who had themselves become dehydrated and were unable to lactate adequately. As the youngest mice became anaemic, the mothers stopped feeding and grooming them. Linnehan drew praise from animal rights campaigners on 1 May, after defying an instruction to destroy a rodent when ultra-fine electrodes implanted in its brain broke loose. "Based on his expertise and professional opinion," said Neurolab Mission Scientist Jerry Homick, "he determined the animal was not in any danger and determined it was appropriate to return that animal to its housing."

THE PACE OF 'BLURROLAB'

By now, the intense pace of the flight and the additional free time given up to care for the sick rodents meant that most of the crew averaged three or four hours of sleep each night. Linnehan started calling it 'Blurrolab', rather than Neurolab, "because I just could never remember what day it was or where we were, we just kept on going". Years later, he described Neurolab as one of the hardest parts of his professional life, not least because, as payload commander, the primary responsibility for the entire science mission fell on his shoulders. "I don't think I'm ever going to work as hard in terms of the time I put in and the midnight oil burned," he said. However, he had nothing but praise for Williams, Buckey and Pawelczyk—"Smart, smart people," he mused, "*Much smarter than me*"—but highlighted that the crew frequently worked 16-hour days, every day.

Nevertheless, despite the neonatal fatalities, the mission was shaping up to be a huge success. "We went into this mission facing a number of challenges," said Homick. "We knew we had a difficult timeline to work with, we knew that we had a number of complex hardware systems…to acquire the data and we knew we were going to be implementing a number of very difficult experimental procedures, using cutting-edge technologies. With all of that in mind, we did expect to achieve a great deal of success with this mission and I'm pleased to report that I think we've exceeded our expectations" (Fig. 13.13).

Although the astronauts averaged just a handful of hours of sleep each night during the rodent crisis, Dave Williams was able to rest fitfully, which aided another of Neurolab's investigations: a series of measurements of brain waves, eye movements, respiration, heart rate, internal body temperature and

Fig. 13.13: Dave Williams is instrumented with elaborate head gear for one of Neurolab's many research investigations.

snoring. As well as improving the performance of astronauts in space, such experiments were expected to support the sleep-wake cycles of workers on Earth. Not only did the hectic workload of the astronauts carry the potential to impair their performance, but so too did the experience of sunrises and sunsets on no less than 16 occasions in each 24-hour period. It was recognised that around 20 percent of astronauts on single-shift missions needed to take sleeping pills, which represented between three and eight times as high a number as the general population on Earth. Linnehan, Williams, Buckey and Pawelczyk also donned body suits and sensor-laden skullcaps before going to sleep in order to monitor their rest patterns. Although Williams admitted that his sleep was reduced, he added that was still able to 'turn over' in his sleep, despite having no pressure points on his body, a purely psychological function.

Two of the medical experiments were provided by Canada. The Visuo-Motor Co-ordination Facility (VCF) studied changes in movement during weightlessness which affected the astronauts' pointing and grasping capabilities, whilst the Role of Visual Cues in Spatial Orientation created 'fake' gravity by applying pressure to the soles of the feed to find out it somehow 'overrode' visual cues and enabled them to readapt to a terrestrial-type environment. Developed by CSA, the VCF projected visual targets onto a screen, to which the crewman grasped and pointed at them and tracked them as they moved, using an instrumented glove. The motor skills thus demonstrated were recorded at various stages of the 16-day mission to evaluate changes in the nervous system as it adapted to microgravity conditions. A rotating chair, the Visual and Vestibular Integration System (VVIS), mounted along the centre aisle of the Spacelab module, was designed to stimulate the astronauts' vestibular systems with spinning and tilting sensations. Performed by all four of the science crew, the chair—which rotated at 45 revolutions per minute—was used six times during the mission, with the astronauts' eyes shielded from external stimuli, giving their nervous systems no visual cues, whilst a video camera, trained on their faces captured their reactions to the sudden motions.

As well as proving somewhat dizzying in nature, a few investigations generated a sting for the crew. One experiment demonstrated an innovative technique known as 'microneurography', whereby a fine needle—about the same diameter as an acupuncture needle—was inserted into a nerve, just below the knee. This allowed nerve signals from the brain to the blood vessels to be measured directly, whilst the cardiovascular system was monitored with the Lower Body Negative Pressure apparatus. Results were expected to aid sufferers of autonomic blood pressure disorders, including 'orthostatic intolerance', an inability to maintain proper blood pressure whilst standing for long periods. To prepare for the microneurography procedure, Linnehan, Williams, Buckey and Pawelczyk spent two months at Vanderbilt University in Nashville, Tennessee, training with the hardware. (In fact, in September 1996, Rhea Seddon was assigned to Vanderbilt's Center for Space Physiology and Medicine as a NASA liaison to evaluate flight equipment and operating procedures.) The insertion of the needle, which all four astronauts typically achieved within about 40 minutes during several runs in orbit, was delicate and difficult. "Finding a nerve is a lot more difficult than finding a vein," said Principal Investigator David Robertson. "This is a very difficult thing to do on Earth and the idea that it can be done in space is a little bit astounding to many people."

The orbiter crew of Searfoss, Altman and Hire participated in several experiments. Hire worked with the Bioreactor Demonstration System (BDS) on the middeck, growing cultures of renal tissue and bone marrow, for possible applications in the treatment of kidney disease, AIDS and other immune-system ailments, as well as for the chemotherapy of cancer patients. The orbiter crew also tended to the very few malfunctions with the Shuttle itself, including a blockage in a wastewater dump line. As STS-90 entered its final stages, it was decided not to extend the mission to 17 days, after reports from the Neurolab team that they had accrued sufficient data and the additional time would be unnecessary. This was a pity, for the conservation efforts of the seven astronauts had added sufficient consumables for a full 36 hours of extra research.

Another limiting factor was the weather at KSC in anticipation of landing opportunities on 3 and 4 May. Although it was expected to be reasonable, forecasts on the 4th were not as good as for the 3rd. Options to land Columbia at Edwards were undesirable, due to the need to remove the animals and perform data collection at the KSC facilities. As it was, Searfoss and Altman guided their ship to a smooth touchdown in Florida at 12:09 p.m. EDT, wrapping up almost 16 full days in orbit.

Even before STS-90 launched, however, there existed a possibility that Neurolab might be reflown, with the same crew and aboard the same orbiter, in August 1998. At a press conference on 15 April, Space Shuttle Program Manager Tommy Holloway noted that engineers were exploring the feasibility of flying the payload again in the late summer months. Further delays to the ISS construction effort had opened up a three-month 'gap' in the manifest, but with the first station flight scheduled for launch on 4 September it was eventually decided to protect that option and the Neurolab reflight was shelved. (Ironically, ISS construction met with further delay and did not ultimately commence until November.) When STS-90's inaugural results were published in April 1999, they offered promising insights into the neurological mechanisms responsible for Alzheimer's disease and epilepsy and the astronauts' cognitive work was expected to lead to diagnoses and rehabilitative measures for brain-injury sufferers. It was also anticipated that the sleep studies would find terrestrial applications for shift workers, the elderly, jetlag sufferers and insomniacs.

Summing up the mission, the Neurolab crew were philosophical about their accomplishment. "If we don't figure out how to stem nerve-muscle degeneration, we're not going to be able to travel to other planets or live in space stations," said Linnehan. "We would face the risk of fractures when returning to Earth." On the other hand, explained Jay Buckey with a wry grin, at least living and working in space allowed him to put his trousers on both legs at the same time…

GAPS IN CAPABILITY

"By 1995, the Spacelab was being phased out," wrote Rhea Seddon in *Go for Orbit*. "There was little science planned for the space station until all the components had been built, launched and constructed in space." Upon her return from SLS-2 in November 1993, Seddon did not anticipate that happening until the early years of the 21st century at the soonest, opening an unconscionably lengthy gap in scientific capability and few opportunities for researchers to fly their experiments. As late as November 2000, in a National Academy of Sciences space studies board paper that emphasised biological and medical research, it was recommended that "NASA should make every effort to mount at least one…life sciences flight in the period between Neurolab and the completion of ISS facilities". Several missions were planned with the commercial Spacehab pressurised laboratory—one in October 1998 and a second (which culminated in Columbia's tragic destruction during re-entry) early in 2003—but for the versatile Spacelab the end was nigh. Budgetary cuts had already precipitated the wholesale cancellation of follow-on SLS, IML, ATLAS and USML missions, the eye-watering costs of Germany's reunification after the collapse of the Berlin Wall put paid to Spacelabs D3 and D4 and a raft of others, including Starlab, Spacelab-E1 and E2 and the unhappy Sunlab had met their own untimely fates. It can never be known what successes or challenges these missions, if flown, might have produced (Fig. 13.14).

Today, Long Module No. 1 (LM1), which began its career with the pioneering Spacelab-1 mission on 28 November 1983, resides in the Udvar-Hazy Center at the Smithsonian's National Air and Space Museum, near Washington, D.C. Following that maiden voyage, the pressurised LM1 flew eight more times: supporting Spacelab-3, SLS-1, Spacelab-D2, IML-2 and the two USML and MSL missions, before being retired to Washington in 1998. The instrument-laden igloo, which flew several pallet-only missions, is today

Fig. 13.14: The Spacelab-Mir pressurised module sits in Atlantis' payload bay during final closeout activities. Note the Orbiter Docking System (ODS) in the forward payload bay.

displayed in Udvar-Hazy's James S. McDonnell Space Hangar. As for the pallets themselves, one (nicknamed 'Elvis', notably used for the TSS flights) sits at the Swiss Museum of Transport in Lucerne, whilst another, which carried a Canadian-built robotic arm and 'hand' to the International Space Station (ISS) in March 2008, resides in the Canada Aviation and Space Museum in Ottawa. And Long Module No. 2 (LM2), fabricated under the Follow-On Production (FOP) contract between NASA and ESA, first flew on West Germany's Spacelab-D1 mission in late 1985. Fittingly, it now sits on display in Germany, until 2010 in the *Bremenhalle* exhibition at Bremen Airport and more recently at the nearby Airbus Defence and Space facility. After Spacelab-D1, this second pressurised module in the fleet logged six more flights, reaching space again on IML-1, Spacelab-J, SLS-2, LMS, Neurolab and—curiously, in June 1995—a mission which literally brought it cheek-to-jowl with a real space station for the first and only time in its history.

Having surveyed some of Spacelab's most significant achievements during its quarter-century of operations, it is difficult to imagine why a facility of such resounding promise should have been abandoned to retirement, at the peak of its success and apparently with such premature haste. But in many minds, not least that of Jeff Hoffman, who might have flown three times with the ASTRO payload, had Challenger not been lost, the facility was never launched often enough to fully exercise its versatility. In summing up her first Spacelab mission on SLS-1, Rhea Seddon opined that the gaps between flights were too long to be scientifically profitable. "SLS-1 was a good beginning," she acquiesced, "but we couldn't do great science when it took over seven years to prepare for a flight and after nine days in space, studying a mere four subjects, the lab had to be dismantled and brought back home." Spacelab's overall flight rate, added John Logsdon, was never high enough to justify its enormous cost. Hopes of eight missions per year came crashing to nought: even at its peak in 1992, the pressurised modules flew barely three times, in other years substantially less. "Very, very inefficient use of hardware, which is certainly not to NASA's credit," added astronaut Joe Allen. "European astronauts had trained to fly…and they made very infrequent flights aboard it. That also was not an efficient use of very skilled persons. Then, NASA set off on this International Space Station and, of course, by then, Spacelab was completely archaic and not used again."

At the same time, operating inside the Shuttle's payload bay had furnished Europe with invaluable experience which led directly to its proposed Columbus pressurised laboratory for the space station. "They were actually launching the Spacelab as a payload in the Shuttle; they had become accustomed to the Shuttle requirements, which are relevant because the Columbus module was going to launch in the Shuttle," remembered Rick Nygren, one-time head of NASA's vehicle integration test team. "That's where the linkage comes. [With] Spacelab, they learned how to work as a payload…and they learned the Shuttle processes in systems requirements. When they went to their contractor to build the Columbus module, they had all of this experience and they could pass it along through the contractual mechanisms to say this is how you're going to have to build it, how you're going to verify what you have to build it to."

A MEANINGFUL ENDEAVOUR

In terms of international co-operation between two former foes, it would be no understatement to assert that STS-71—the United States' hundredth human-carrying space mission—was the singular most important Shuttle flight of the 1990s. For the first time, the reusable orbiter accomplished what it had been designed to do: it docked with an Earth-circling space station and exchanged crew members of different nationalities. It supported shared research and its technical and human success facilitated the incremental steps which followed and guided the United States and Russia onto a new path of space collaboration which would endure for decades. The first docking between Shuttle Atlantis and the Mir space station in June 1995 was a direct stepping stone to building today's ISS and its ramifications can still be heard in the successful partnership which (though severely strained following Russia's invasion of Ukraine) continues to endure into the third decade of the 21st century. In the words of NASA Administrator Dan Goldin, the mission of STS-71 heralded "a new era of friendship and co-operation between our two countries". And it was entirely fitting that Spacelab (in the form of LM2) would play a direct role in its success.

Yet the notion of a joint U.S./Russian Shuttle/station mission was far older. As early as May 1975, NASA made its first proposal to the Soviet Union to fly a cosmonaut aboard one of the reusable orbiters, which were then scheduled to begin flying later in the decade. A rendezvous and docking with a Salyut station and participation by a cosmonaut on a Spacelab flight were discussed, as noted by *Time* magazine in August 1975. A little over a year later, a series of U.S./Soviet talks at NASA Headquarters established a "meeting of minds" on future co-operation, with two principal foci: a scientific venture involving Shuttle/Salyut or the development of "a space platform…bilaterally or multi-laterally". In May 1977, NASA Acting Administrator Alan Lovelace and Anatoli Alexandrov of the Soviet Academy of Sciences explored the Shuttle/

Salyut option in greater depth and produced a document, ponderously titled *Objectives, Feasibility and Means of Accomplishing Joint Experimental Flights of a Long-Duration Station of the Salyut Type and a Reusable Shuttle Spacecraft*. Ongoing meetings over the next year or two raised a glimmer of hope that it might happen and in April 1978 *Flight International* suggested that a rendezvous might take place as soon as 1981, followed by a docking at some point thereafter. Unfortunately, these plans came to nought when the deteriorating geopolitical situation in the late 1970s and early 1980s (particularly the Soviet Union's invasion of Afghanistan in December 1979) led to a rapid cooling of relations into the succeeding decade.

Not until the early 1990s were serious inroads into collaboration next made. The first formal planning for a Shuttle/Mir mission began in July 1991, when U.S. President George H.W. Bush and Soviet General Secretary Mikhail Gorbachev met for two days in Moscow and agreed to fly a cosmonaut aboard the orbiter and an astronaut aboard the space station for three months. A year later, planning expanded to include a single Shuttle/Mir docking mission, but by the time U.S. Vice President Al Gore and Russian Prime Minister Viktor Chernomyrdin met in December 1993, flesh was added to the bones of up to nine flights in 1995-1997. These flights would deliver equipment and supplies and exchange crew members, including a succession of long-duration NASA astronauts. In April 1994, Harry Holloway, NASA's associate administrator for the Office of Life and Microgravity Sciences and Applications, outlined plans for the Spacelab long module to fly on five Shuttle/Mir missions "to conduct priority research in fluid physics and combustion" (Fig. 13.15).

In June 1994, the core of the STS-71 crew was announced. In command was Robert 'Hoot' Gibson, chief of the astronaut corps and veteran of Spacelab-J, joined by Spacelab-D2's Charlie Precourt as pilot. Payload commander was USML-1's Ellen Baker, together with engineer Greg Harbaugh and seasoned Spacelab veteran Bonnie Dunbar, the first person to fly three times with the pressurised module. Dunbar's assignment to STS-71 was interesting in itself, for she had earlier been selected to back up fellow astronaut Norm Thagard in his training for a three-month Mir stay, beginning in March 1995. Plans called for STS-71 to transport a pair of cosmonauts up to the space station and return home with Thagard and two other cosmonauts, a unique situation in which Atlantis would launch with a crew of seven and land with eight. As circumstances transpired, Dunbar was unavailable for STS-71 training until after Thagard's launch, leaving her an unenviable three months to prepare. "It was my third Spacelab flight and so I understood the systems," she told the NASA oral historian. "The experiments were either the same as what we had on Mir—because we were putting Norm and the two

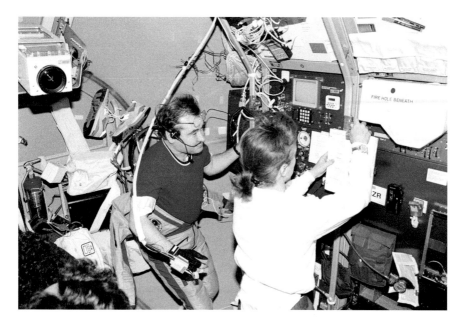

Fig. 13.15: Gennadi Strekalov and Bonnie Dunbar are pictured at work during Spacelab-Mir's many medical activities.

cosmonauts into the Spacelab and doing medical tests on them—or they were experiments I had done before. It was actually very easy to train for it."

So it was that on the morning of 27 June 1995, seven astronauts and cosmonauts departed the KSC crew quarters, bound for the launch pad. Leading the throng were Gibson and Russia's Anatoli Solovyov, the latter teamed with fellow cosmonaut Nikolai Budarin and targeting a multi-month stay aboard Mir. Atlantis roared aloft at 2:32 p.m. EDT, delivering her seven crewmembers and a unique payload, dubbed 'Spacelab-Mir', into the Shuttle's lowest orbit, with an apogee of 180 miles (290 kilometres) and a perigee of 98 miles (157 kilometres). Over the next two days, Gibson, Precourt and Harbaugh manoeuvred their ship across the 8,000-mile (12,960-kilometre) distance to reach Mir, closing on their quarry at an initial rate of a thousand miles (1,630 kilometres) with each orbit. Meanwhile, Baker and Dunbar set to work activating Spacelab-Mir, whose pressurised tunnel linking the middeck to the module also included a large Orbiter Docking System (ODS) to connect the spacecraft with the space station. Early on the 29th, Gibson and Precourt brought the Shuttle in to a smooth docking at Mir and for the first and only time a Spacelab pressurised module came face to face with a space station.

"Houston, Atlantis," Gibson triumphantly radioed, "we have capture."

"Copy, capture," replied CAPCOM Dave Wolf. "Congratulations Space Shuttle Atlantis and Space Station Mir. Our new era has begun."

After docking, a lengthy period of almost two hours ensued as pressurisation and leak checks were conducted. In his oral history, Precourt remembered floating into the ODS and spotting the Mir crew—Norm Thagard and his Russian crewmates Vladimir Dezhurov and Gennadi Strekalov—through the porthole. "Their hatch is already open and we're doing pressure checks with our hatch," he recalled. "And our hatch opens last, to physically give us access, so you can look through this little porthole and wave to the guys on the other side and you can see that they're really antsy for us to open our hatch." Finally, hatches were opened and in a highly symbolic gesture Gibson and Dezhurov shook hands and exchanged greetings at the interface between Atlantis and Mir.

Precourt's first view of Thagard was a comical one. "Norm," the newcomers shouted, "you guys are *upside down*!"

"Naw," retorted Thagard. "*You* guys are upside down!"

For the next five days, symbolism and science were close bedfellows, as collaboration gave way to meaningful medical research inside Spacelab-Mir. Specimens from Thagard's three months aboard Mir were transferred into the module, including disks and cassettes, over a hundred urine and saliva samples, 30 blood-filled vials and 20 surface swabs were transferred to the Shuttle for return to Earth. And Dezhurov, Strekalov and Thagard were themselves subjected to medical tests inside Spacelab-Mir, under the watchful gaze of Baker. Six metabolic experiments focused on a range of physiological responses in the men's bodies, which expanded on studies already begun aboard Mir to examine human metabolism and endocrinology and determine how fluids redistributed themselves around the human body. The long-duration crewmen participated in efforts to understand whether prolonged exposure to the microgravity environment affected their ability to mount an antibody response and if their immune cells were altered in any way. Other experiments involved the use of both Russian and U.S. lower-body negative pressure devices to assess their usefulness as countermeasures to the kind of orthostatic intolerance frequently reported by spacefarers upon their return to terrestrial gravity. As part of this research, Dezhurov, Strekalov and Thagard would return to Earth in a reclining position, aboard custom-moulded recumbent seats in Atlantis' middeck, with changes in their heart rates, blood pressure, voices and posture being continuously monitored throughout re-entry and landing (Fig. 13.16).

As part of ongoing neurosensory investigations, the crew measured muscle tone, strength and endurance by electromyography and utilisation of oxygen

Fig. 13.16: Norm Thagard (lower left) and Bonnie Dunbar (right) jointly became the first Americans to fly three times with the pressurised module, whilst Ellen Baker (left), as the mission's payload commander, was making her second Spacelab mission.

during one to two hours of daily walking or running on a treadmill, together with other exercise sessions. "We had a bicycle ergometer there and a gas-exchange system," remembered Dunbar. "It's just like taking a treadmill test here on the ground. Then looking at the efficiency of your body. You do have degradation over time. How much does their cardiovascular system degrade? And you want to get those measurements before they get down on the ground and start to reacclimate." Years later, Dunbar recalled spending a sizeable portion of her time working with the three long-duration veterans inside Spacelab-Mir. "It was full-day operations, that's what I did most of the day," she said. "They weren't all three in there at a time, because there were still handover activities going on [aboard] Mir, so we had the old crew and Norm in there most of the time."

But the mission afforded an entirely different setup from her prior experience with Spacelab-D1 in late 1985 and USML-1 in the summer of 1992. Notwithstanding this past knowledge, Dunbar's three months of training for STS-71 was rushed. "I wasn't integrated into the crew that well," she said. "They'd been training together for, what, nine months, and here I kind of step in. It was professionally integrated, but there's a certain amount of bonding that occurs up front that is more informal and I don't think I ever got that way. I wasn't in the in-jokes, for example. You go from 13 months in Russia into…the last three months of a Shuttle training flow, it is pretty much running the whole time."

Microbial samples were taken from both Mir and the Shuttle, as well as specimens from the crew themselves, in order to determine if the closed environment of a spacecraft or space station affected microbial physiology and its interaction with humans in orbit. Muscle co-ordination and mental agility were also monitored. The Spacelab module also supported a variety of other experiments. It provided a means to return the pre-fertilised Japanese quail eggs to Earth and deliver new sensors to Mir for the station's on-board greenhouse. A series of several hundred protein crystal growth investigations, frozen in a thermos-bottle-like vacuum dewar, were delivered to Mir for the next several months.

Although she never flew to the ISS, Dunbar compared the sprawling orbital outpost favorably to the Spacelab that was her home from home for three missions and more than four weeks between the mid-1980s and mid-1990s. "The International Space Station is more similar to Spacelab," she remarked to a NASA oral historian. "Those are not overly big volumes. In fact, they're still significantly smaller than the Skylab work volumes. Those are very comfortable volumes to work."

Comfortable, indeed, as Russia's Gennadi Strekalov discovered when he floated inside Spacelab-Mir for the first time. At the time, one of the module's research racks had been removed, leaving a small open volume against one of the module's walls. The crew positioned some foot-loops inside it. Strekalov went inside and pulled on a headset to listen to some music. "Tell me when you're ready to test me," he told Dunbar. "I really like it in here. It's light, it's roomy. This is very comfortable. In fact, I will sleep until you're ready for me." Whenever Dunbar or Baker needed Strekalov for anything, they tapped him lightly on the shoulder. "I think that the room is an advantage," Dunbar said of Spacelab. "It's not a disadvantage. Of course, the hatches need to be as big as they are to move our racks back and forth and that's what's going to make our station a real research laboratory…the ability to move things back and forth."

Moving from Spacelab-Mir and Atlantis' middeck often proved difficult, for as previous astronauts had described, it was virtually impossible for two crewmembers to pass each other in the relative narrowness of the tunnel. On one occasion, Dunbar came face-to-face with Gibson as one departed the module and the other headed towards it.

"Oh, oh, collision," Dunbar joked. "Signal left or right?"

"Ships passing in the night," retorted Gibson. "Guess we should pass on the right, shouldn't we?"

"Yeah."

Early on 30 June, with the Russian tricolour and Stars and Stripes as a backdrop in the Spacelab module, the crews exchanged gifts, including the ceremonial joining of a halved pewter medallion, which bore a relief image of the docked Shuttle-Mir combination. A 1/200-scale model of the two spacecraft was also joined, with the intention that both gifts would be presented to U.S. and Russian heads of state after the mission. Furthermore, a proclamation was signed by all ten astronauts and cosmonauts, certifying the date and time of docking, which declared that "The success of this endeavour demonstrates the desire of these two nations to work co-operatively to achieve the goal of providing tangible scientific and technical rewards that will have far-reaching effects to all people of the planet Earth" (Fig. 13.17).

Fig. 13.17: The Spacelab-Mir pressurised module is pictured inside Atlantis payload bay as the Shuttle undocks from Mir on 4 July 1995. This view was captured by cosmonauts Anatoli Solovyov and Nikolai Budarin, station-keeping in the Soyuz TM-21 spacecraft.

It was already intended that Solovyov and Budarin would board and undock the Soyuz TM-21 spacecraft from Mir about 15 minutes prior to Atlantis' own separation on 4 July, in order to capture still and video imagery. Shortly before the crews parted company, the STS-71 astronauts gifted Solovyov and Budarin flight pins, watches, fresh fruit and big, soft Mexican tortillas. After training with Dunbar as Thagard's backup for more than a year, Solovyov playfully pulled her into his arms, just prior to hatch closure. Which side of the hatch would she like to stay on, he jokingly asked. Dunbar smiled. As much as she wanted to fly a long-duration mission on Mir, her place on this flight was very much on the Shuttle.

Early on Independence Day, 4 July 1995, Soyuz MS-21 with Solovyov and Budarin undocked from Mir, to be followed by Atlantis herself a mere 14 minutes later. As the two cosmonauts fired off photograph after photograph of the event, they cannot have failed to be astounded by its historical significance. But for the purposes of this book, there was another significant point of import about the STS-71 undocking. It marked the only time in Spacelab's history that a pressurised module in the Shuttle's payload bay was photographed in its entirety during the course of a space mission. Returning safely to Earth on 7 July, Thagard became the United States' most experienced astronaut, having flown for 115 days, in addition to his multiple prior Shuttle missions. With Dunbar, he also jointly became one of the first two astronauts to have flown as many as three times with the pressurised module. After his two earlier Spacelab flights, it had taken him about a week to restore his pre-flight strength and regain his 'Earth-legs'; after more than three months aboard Mir, this basic physical readaptation took nearly a week. Not long after landing, Thagard went jogging in Houston with Precourt and Baker. "It was the hardest three miles I ever did," the physician-astronaut said later. "But I did it!"

Spacelab-Mir, though perhaps one of the pressurised module's most visible missions, was not its last; five more research missions would occur over the next few years, with LM1 returning to Earth for the final time on 17 July 1997 and LM2 doing likewise after the Neurolab flight on 3 May 1998. The pallets survived somewhat longer, though not for 'scientific' Spacelab missions, but rather to transport hardware into orbit for the servicing of Hubble or construction and maintenance of the space station. These hardware pieces included segments of the station's truss backbone in October 2000, the Quest airlock in July 2001 and the Canadarm2 and Dextre robotic manipulators in March 2008. The final Hubble servicing and the swansong outing of a Spacelab pallet came in May 2009.

Four decades since its maiden voyage, and a half-century since its conception, Spacelab's hardware now exists as a suite of museum-pieces at various locations around the world, relics of a bygone era. Yet in many ways, although its 1970s-era potential—like that of the Shuttle itself—went unrealised, and many promising missions were lost, Spacelab cemented its credentials as an exercise in international co-operation and crucially enabled humans to learn how to conduct real and meaningful science in a weightless environment, with around 800 discrete experiments conducted. "Spacelab served as a precursor to the space station, because it taught scientists how to design, integrate and operate experiments in an orbiting laboratory," said Rick Rodriguez, a former Spacelab simulation engineer at MSFC. "It provided a plug-in structure for experiments similar to what is available in ground laboratories, that provide power, communication, cooling, vacuum and other resources needed. Scientists learned how to design experiments that could be integrated into Spacelab. They also learned how to design experiments that could withstand being launched into space, operated in microgravity and returned to Earth."

Several Spacelab veterans progressed from this miniature space station to live for long periods aboard fully fledged space stations, notably Ulf Merbold, Norm Thagard, Shannon Lucid, John Blaha, Jerry Linenger and Dave Wolf on Russia's Mir and Susan Helms, Dan Bursch, Carl Walz, Ken Bowersox, Leroy Chiao, Bill McArthur, Mike Lopez-Alegria, Bob Thirsk and Cady Coleman on the ISS. And ATLAS veteran Mike Foale logged long-duration stays on both Mir and the ISS. In the words of Coleman, who flew USML-2, almost flew MSL-1 and went on to log a six-month ISS increment between December 2010 and May 2011, the Spacelab experience represented nothing less than "a practice session" for the space stations which followed. And the ISS, which has for a quarter-century orbited high above our heads and contributed immeasurably not only to science, technology and exploration, but also to world peace and harmony, owes its enormous success in no small part to Spacelab.

Bibliography

Books and Memoirs

'Aeronautical and Astronautical Events of 1962: Report of the National Aeronautics and Space Administration to the Committee on Science and Astronautics, U.S. House of Representatives, Eighty-Eighth Congress, First Session.' Washington, D.C.: U.S. Government Printing Office, June 1963

'Aeronautics and Astronautics, 1963: Chronology on Science, Technology and Policy' (1964). Scientific and Technical Information Division, NASA Headquarters, Washington, D.C.

'Aeronautics and Astronautics, 1964: Chronology on Science, Technology and Policy' (1965). Scientific and Technical Information Division, NASA Headquarters, Washington, D.C.

'Aeronautics and Astronautics, 1965: Chronology on Science, Technology and Policy' (1966). Scientific and Technical Information Division, NASA Headquarters, Washington, D.C.

'Aeronautics and Astronautics, 1966: Chronology on Science, Technology and Policy' (1967). Scientific and Technical Information Division, NASA Headquarters, Washington, D.C.

'Aeronautics and Astronautics, 1967: Chronology on Science, Technology and Policy' (1968). Scientific and Technical Information Division, NASA Headquarters, Washington, D.C.

'Aeronautics and Astronautics, 1968: Chronology on Science, Technology and Policy' (1969). Scientific and Technical Information Division, NASA Headquarters, Washington, D.C.

'Aeronautics and Astronautics, 1969: Chronology on Science, Technology and Policy' (1970). Scientific and Technical Information Division, NASA Headquarters, Washington, D.C.

'Aeronautics and Astronautics, 1970: Chronology on Science, Technology and Policy' (1972). Scientific and Technical Information Division, NASA Headquarters, Washington, D.C.

'Aeronautics and Astronautics, 1971: Chronology on Science, Technology and Policy' (1972). Scientific and Technical Information Division, NASA Headquarters, Washington, D.C.

'Aeronautics and Astronautics, 1972: Chronology on Science, Technology and Policy' (1974). Scientific and Technical Information Division, NASA Headquarters, Washington, D.C.

'Aeronautics and Astronautics, 1973: Chronology on Science, Technology and Policy' (1975). Scientific and Technical Information Division, NASA Headquarters, Washington, D.C.

'Aeronautics and Astronautics, 1978: Chronology on Science, Technology and Policy' (1986). Scientific and Technical Information Division, NASA Headquarters, Washington, D.C.

'Aeronautics and Astronautics, 1979-1984: A Chronology' (1990). Scientific and Technical Information Division, NASA Headquarters, Washington, D.C.

'Aeronautics and Astronautics, 1985: A Chronology' (1988). Scientific and Technical Information Division, NASA Headquarters, Washington, D.C.

Apt, Jay, Helfert, Michael and Wilkinson, Justin (1996) *Orbit*. National Geographic Society

Brun, Nancy L. (1977) *Aeronautics and Astronautics, 1974: A Chronology*. The NASA History Series: Scientific and Technical Information Branch, NASA Headquarters, Washington, D.C.

Brun, Nancy L. and Ritchie, Eleanor H. (1979) *Aeronautics and Astronautics, 1975: A Chronology*. The NASA History Series: Scientific and Technical Information Branch, NASA Headquarters, Washington, D.C.

Brun, Nancy L. (1984) *Aeronautics and Astronautics, 1976: A Chronology*. The NASA History Series: Scientific and Technical Information Branch, NASA Headquarters, Washington, D.C.

Bugos, Glenn E. (2010) *Atmosphere of Freedom: 70 Years at the NASA Ames Research Center*. NASA History Office, Washington, D.C.

Cassutt, Michael (2018) *The Astronaut Maker: How One Mysterious Engineer Ran Human Spaceflight for a Generation*. Chicago, Illinois: Chicago Review Press Inc.

Cleary, Mark C. (2012) *The Cape: Military Space Operations 1971-1992*. Cape Canaveral, Florida: 45th Space Wing History Office

Cooper, Henry S.F., Jr. (1987) *Before Lift-Off: The Making of a Space Shuttle Crew*. Baltimore, Maryland: The Johns Hopkins University Press

Croft, Melvin and Youskauskas, John (2019) *Come Fly With Us: NASA's Payload Specialist Program*. Lincoln, Nebraska: University of Nebraska Press

Dethloff, Henry C. (1993) *Suddenly, Tomorrow Came: A History of the Johnson Space Center*. NASA Lyndon B. Johnson Space Center, Houston, Texas: The NASA History Series

Dunar, Andrew J. and Waring, Stephen P. (1999) *Power to Explore: A History of Marshall Space Flight Center 1960-1990*. NASA History Office, Washington, D.C.: Office of Policy and Plans

Froehlich, Walter (1983) *Spacelab: An International Short-Stay Orbiting Laboratory*. NASA Headquarters, Washington, D.C.

Gawdiak, Ihor Y., Miro, Ramon J. and Stueland, Sam (1997) 'Astronautics and Aeronautics, 1986-1990: A Chronology'. NASA Headquarters, Washington, D.C.: Office of Policy and Plans

Gawdiak, Ihor Y. and Shetland, Charles (2000) 'Astronautics and Aeronautics, 1991-1995: A Chronology'. NASA Headquarters, Washington, D.C.: Office of Policy and Plans

Hilmers, David C. (2015) *Man on a Mission*. Grand Rapids, Michigan: Zonderkidz

Iliff, Kenneth W. and Peebles, Curtis L. (2004) *From Runway to Orbit: Reflections of a NASA Engineer*. NASA History Office, Washington, D.C.: The NASA History Series

Ivey, William Noel and Lewis, Marieke (2010) *Astronautics and Aeronautics: A Chronology, 2001-2005*. NASA Headquarters, Washington, D.C.: NASA History Division

Janson, Bette R. and Ritchie, Eleanor H. (1990) *Aeronautics and Astronautics, 1979-1984: A Chronology*. The NASA History Series: Scientific and Technical Information Branch, NASA Headquarters, Washington, D.C.

Jenkins, Dennis R. (2001) *Space Shuttle: The History of the National Space Transportation System*. Hinckley, England: Midland Publishing

Krige, J. and Russo, A. (2000) *A History of the European Space Agency 1958-1987 Vol. 1: The Story of ESRO and ELDO, 1958-1973*. Paris, France: European Space Agency

Lenehan, Anne E. (2004) *Story: The Way of Water*. New South Wales, Australia: The Communications Agency

Lewis, Marieke and Swanson, Ryan (2009) *Astronautics and Aeronautics: A Chronology, 1996-2000*. NASA Headquarters, Washington, D.C.: NASA History Division

Logsdon, John M. (1996) The development of international space co-operation. In Logsdon, John M. (ed.) *Exploring the Unknown: Selected Documents in the History of the U.S. Civil Space Program Vol II: External Relationships*. NASA History Series, Washington, D.C.

Logsdon, John M. (ed.) with Day, Dwayne A. and Launius, Roger D. (1996) *Exploring the Unknown: Selected Documents in the History of the U.S. Civil Space Program Vol II: External Relationships*. NASA History Series, Washington, D.C.

Logsdon, John M. (1998) *Together in Orbit: The Origins of International Participation in the Space Station*. NASA History Office, Washington, D.C.: Monographs in Aerospace History

Logsdon, John M. (ed.) with Williamson, Ray A., Launius, Roger D., Acker, Russell J., Garber, Stephen J. and Friedman, Jonathan L. (1999) *Exploring the Unknown: Selected Documents in the History of the U.S. Civil Space Program Vol IV: Accessing Space*. NASA History Series, Washington, D.C.

Logsdon, John M. (ed.) with Snyder, Amy Paige, Launius, Roger D., Garber, Stephen J. and Newport, Regan Anne (2001) *Exploring the Unknown: Selected Documents in the History of the U.S. Civil Space Program Vol V: Exploring the Cosmos*. NASA History Series, Washington, D.C.

Logsdon, John M. (ed.) with Garber, Stephen J., Launius, Roger D. and Williamson, Ray A. (2004) *Exploring the Unknown: Selected Documents in the History of the U.S. Civil Space Program Vol VI: Space and Earth Science*. NASA History Series, Washington, D.C.

Lord, Douglas R. (1987) *Spacelab: An International Success Story*. Scientific and Technical Information Division, NASA Headquarters, Washington, D.C.

Muenger, Elizabeth A. (1985) *Searching the Horizon: A History of Ames Research Center 1940-1976*. NASA History Series, Washington, D.C.

Neal, Valerie, McMahan, Tracy and Dooling, Dave (1988) *Science in Orbit: The Shuttle and Spacelab Experience 1981-1986*. NASA Marshall Space Flight Center (MSFC), Huntsville, Alabama

Parazynski, Scott with Flory, Susy (2017) *The Sky Below*. New York: Little A

Portree, David S.F. and Treviño, Robert C. (1997) *Walking to Olympus: An EVA Chronology*. NASA History Office, Washington, D.C.: Monographs in Aerospace History

Ritchie, Eleanor H. (1984) *Aeronautics and Astronautics, 1976: A Chronology*. The NASA History Series: Scientific and Technical Information Branch, NASA Headquarters, Washington, D.C.

Ritchie, Eleanor H. (1986) *Aeronautics and Astronautics, 1977: A Chronology*. The NASA History Series: Scientific and Technical Information Branch, NASA Headquarters, Washington, D.C.

Rottner, Renee M. (2017) *Making the Invisible Visible: A History of the Spitzer Infrared Telescope Facility (1971-2003)*. NASA History Division, Washington, D.C.: Monographs in Aerospace History

Sebesta, Lorenza (1997) *Spacelab in Context*. Italy: Facolta di Scienze Politiche, Universita degli Studi Bologna, Forli Camp

Seddon, Rhea (2015) *Go for Orbit*. Murfreesboro, Tennessee: Your Space Press

Shapland, David and Rycroft, Michael (1984) *Spacelab: Research in Earth Orbit*. Cambridge, England: Cambridge University Press

Shayler, David J. and Burgess, Colin (2007) *NASA's Scientist-Astronauts*. Chichester, UK: Springer-Praxis

'STS Flight Assignment Baseline' (1977). STS Utilization and Planning Office, NASA Lyndon B. Johnson Space Center, Houston, Texas

Sullivan, Kathryn D. (2019) *Handprints on Hubble: An Astronaut's Story of Invention*. Cambridge, Massachusetts. The MIT Press

Williams, Dr. Dave (2018) *Defying Limits: Lessons From the Edge of the Universe*. Toronto: Simon & Schuster Canada

Young, John W. with Hansen, James R. (2012) *Forever Young: A Life of Adventure in Air and Space*. Gainesville, Florida: University Press of Florida

Reports

Acton, Loren (1985) *Max '91: The Active Sun*. Solar Physics Division of the American Astronomical Society & National Aeronautics and Space Administration

'Alignment Control Study for the Solar Optical Telescope: Final Report.' Ball Brothers Research Corp., Boulder, Colorado, October 1976

'ASSESS II Spacelab Simulation.' NASA Ames Research Center, Moffett Field, California, December 1977

Assured Access to Space: Hearings Before the Subcommittee on Space Science and Applications of the Committee on Science and Technology, House of Representatives. Ninety-Ninth Congress, Second Session (1987). Washington, D.C.: U.S. Government Printing Office

Bareiss, L.E., Payton, R.M. and Papazian, H.A. (1986) *Shuttle/Spacelab Contamination Environment and Effects Handbook*. Denver, Colorado: Martin Marietta Aerospace Denver Division

Catura, R.C., Acton, L.W., Brown, W.A., Gilbreth, C.W., Springer, L.A., Vieira, J.R., Culhane, J.L., Mason, I.W., Siegmund, O. and Patrick, T.J. (1982) *High resolution large area modular array of reflectors (LAMAR) Wolter type I X-ray telescope for Spacelab*. NASA Headquarters, Washington, D.C.

Dabbs, J.R., Tandberg-Hanssen and Hudson, H.S. (1982) *The Pinhole/Occulter Facility Executive Summary*. NASA Marshall Space Flight Center, Huntsville, Alabama

Eaton, Larry R. and Neste, Sherman L. (1977) *Atmospheric Cloud Physics Laboratory (ACPL) Optical Motion Control Study: Final Report*. Philadelphia, Pennsylvania: General Electric Co.

Gorenstein, Paul (1996) *The Large Area Modular Array of Reflectors (LAMAR) Final Report*. Cambridge, Massachusetts: Smithsonian Institution Astrophysical Observatory

Harrington, James C. *Spacelab Program Preparations for First Flight and Projected Utilization* (1983). The Space Congress Proceedings, 1

Hudson, Hugh S. (1986) *The Concept of the Pinhole/Occulter Facility*. La Jolla, California: University of California at San Diego

Interim Report of the Astronomy Spacelab Payload Study, Astronomy Spacelab Payloads Project. NASA Headquarters, Washington, D.C.: NASA History Office, July 1975

Investigation of the Challenger Accident (Vol. 1) Hearings Before the Committee on Science and Technology, House of Representatives. Ninety-Ninth Congress, Second Session (June 1986. Washington, D.C.: U.S. Government Printing Office

Jordan, S.D. (1981) *The Solar Optical Telescope (SOT)*. In *Space Science Reviews*, Vol. 29, Issue 4, pp 333-340

Lemke, D., Grewing, M., Offermann, D., Drapatz, S. and Klipping, G. (1982) *GIRL: The German Infrared Laboratory for Spacelab*. In *Advances in Space Research Vol. 2, Issue 4*, pp 123-130

NASA Mission Planning and Operations Team Report (1986). In *Report of the Presidential Commission on the Space Shuttle Challenger Accident Vol. 2. Appendix J*. NASA Lyndon B. Johnson Space Center, Houston, Texas

Oliver, Randall G. and Moye, John E., *Paper Session I-A - Starlab Overview* (1990). The Space Congress Proceedings 15

Portree, David S.F. (2012) *What Shuttle Should Have Been: The October 1977 Flight Manifest*. Wired.com. Accessed August 2023

Pounds, Kenneth A. (1980) *Large Area Modular Array of Reflectors (LAMAR)*. In *Journal of the Washington Academy of Sciences* Vol. 71, No. 2, Proceedings of the Uhuru Memorial Symposium: The Past, Present and Future of X-ray Astronomy. Greenbelt, Maryland: NASA Goddard Space Flight Center (GSFC)

Pressures on the System (1986). In *Report of the Presidential Commission on the Space Shuttle Challenger Accident Vol. 1 Ch. 8.* NASA Lyndon B. Johnson Space Center, Houston, Texas

Progress Report: Co-ordinated Study of Solar-Terrestrial Payloads on Space Station (1985) NASA Marshall Space Flight Center, Huntsville, Alabama

Schuiling, R.L. *Paper Sessions IV-B – Shuttle and Space Station Scientific Payloads: Their Role in the Next Generation* (1989). The Space Congress Proceedings 3

SHEAL II: NASA's Shuttle High Energy Astrophysics Laboratory (1988). Greenbelt, Maryland: NASA Goddard Space Flight Center (GSFC)

Shepherd, Gordon, Gault, William A. and Miller, D.W. (1985) *WAMDII: Wide-Angle Michelson Doppler Imaging Interferometer for Spacelab*. In *Applied Optics* 24:11

Spacelab Flight Division Missions (1983) NASA Headquarters, Washington, D.C.: Shuttle Payloads Engineering Division

Space Task Group (1969) *The Post-Apollo Space Program: Directions for the Future*. NASA Headquarters: History Office

Tarbell, Theodore D. and Title, Alan M. (1992) *A Solar Magnetic and Velocity Field Measurement System for Spacelab-2: The Solar Optical Universal Polarimeter (SOUP)*. Bethesda, Maryland: Lockheed Missiles and Space Co.

Torr, Marsha R. (1986) *A Possible Glow Experiment for the EOM-1/2 Mission*. Logan, Utah: Utah State University

VanderVoort, R.J. (1985) *Modeling and Analysis of Pinhole Occulter Experiment*. Clearwater, Florida: Honeywell Space & Strategic Avionics Division

WAMDII: The Wide Angle Michelson Doppler Imaging Interferometer (1988). Greenbelt, Maryland: NASA Goddard Space Flight Center (GSFC)

Werner, Michael W. and Witteborn, Fred C. (1982) *SIRTF – The Shuttle Infrared Telescope Facility*. In *Proceedings of the Second ESO Infrared Workshop*, Garching, West Germany

News Releases

'Space Station Design Study Release.' NASA Manned Spacecraft Center News Release, 28 April 1969

'MSC Establishes a Space Station Task Group.' NASA Manned Spacecraft Center News Release, 14 May 1969

'Study of Reusable Space Tug Requested from Aerospace Industry.' NASA Manned Spacecraft Center News Release, 6 April 1970

'North American Rockwell Awarded Contract for Study of Space Tug.' NASA Manned Spacecraft Center News Release, 4 June 1970

'Messerschmitt and British Aircraft Corp. to Join With U.S. Firms in Design Studies of Space Shuttle.' NASA Manned Spacecraft Center News Release, 8 September 1970

'Study of Space Station Assembly in Earth Orbit.' NASA Manned Spacecraft Center News Release, 12 November 1970

'North American Rockwell Requested to Study Modular Space Station.' NASA Manned Spacecraft Center News Release, 15 December 1970

'Contract Extensions for Earth-Orbital Space Station Phase B Contract With McDonnell Douglas and North American.' NASA Manned Spacecraft Center News Release, 5 March 1971

'Nixon/Fletcher Shuttle Statements.' NASA Manned Spacecraft Center News Release, 5 January 1972

'MSC Releases Request for Shuttle Payload Study.' NASA Manned Spacecraft Center News Release, 7 March 1972

'TRW Wins Shuttle Payload Study Contracts.' NASA Manned Spacecraft Center News Release, 6 June 1972

'Astronauts Chapman and England Resign.' NASA Manned Spacecraft Center News Release, 14 July 1972

'NASA/NR Sign Shuttle Letter Contract.' NASA Manned Spacecraft Center News Release, 9 August 1972

'Europe to Develop Sortie Lab.' NASA Manned Spacecraft Center News Release, 22 January 1973

'Four Firms to Study Space Tug Systems.' NASA Manned Spacecraft Center News Release, 14 February 1973

'Europe to Build Spacelab for U.S. Reusable Space Shuttle.' NASA Lyndon B. Johnson Space Center News Release, 24 September 1973

'ESRO Meeting.' NASA Lyndon B. Johnson Space Center News Release, 22 July 1974

'Dr. Robert A. Parker New Chief, Astronaut Office S&AD.' NASA Lyndon B. Johnson Space Center News Release, 15 August 1974

'Spacelab Simulation Underway at JSC.' NASA Johnson Space Center News Release, 1 October 1974

'NASA Seeks Proposals on Space Station Study.' NASA Lyndon B. Johnson Space Center News Release, 11 December 1975

'Spacelab Simulation.' NASA Lyndon B. Johnson Space Center News Release, 23 January 1976

'Dr. Robert S. Clark Member of Shuttle Team.' NASA Lyndon B. Johnson Space Center News Release, 30 January 1976

'Dr. Charles F. Sawin Member of Shuttle Team.' NASA Lyndon B. Johnson Space Center News Release, 30 January 1976

'Spacelab Simulation 'Crew' Undergoes Medical Tests.' NASA Lyndon B. Johnson Space Center News Release, 1 December 1976

'Spacelab Network Simulation Starts at JSC.' NASA Lyndon B. Johnson Space Center News Release, 13 May 1977

'First Shuttle Payload to Investigate Earth Resources.' NASA Lyndon B. Johnson Space Center News Release, 7 September 1977

'Astronaut William E. Thornton Nominated for Spacelab Crew Position.' NASA Lyndon B. Johnson Space Center News Release, 5 December 1977

'Singer Company to Build Spacelab Simulator.' NASA Lyndon B. Johnson Space Center News Release, 1 February 1978

'Mission Specialists for Spacelab-1 Named at JSC.' NASA Lyndon B. Johnson Space Center News Release, 1 August 1978

'England Returns to Astronaut Program.' NASA Lyndon B. Johnson Space Center News Release, 7 June 1979

'Two Europeans Accepted for Space Shuttle Mission Specialist Training.' NASA Lyndon B. Johnson Space Center News Release, 7 July 1980

'STS-2 Crew Selected.' NASA Lyndon B. Johnson Space Center News Release, 23 April 1981

'Columbia's Second Crew, Mission Control Simulate First Two Flight Days.' NASA Lyndon B. Johnson Space Center News Release, 6 July 1981

'OSTA-1 Mission.' NASA Lyndon B. Johnson Space Center News Release, 10 September 1981

'First Spacelab-1 Hardware Shipped.' NASA Lyndon B. Johnson Space Center News Release, 9 October 1981

'Payload Operations Control Center (POCC).' NASA Lyndon B. Johnson Space Center News Release, 21 March 1982

'Three Shuttle Crews Announced.' NASA Lyndon B. Johnson Space Center News Release, 19 April 1982

'NASA Awards OSTA-3 Payload Contract to Rockwell Space Operations.' NASA Lyndon B. Johnson Space Center News Release, 11 May 1982

'Crewmembers Named for STS 13, Spacelab 2 and Spacelab 3.' NASA Lyndon B. Johnson Space Center News Release, 18 February 1983

'Second Tracking and Data Relay Satellite (TDRS) Deleted from STS-8 Manifest.' NASA Lyndon B. Johnson Space Center News Release, 27 May 1983

'STS Flight Assignments.' NASA Lyndon B. Johnson Space Center News Release, 17 November 1983

'Flight Control of STS-9.' NASA Lyndon B. Johnson Space Center News Release, 18 November 1983

'Extension of STS-9 Mission By One Day.' NASA Lyndon B. Johnson Space Center News Release, 3 December 1983

'51D, 61D Crew Announced.' NASA Lyndon B. Johnson Space Center News Release, 2 February 1984

'51K Crew Announcement.' NASA Lyndon B. Johnson Space Center News Release, 14 February 1984

'NASA Announces Crew Members for Future Space Shuttle Flights.' NASA Lyndon B. Johnson Space Center News Release, 7 June 1984

'NASA Announces Flight Crew Assignments and Changes.' NASA Lyndon B. Johnson Space Center News Release, 22 October 1984

'Space Shuttle Payload Flight Assignments.' Customer Services Division, NASA Headquarters, Washington, D.C., March 1985

'NASA Changes 51B Landing Site to Edwards Air Force Base.' NASA Lyndon B. Johnson Space Center News Release, 24 April 1985

'Space Shuttle Payload Flight Assignments.' Customer Services Division, NASA Headquarters, Washington, D.C., June 1985

'Flight Control of Shuttle Mission 51F.' NASA Lyndon B. Johnson Space Center News Release, 10 June 1985

'NASA Names Crews for Upcoming Space Shuttle Flights.' NASA Lyndon B. Johnson Space Center News Release, 19 September 1985

'Space Shuttle Payload Flight Assignments.' Customer Services Division, NASA Headquarters, Washington, D.C., November 1985

'NASA Switches Earth Observation Mission and Telescope Launches.' NASA George C. Marshall Space Center, Huntsville, Alabama, 6 January 1986

'Flight Data File Crew Activity Plan STS-61K.' Mission Operations Directorate, Operations Division, NASA Lyndon B. Johnson Space Center, Houston, Texas, 17 February 1986

'Astronaut Owen K. Garriott Leaves NASA.' NASA Lyndon B. Johnson Space Center News Release, 11 June 1986

'Astronaut Group Provides Interface With Space Shuttle Customers.' NASA Lyndon B. Johnson Space Center News Release, 29 February 1988

'Payload Flight Assignments: NASA Mixed Fleet.' Office of Space Flight, NASA Headquarters, Washington, D.C., August 1988

'Four New Shuttle Crews Named.' NASA Lyndon B. Johnson Space Center News Release, 30 November 1988

'Partial Shuttle Crew Assignments Announced.' NASA Lyndon B. Johnson Space Center News Release, 29 June 1989

'Space Shuttle Crew Members Named to DoD, Life Sciences Missions.' NASA Lyndon B. Johnson Space Center News Release, 24 February 1989

'Astronauts Named to Space Science Missions.' NASA Lyndon B. Johnson Space Center News Release, 5 April 1989

'McBride to Leave NASA; Brand Named Commander of STS-35.' NASA Lyndon B. Johnson Space Center News Release, 24 April 1989

'Partial Shuttle Crew Assignments Announced.' NASA Lyndon B. Johnson Space Center News Release, 29 June 1989

'Astronauts Named to Shuttle Crews.' NASA Lyndon B. Johnson Space Center News Release, 29 September 1989

'Payload Flight Assignments: NASA Mixed Fleet.' Office of Space Flight, NASA Headquarters, Washington, D.C., January 1990

'Astronaut Crew Named to International Microgravity Mission'. NASA Lyndon B. Johnson Space Center News Release, 2 January 1990

'NASA Announces Payload Specialists for Spacelab IML-1 Mission.' NASA Headquarters News Release, 19 January 1990

'Science Payload Commanders Named; Sonny Carter Replaces Mary Cleave on IML-1. NASA Lyndon B. Johnson Space Center News Release, 25 January 1990

'NASA and Rockwell Sign Agreement for EDO Pallet.' NASA Headquarters News Release, 2 April 1990

'Shuttle Crews Named for 1991 Missions.' NASA Lyndon B. Johnson Space Center News Release, 24 May 1990

'Shuttle Crew Commanders Reassigned.' NASA Lyndon B. Johnson Space Center News Release, 9 July 1990

'NASA Selects Microgravity Mission Payload Specialists.' NASA Headquarters News Release, 6 August 1990

'Bonnie Dunbar Named Payload Commander for USML-1.' NASA Lyndon B. Johnson Space Center News Release, 13 September 1990

'NASA Names Space Shuttle Hydrogen Leak Investigation Team.' NASA Headquarters News Release, 19 September 1990

'NASA Announces Crew Members for Future Space Shuttle Flights.' NASA Lyndon B. Johnson Space Center News Release, 19 December 1990

'NASA Issues Modifications to Shuttle Manifest.' NASA Headquarters News Release, 20 March 1991

'Shuttle Crew Assignments Announced.' NASA Lyndon B. Johnson Space Center News Release, 19 April 1991

'Astronaut Mary Cleave Joins NASA Environmental Project.' NASA Headquarters News Release, 1 May 1991

'NASA Selects Payload Specialists for Spacelab Mission.' NASA Headquarters News Release, 1 May 1991

'ASTRO Mission to Refly.' NASA Headquarters News Release, 19 May 1991

'New Crew Transport Vehicle (CTV) Allows More Timely Orbiter Egress.' NASA Lyndon B. Johnson Space Center News Release, 28 May 1991

'NASA Announces Crew Members for Future Shuttle Flights.' NASA Lyndon B. Johnson Space Center News Release, 22 August 1991

'NASA Appoints Spacelab Payload Specialist.' NASA Headquarters News Release, 10 September 1991

'Early Results from Life Sciences Mission Show New Discoveries.' NASA Headquarters News Release, 19 September 1991

'Payload Specialists for Tethered Satellite Mission Named.' NASA Lyndon B. Johnson Space Center News Release, 26 September 1991

'Dr. M. Rhea Seddon Named Payload Commander for STS-58 Spacelab Life Sciences-2 (SLS-2).' NASA Lyndon B. Johnson Space Center News Release, 23 October 1991

'Payload Crew Named for Spacelab Life Sciences-2 (STS-58) Mission.' NASA Lyndon B. Johnson Space Center News Release, 6 December 1991

'Crew Assignments Announced for Future Shuttle Missions.' NASA Lyndon B. Johnson Space Center News Release, 21 February 1992

'Space Shuttle Crew Assignments Announced.' NASA Lyndon B. Johnson Space Center News Release, 16 March 1992

'Unexpected Results from Life Science Mission.' NASA Headquarters News Release, 24 July 1992

'Tethered Satellite Investigation Underway.' NASA Lyndon B. Johnson Space Center News Release, 11 August 1992

'Spacelab Studies Feature Eggs and Space Motion Sickness.' NASA Headquarters News Release, 13 August 1992

'Crew Assignments Announced for STS-58 and STS-61.' NASA Lyndon B. Johnson Space Center News Release, 27 August 1992

'Richard J. Hieb Named Payload Commander for STS-66 (IML-2) Mission.' NASA Lyndon B. Johnson Space Center News Release, 2 September 1992

'Mission Specialists Named for STS-65 (IML-2) Mission.' NASA Lyndon B. Johnson Space Center News Release, 27 October 1992

'Payload Specialist Selected for Second Life Sciences Mission STS-58 (SLS-2).' NASA Lyndon B. Johnson Space Center News Release, 29 October 1992

'STS-55 Experiment Focuses on Lymphocyte Activation.' NASA Lyndon B. Johnson Space Center News Release, 3 February 1993

'STS-62 and STS-59 Space Shuttle Crew Assignments Announced.' NASA Lyndon B. Johnson Space Center News Release, 5 March 1993

'STS-56 Astronauts to Document Earth Observations.' NASA Lyndon B. Johnson Space Center News Release, 7 April 1993

'STS-55 Astronauts to Document Earth Observations'. NASA Lyndon B. Johnson Space Center News Release, 26 April 1993

'Astronaut Seddon Injured During Training.' NASA Lyndon B. Johnson Space Center News Release, 4 May 1993

'Payload Commanders Named for Future Shuttle Missions.' NASA Lyndon B. Johnson Space Center News Release, 3 August 1993

'Crew Members Selected for STS-65.' NASA Lyndon B. Johnson Space Center News Release, 23 August 1993

'Shuttle Mission to Conduct Scientific First.' NASA Headquarters News Release, 28 September 1993

'Crew Selected for Space Shuttle Mission STS-68.' NASA Lyndon B. Johnson Space Center News Release, 28 October 1993

'Grunsfeld Named to STS-67 ASTRO-2 Crew.' NASA Lyndon B. Johnson Space Center News Release, 28 October 1993

'NASA Names Space Shuttle Mission STS-64 Crew.' NASA Lyndon B. Johnson Space Center News Release, 16 November 1993

'NASA and Russian Space Agency Sign Agreement for Additional Space Shuttle/Mir Missions.' NASA Headquarters News Release, 16 December 1993

'Astronauts Selected for Atlantis' STS-66 Mission.' NASA Lyndon B. Johnson Space Center News Release, 10 January 1994

'Crew Selected for STS-67 Astronomy Mission.' NASA Lyndon B. Johnson Space Center News Release, 10 January 1994

'Astronaut Linenger Joins STS-64 Crew.' NASA Lyndon B. Johnson Space Center News Release, 28 February 1994

'Reflight of Tethered Satellite System Confirmed.' NASA Lyndon B. Johnson Space Center News Release, 10 March 1994

'Payload Commander, Mission Specialist Named to STS-73.' NASA Lyndon B. Johnson Space Center News Release, 17 March 1994

'STS-59 Launch Rescheduled to April 8th.' NASA Lyndon B. Johnson Space Center News Release, 4 April 1994

'Crew Named for First Space Shuttle, Mir Docking Mission.' NASA Lyndon B. Johnson Space Center News Release, 3 June 1994

'NASA Selects Payload Specialists for Spacelab Mission.' NASA Lyndon B. Johnson Space Center News Release, 20 June 1994

'NASA Selects Scientists for Neurolab Shuttle Mission.' NASA Lyndon B. Johnson Space Center News Release, 22 June 1994

'Chang-Díaz Named Payload Commander for TSS Reflight.' NASA Lyndon B. Johnson Space Center News Release, 25 August 1994

'Space Shuttle Mission STS-64 Press Kit.' NASA Headquarters, Washington, D.C., September 1994

'Guidoni Named to Crew of Tethered Satellite Reflight.' NASA Lyndon B. Johnson Space Center News Release, 12 October 1994

'NASA Radar Observes Erupting Volcano on the Ring of Fire.' NASA Headquarters News Release, 24 October 1994

'Commander, Pilot and Flight Engineer Named to STS-73 Crew.' NASA Lyndon B. Johnson Space Center News Release, 18 November 1994

'New Spacelab Science Mission to Fly in 1996.' NASA Lyndon B. Johnson Space Center News Release, 13 December 1994

'Space Shuttle Crew Selected for Tethered Satellite Mission.' NASA Lyndon B. Johnson Space Center News Release, 27 January 1995

'Mission and Payload Specialists Named to Life, Microgravity Flight.' NASA Lyndon B. Johnson Space Center News Release, 8 May 1995
'ASTRO-2 Provides First Definitive Detection of Primordial Helium.' NASA Lyndon B. Johnson Space Center News Release, 12 June 1995
'Commander, Pilot Round Out STS-78 Crew.' NASA Lyndon B. Johnson Space Center News Release, 6 October 1995
'NASA Selected Payload Specialists for Shuttle Mission.' NASA Headquarters News Release, 29 January 1996
'Trainees Selected for Shuttle Neurolab Mission.' NASA Headquarters News Release, 4 April 1996
'Commander, Pilot, Flight Engineer Round Out STS-83 Crew.' NASA Headquarters News Release, 31 May 1996
'Tethered Satellite Investigation Report is Released.' NASA Lyndon B. Johnson Space Center News Release, 4 June 1996
'Future Topographic Radar Shuttle Mission Will Map 80 Percent of Earth.' NASA Headquarters News Release, 15 July 1996
'Crew Members Named to Life Science Mission.' NASA Headquarters News Release, 12 August 1996
'Seddon to Support Life Science Investigations at Vanderbilt.' NASA Headquarters News Release, 10 September 1996
'Astronaut Cady Coleman Begins Training as Backup Mission Specialist for STS-83.' NASA Lyndon B. Johnson Space Center News Release, 18 February 1997
'Commander, Pilot, Flight Engineer Round Out STS-90 Crew.' NASA Lyndon B. Johnson Space Center News Release, 18 April 1997
'Microgravity Science Laboratory Mission Set for July.' NASA Lyndon B. Johnson Space Center News Release, 25 April 1997
'Payload Specialists Selected for Future Shuttle Mission.' NASA Lyndon B. Johnson Space Center News Release, 28 April 1997
'Crew Members Named for Earth-Mapping Mission.' NASA Lyndon B. Johnson Space Center News Release, 26 October 1998

Oral Histories

Alexander, S. Jean (1998) *NASA Johnson Space Center Oral History Project*
Allen, Joseph P. (2004) *NASA Johnson Space Center Oral History Project*
Blaha, John E. (2004) *NASA Johnson Space Center Oral History Project*
Bluford, Guion S. (2004) *NASA Johnson Space Center Oral History Project*
Bolden, Charles F. (2004) *NASA Johnson Space Center Oral History Project*
Brand, Vance D. (2002) *NASA Johnson Space Center Oral History Project*
Brooks, Melvin F. (2000) *NASA Johnson Space Center Oral History Project*
Cabana, Robert D. (2015) *NASA Johnson Space Center Oral History Project*

Charles, John B. (2015) *NASA Johnson Space Center Oral History Project*
Chiao, Leroy (2011) *NASA Johnson Space Center Oral History Project*
Chrétien, Jean-Loup (2002) *NASA Johnson Space Center Oral History Project*
Cleave, Mary L. (2002) *NASA Johnson Space Center Oral History Project*
Crippen, Robert L. (2006) *NASA Johnson Space Center Oral History Project*
Dalton, Bonnie P. (2002) *NASA Johnson Space Center Oral History Project*
Duffy, Brian (2004) *NASA Johnson Space Center Oral History Project*
Dunbar, Bonnie J. (1998 and 2005) *NASA Johnson Space Center Oral History Project*
Engle, Joseph H. (2004) *NASA Johnson Space Center Oral History Project*
Fabian, John M. (2006) *NASA Johnson Space Center Oral History Project*
Frutkin, Arnold W. (2002) *NASA Johnson Space Center Oral History Project*
Fullerton, C. Gordon (2002) *NASA Johnson Space Center Oral History Project*
Garriott, Owen K. (2000) *NASA Johnson Space Center Oral History Project*
Gibson, Robert L. 'Hoot' (2018) *NASA Johnson Space Center Oral History Project*
Gregory, Frederick D. (2004) *NASA Johnson Space Center Oral History Project*
Hale, N. Wayne (2008) *NASA Johnson Space Center Oral History Project*
Hanley, Jeffrey M. (2016) *NASA Johnson Space Center Oral History Project*
Hart, Terry J. (2003) *NASA Johnson Space Center Oral History Project*
Hartsfield, Henry W. (2001) *NASA Johnson Space Center Oral History Project*
Hoffman, Jeffrey A. (2009) *NASA Johnson Space Center Oral History Project*
Horowitz, Scott J. (2007) *NASA Johnson Space Center Oral History Project*
Kennel, Charles F. (2002) *NASA Johnson Space Center Oral History Project*
Kleinknecht, Kenneth S. (1998 and 2000) *NASA Johnson Space Center Oral History Project*
Leestma, David C. (2002) *NASA Johnson Space Center Oral History Project*
Lenoir, William B. (2004) *NASA Johnson Space Center Oral History Project*
Lind, Don L. (2005) *NASA Johnson Space Center Oral History Project*
Lounge, John M. (2008) *NASA Johnson Space Center Oral History Project*
Lousma, Jack R. (2001) *NASA Johnson Space Center Oral History Project*
Lucas, William R. (2010) *NASA Johnson Space Center Oral History Project*
McArthur, William S. (2017) *NASA Johnson Space Center Oral History Project*
McBride, Jon A. (2012) *NASA Johnson Space Center Oral History Project*
Myers, Dale D. (1998) *NASA Johnson Space Center Oral History Project*
Nagel, Steven R. (2002) *NASA Johnson Space Center Oral History Project*
Nygren, Richard W. (2006) *NASA Johnson Space Center Oral History Project*
O'Connor, Bryan D. (2006 and 2007) *NASA Johnson Space Center Oral History Project*
Oswald, Stephen S. (2008) *Space Shuttle Program Tacit Knowledge Capture Oral History Project*
Parker, Robert A.R. (2002) *NASA Johnson Space Center Oral History Project*
Readdy, William F. (1998) *NASA Johnson Space Center Oral History Project*
Reyes, Raul E. (1998) *NASA Johnson Space Center Oral History Project*
Rice, William E. (2004) *NASA Johnson Space Center Oral History Project*
Richards, Richard N. (2006) *NASA Johnson Space Center Oral History Project*

Ride, Sally K. (2002) *NASA Johnson Space Center Oral History Project*
Ross, Jerry L. (2003 and 2004) *NASA Johnson Space Center Oral History Project*
Rotter, Henry A. (2009) *NASA Johnson Space Center Oral History Project*
Schwinghamer, Robert J. (2010) *Space Shuttle Recordation Oral History Project*
Seddon, M. Rhea (2010 and 2011) *NASA Johnson Space Center Oral History Project*
Shaw, Brewster H. (2002) *NASA Johnson Space Center Oral History Project*
Shriver, Loren J. (2002) *NASA Johnson Space Center Oral History Project*
Stonesifer, John C. (2001) *NASA Johnson Space Center Oral History Project*
Sullivan, Kathryn D. (2007 and 2008) *NASA Johnson Space Center Oral History Project*
Thagard, Norman E. (1998) *NASA Johnson Space Center Oral History Project*
Truly, Richard H. (2003) *NASA Administrators Oral History Project*
Wilcutt, Terrence W. (2015) *NASA Johnson Space Center Oral History Project*

Index

A

Abbey, George, 58, 82, 107, 139, 166, 227, 353
ABLE Deployable Articulated Mast (ADAM), 253
Abort to Orbit (ATO), 84, 85, 102, 104, 105
Active Cavity Radiometer (ACR), 62, 206, 218
Acton, Loren, 86, 87, 95–99, 103, 107
Advanced Gradient Heating Furnace (AGHF), 333
Advanced Protein Crystallisation Facility (APCF), 318, 332
Agenzia Spaziale Italiana (ASI), 239, 251, 274, 278, 283, 332
Agnew, Spiro, 2
Airborne Science/Spacelab Experiment System Simulation (ASSESS), 44–46, 56, 96
albino rats, 113, 114
Alexander, Carter, 47
Allen, Andy, 275, 277–279, 283, 284, 287
Allen, Joe, 390
Altman, Scott, 378, 379, 382, 387, 388
American Flight Echocardiograph (AFE), 374
Ames Double Rack (ADR), 113
Ames Research Center (ARC), 44, 47, 113, 123, 191, 295
Ames Single Rack (ASR), 113
Anders, Bill, 12
Animal Enclosure Module (AEM), 362
Anthrorack, 156, 157
Applied Research on Separation Methods Using Space Electrophoresis (RAMSES), 320
Apt, Jay, 237, 238, 262, 268–270
Aquatic Animal Experiment Unit (AAEU), 317, 319
ASTRO-1, 98, 165, 167–184, 186, 187, 276
ASTRO-2, 167, 181, 187, 222, 223
ASTRO-3, 167, 181–183
Astroculture, 312, 326
ATLAS-1, 202–212, 214–218, 222, 223
ATLAS-2, 204, 206, 216–220, 222
ATLAS-3, 204, 220–224

Atmospheric Cloud Physics Laboratory (ACPL), 24, 25
Atmospheric Emissions Photometric Imager (AEPI), 62, 196, 207
Atmospheric Laboratory for Applications and Science (ATLAS), 196, 203–204, 207, 208, 217, 220–222, 224, 389, 399
Atmospheric Lyman-Alpha Emissions (ALAE), 196, 204, 205
Atmospheric, Magnetospheric and Plasmas in Space (AMPS), 25
Atmospheric Trace Molecule Spectroscopy (ATMOS), 110, 111, 114, 125, 126, 204–205, 215, 218, 219, 223
Atomic Oxygen Exposure Tray (AOET), 153
aurora, 106, 126, 148, 215, 219
Autogenic Feedback Training (AFT), 114, 132, 266
Auxiliary Power Unit (APU), 81, 244

B

Baker, Ellen, 304, 306, 310, 311, 314, 392–396, 398
Baker, Mike, 243, 245, 246
Ball Aerospace, 300
Bantle, Jeff, 313
Baroreflex, 156, 162, 359
Bartoe, John-David, 86, 87, 96, 99, 100, 102, 104, 107, 188
Bechis, Kenneth, 300, 301
Beggs, James, 81
Bignier, Michel, 34, 35, 39, 41, 43, 89, 91
Biliu, David, 46
Biolabor, 156, 157
Biorack, 140, 145, 295, 317, 318
Bioreactor Demonstration System (BDS), 387
Biostack, 65, 153, 295, 316
Biotelemetry System (BTS), 113, 122
Biowissenschaften, 140
Blom, Ronald, 226, 233
Bluford, Guy, 137–139, 142, 143, 145, 148, 161, 270
Boeing, 72
Boesen, Denny, 300, 301
Bolden, Charlie, 208, 209, 211, 212
Bondar, Roberta, 291, 292, 294–296, 299
Bondi, Herman, 8, 38
Boudreaux, Marc, 330
Bowersox, Ken, 303, 304, 306, 307, 310, 312, 324, 325, 399
Brady, Chuck, 331, 334–337
Brandenstein, Dan, 150, 277, 293, 303
Brand, Vance, 174–177, 180, 200, 201, 354, 356
Branscome, Darrell, 282
Bredt, James, 24
Bridges, Roy, 83, 84, 86, 87, 97, 103, 104, 106, 107, 117, 346
Briscoe, Lee, 100
British Aircraft Corporation (BAC), 11
Broad Band X-ray Telescope (BBXRT), 170–173, 177–180, 196
Brooks, Mel, 38, 58, 66
Brown, Curt, 220, 221, 262, 268–272
Brümmer, Renate, 152, 153
Bubble, Drop and Particle Unit (BDPU), 320, 334
Buchli, Jim, 137, 138, 147, 148
Buckey, Jay, 371, 372, 379, 382, 384–388
Budarin, Nikolai, 393, 397, 398
Bundesministerium für Forschung und Technologie (BMFT), 140
Burgess, Colin, 47, 64, 102
Bursch, Dan, 243, 245, 247, 399
Bush, George H.W., 67, 68, 374, 379, 392

C

Cabana, Bob, 315, 316, 318, 319, 321
Cameron, Ken, 216–219
Canadian Space Agency (CSA), 290, 295, 315, 332, 380, 386
Carter, Sonny, 291, 293
Casper, John, 303
Cassutt, Michael, 57, 58
Castle, Bob, 177
Causse, Jean-Pierre, 34
Centre National d'Études Spatiales (CNES), 34, 37, 40, 152, 290, 298, 315, 332, 379
Chandler, Doris, 44
Chang-Díaz, Franklin, 276–278, 281, 283, 284, 286
Chappell, Rick, 56, 79, 201, 208, 210
Charles, John, 118
Cheli, Maurizio, 283, 287
Chiao, Leroy, 315–318, 321, 399
Chilton, Kevin, 224, 237, 238, 242
Chrétien, Jean-Loup, 57
Clapp, Nicholas, 225
Clark, Robert, 46, 47
Cleave, Mary, 291
Clervoy, Clervoy, 220, 221
Clifford, Rich, 237, 238, 240, 244
Coca Cola/Pepsi, 97
Cockrell, Ken, 216
Coleman, Catherine 'Cady', 323, 324, 327–329, 340, 399
Colombo, Guiseppe, 273, 274
Columbus, Christopher, 289, 314
Columbus module, 390–391
Combustion Module-1 (CM-1), 342, 349
Common Market, 7
Concept Verification Test (CVT), 44
Contamination Monitor Package (CMP), 50
Convair 990, 14, 44–46

Coronal Helium Abundance Spacelab Experiment (CHASE), 93, 99–101, 188
Cosmic Ray Nuclei Experiment (CRNE), 89, 91, 92, 98, 171, 195
Craft, Harry, 69
Crew Telesupport Equipment (CTE), 156
Crew Transport Vehicle (CTV), 369
Crippen, Bob, 27, 227, 229, 231, 234, 235
Critical Point Facility (CPF), 298, 319, 320
Croft, Melvin, 58, 93, 124, 137, 210
Crouch, Roger, 291, 339, 340, 344–346, 349, 352
Crystal Growth Furnace (CGF), 308, 309, 328, 329
Culbertson, Philip, 17, 39
Curien, Huberty, 40

D

d'Allest, Frederic, 37
Dabbs, J.R., 194
Dalton, Bonnie, 114, 116, 117, 133, 354, 361
Dark Sky, 195
Data Display Unit (DDU), 90, 93, 104, 177, 179, 212
Davidsen, Art, 180
Davis, Jan, 261, 262, 264, 265, 267, 269, 271
Deloffre, Bernard, 34
Delorme, Jean, 5
DeLucas, Larry, 303, 304, 306–308, 311, 313
Demas, Louis, 189
Deployer Core Equipment and Satellite Core Equipment (DCORE/SCORE), 274

Deutsche Agentur für Raumfahrtangelegenheiten (DARA), 152, 239, 252, 290, 315, 380
Deutsche Forschungsanstalt für Luft-und Raumfahrt (DFVLR), 140, 142, 152
Deutsches Zentrum für Luft-und Raumfahrt (DLR), 152
Dezhurov, Vladimir, 394
Diffusion-Controlled Crystallisation Apparatus for Microgravity (DCAM), 326
Diller, George, 317
Dodeck, Hauke, 152, 153
Dornier, 25, 88–90, 239
Downey, Patton, 332, 339
Drop Dynamics Module (DDM), 113, 119, 127–129
Drop Physics Module (DPM), 311, 312, 328
Droplet Combustion Apparatus (DCA), 342
Droplet Combustion Experiment (DCE), 330, 342, 350
Duffy, Brian, 208, 209, 211, 212, 216
Dunar, Andrew, 15, 16
Dunbar, Bonnie, 137–140, 142, 143, 146, 147, 149, 303–307, 309–313, 392, 393, 395, 396, 398
Dunlap, Alex, 379
Duque, Pedro, 331
Durrance, Sam, 167–169, 174, 177–179, 181–184
Dye, Paul, 259
Dynamic Environment Measurement System (DEMS), 113
Dynamics of Rotating and Oscillating Free Drops (DROP), 113, 119

E
Earth Observation Mission (EOM), 199–203, 206, 208
Earth Observing System (EOS), 220
Earth Radiation Budget Satellite (ERBS), 227, 229, 231
Edelson, Burt, 190
Edwards Air Force Base, 3, 27, 79, 82, 106, 148, 287, 299, 368
Elachi, Charles, 252, 253
Electromagnetic Containerless Processing Facility (TEMPUS), 320, 344, 345, 351
Energetic Ion Mass Spectrometer (EIMS), 196
Energetic Neutral Atom Precipitation (ENAP), 96, 207
England, Tony, 86, 96, 107
Engle, Joe, 27, 31–34
Europa rocket, 9
European Launcher Development Organisation (ELDO), 5, 7, 9, 11, 18, 21
European Space Agency (ESA), v, 2, 21, 75
European Space Operations Centre (ESOC), 322
European Space Research Organisation (ESRO), 2, 5, 7, 8, 16–22, 26, 34, 36, 38, 88
European Space Technology Centre (ESTEC), 16, 62, 141
EXPRESS rack, 350
Extended Duration Orbiter (EDO), 182, 183, 290, 291, 301–303, 313–315, 322, 356, 377
External Tank (ET), 2, 3, 31, 71, 73, 76, 83, 174, 381
Extravehicular Activity (EVA), 64, 65

Fabian, John, 353, 354, 356
Far Ultraviolet Space Telescope (FAUST), 62, 206
Favier, Jean-Jacques, 316, 331, 334, 335, 337, 338
Feature Identification and Location Experiment (FILE), 30, 33, 230, 232
Fein, Juergen, 46
Fettman, Marty, 371–376, 378
Fichtl, George, 111, 123
Finke, Wolfgang, 40
Fischer, Craig, 57
Fisk, Lennard, 182, 190, 210, 372
Flanigan, Peter, 12
Fletcher, James, 1, 2, 20–22, 189, 354
Fluids Experiment System (FES), 298
Foale, Mike, 206, 208, 211, 215, 216, 399
Frazier, Don, 310
Frimout, Dirk, 201, 206, 208–212, 216
Froehlich, Walter, 57
Frosch, Robert, 35, 38, 39, 58
Frutkin, Arnold, 5, 8, 10
Fullerton, Gordon, 48, 50–54, 83–88, 94, 96, 102–104, 106, 107
Furrer, Reinhard, 137–139, 141, 148, 149

G

Gaffney, Drew
 and central venous catheter experiment, 359
Galactic Ultra-Wide-Angle Schmidt System Camera (GAUSS), 154, 156, 158
Gardner, Guy, 174, 175, 177, 180
Garneau, Marc, 227

Garriott, Owen, 55, 56, 58–60, 64, 65, 67, 69, 70, 74, 76, 78–80, 82, 199, 200, 202, 203, 206
Gas Analyser Mass Spectrometer (GAMS), 364, 377
Gatland, Kenneth, 13, 21
General Electric, 25, 77
General Purpose Workstation (GPWS), 123, 124, 361, 362, 382, 383
Generic Bioprocessing Apparatus (GBA), 312, 313
Geophysical Fluid Flow Cell (GFFC), 324, 325
Geophysical Fluid Flow Experiment (GFFE), 113, 129, 130, 324, 325
German Infrared Radiation Laboratory (GIRL), 159
German Space Operations Centre (GSOC), 142, 157
Gernhardt, Mike, 340
Gibson, Robert 'Hoot,' 262, 264–266, 268–272, 277, 282, 356, 392–394, 396, 397
Gibson, Roy, 26, 34–36, 38
Glovebox, 140, 265, 295, 305, 307, 309, 311, 312, 325, 329, 346
Goddard Space Flight Center (GSFC), 31, 49, 50, 167, 170, 171, 173, 178, 180, 190, 192, 193, 196, 207, 217
Godwin, Linda, 236–238, 241, 242
Goldin, Dan, 283, 316, 374, 391
Goodyear, 63
Gorie, Dom, 254–256, 260, 349
Grabe, Ron, 291, 293, 294
Gravitational Plant Physiology Facility (GPPF), 295
gravity gradient attitude, 94, 111, 112, 127, 147, 310, 324, 332
Gregory, Bill, 182, 184

Gregory, Fred, 117, 119–125, 127, 131, 132
Griggs, Dave, 96, 97, 201, 354, 356, 359
Grille Spectrometer, 60, 73, 207, 209
Griner, Carolyn, 44
Grossi, Mario, 273, 274
Grunsfeld, John, 182, 184, 185
Guidoni, Umberto, 278, 283, 285, 287
Gull, Ted, 178, 180
Gutierrez, Sid, 237, 238, 242, 359, 369

H

Hale, Wayne, 81
Halley's Comet, 98, 165, 170, 175
Halsell, Jim, 315–317, 321, 340, 343, 344, 346–348, 352
Hammond, Blaine, 249
Hanley, Jeff, 281–283
Harbaugh, Greg, 392, 393
Harrington, James, 43
Harris, Bernard, 152, 153, 155, 156, 160–163
Hart, John, 129, 325
Hartsfield, Hank, 120, 137, 138, 143, 145–147, 202
Hart, Terry, 139
Harvey, Jack, 96
Hauck, Rick, 34
Heflex Bioengineering Test (HBT), 28, 31
Helms, Susan, 249, 331, 333, 334, 338, 399
Henize, Karl, 44–46, 85–87, 96, 101, 103, 107
Henricks, Tom, 152, 155, 156, 160, 161, 331, 332, 337, 339
herpes, 116, 117, 354
Hieb, Rick, 315–320
High-Packed Digital Television Technical Demonstration (HI-PAC), 327

High-Resolution Telescope and Spectrograph (HRTS), 93, 96, 99–101, 187
Hilmers, Dave, 234, 293, 294, 297–299
Hire, Kay, 379, 382, 387
Hocker, Alexander, 2, 17, 18, 20–22, 26
Hoffman, Jeff, 166–169, 174–177, 179, 181, 182, 273, 276–278, 280–288, 390
Holloway, Harry, 373, 374, 392
Holloway, Tommy, 347, 388
Holographic Optical Laboratory (HOLOP), 154, 155
Holt, Glynn, 324, 328
Homick, Jerry, 385
Hopkins Ultraviolet Telescope (HUT), 169, 170, 172, 176, 177, 179, 184–187
Hornig, Donald, 7
Horowitz, Scott, 283, 284
Howard, Jenny, 84–86
Hubble Space Telescope (HST), 4, 23, 201, 240
Hudson, Hugh, 194
Hughes-Fulford, Millie, 353, 355–357, 359, 361, 363, 366, 368, 369
Humphrey, Hubert, 6

I

igloo, 29, 78, 91–93, 159, 167, 172, 181, 182, 188, 194, 195, 200, 202–204, 210, 218, 220, 389
Image Motion Compensation System (IMCS), 169, 185, 187, 191
Imaging Spectrometric Observatory (ISO), 60, 207
Inertial Upper Stage (IUS), 72
Instrument Pointing System (IPS), 82, 87–93, 98–101, 106, 111, 159, 165–169, 172, 173, 177–179, 182, 184–188, 190–195

Interface Configuration Experiment (ICE), 311
Intergovernmental Agreement, 20, 36, 38, 39, 42, 135
International Microgravity Laboratory-1 (IML-1), 290–296, 298, 299, 315, 318, 319, 339, 390
International Microgravity Laboratory-2 (IML-2), 290, 314–317, 319–323, 331, 332, 339, 343, 344, 379, 389
International Space Station (ISS), v, vi, 4, 23, 135, 253, 266, 305, 326, 332, 335, 341, 346, 350, 384, 388–391, 396, 399
Investigators Working Group (IWG), 56, 57, 95, 111, 119, 194, 262, 267
IONS experiment, 114
Ivins, Marsha, 273, 277, 278, 284

J

Jayroe, Robert, 187
Jemison, Mae, 262, 267–269, 271
Jernigan, Tammy, 182, 184, 186, 357, 359, 367, 369
Jet Propulsion Laboratory (JPL), 29, 46, 62, 94, 113, 119, 170, 204, 206, 226, 229, 239, 242, 243, 252, 253, 303, 305, 324
Johnson, Lyndon, 6, 7
Johnson Space Center (JSC), 15, 16, 23, 24, 38, 46, 47, 57, 58, 64, 67, 82, 102, 114, 119, 139, 166, 168, 173, 178, 185, 227, 229, 230, 236, 253, 287, 293, 301, 340, 353
Johnston, Mary Helen, 44, 119, 128
Joint Spacelab Working Group (JSLWG), 35, 39
Jones, Brendan, 7
Jones, Tom, 236, 237, 240–247
Jordan, Stuart, 193

K

Kavandi, Janet, 254, 255, 258
Kelso, Rob, 347
Kennedy, John, 7
Kennedy Space Center (KSC), 19, 27, 28, 31, 37, 55, 68, 75, 84, 89, 93, 110, 114, 117, 125, 137, 140, 163, 172, 173, 183, 187, 201, 207, 214, 216, 219, 223, 228, 234, 236, 240, 242, 247, 249, 251, 259, 269, 272, 282, 313, 322, 330, 336, 339, 346, 348, 388
Kennel, Charles, 252, 253
Kerwin, Joe, 57
Kissinger, Henry, 13
Kleinknecht, Kenny, 35, 36
Klyuchevskaya Sopka eruption, 246
Koszelak, Stan, 268
Kraft, Chris, 16, 32, 166
Kramp, Klaus, 46
Kregel, Kevin, 254–256, 258, 260, 331, 332, 334

L

Lacefield, Cleon, 84, 85
LaComb, Maureen, 300, 301
Laminar Soot Processes (LSP), 342, 349, 350
Lamonds, Harold, 119
Lampton, Mike, 57, 59, 67, 200, 201, 206, 208–210
Large Area Modular Array of Reflectors (LAMAR), 171, 172
Large Format Camera (LFC), 154, 230, 232
Large Isothermal Furnace (LIF), 265, 321, 344, 345, 351

424 Index

Lawrence, Wendy, 182, 184
Leavitt, William, 11
Lee, Jack, 16–18, 35–37
Lee, Mark, 249, 250, 261, 262, 264, 267, 269, 270, 272
Leestma, Dave, 167, 168, 174, 208, 211–213, 227, 228, 231–234
Lefèvre, Théo, 17, 18
Leibacher, John, 189
Leinbach, Mike, 175, 361
Lenoir, Bill, 139
Leslie, Fred, 324, 325, 327–329
Lester, Roy, 101
Lichtenberg, Byron, 55–60, 65, 69, 70, 74, 78, 82, 200, 201, 206, 208, 209, 211, 213
Lidar In-Space Technology Experiment (LITE), 248–251
Lieberman, Henry, 6
Life and Microgravity Spacelab (LMS), 330, 332, 333, 335–339, 379, 390
Lind, Don, 109, 110, 114, 116, 117, 119, 120, 122, 124–126, 131, 132
Linenger, Jerry, 249, 399
Linnehan, Rick, 331, 335, 337, 379, 384–388
Linteris, Greg, 339, 341, 342, 349
Liquid Phase Sintering (LPS), 345, 351
Lockheed, 62, 93, 96, 113, 187, 189, 193, 196, 207
Logsdon, John, 9, 13, 14, 17, 81, 390
Lopez-Alegria, Mike, 324, 399
Lord, Douglas, 16, 17, 21, 22, 34, 65, 68, 72, 73, 88, 89, 110
Lounge, Mike, 174, 175, 177, 179
Lousma, Jack, 48, 50–54
Lucas, Bill, 13, 16, 36, 213
Lucid, Shannon, 371, 374, 399

M

Magellan, Ferdinand, 289
Magnetic Field Experiment for TSS Missions (TEMAG), 274, 287
Magnetospheric Multiprobes (MMP), 196
Malerba, Franco, 57, 276, 278, 284
Marshall Space Flight Center (MSFC), 11–18, 24, 25, 35, 36, 42, 44, 56–58, 66, 69, 82, 95, 106, 111, 113, 114, 119, 130, 166, 169, 171, 173, 175, 178, 184, 185, 193, 194, 199, 201, 217, 267, 274, 290, 301, 310, 324, 330, 332, 342, 357, 399
Martin Marietta, 25, 26
Materials Experiment Assembly (MEA), 141
Materials Science Autonomous Payload (MAUS), 153
Materials Science Double Rack, 66
Materials Science Experiment Double Rack for Experiment Modules and Apparatus (MEDEA), 141, 144, 154
Matthiesen, David, 324, 328
McArthur, Bill, 372, 399
McBride, Jon, 167–169, 174, 227, 231
McCloy, John, 5
McDonnell Douglas, 4, 24, 25, 64, 68, 113, 390
McMonagle, Don, 220, 221, 223, 224
McNamara, Robert, 6, 7
Meade, Carl, 249, 250, 303, 304, 309, 312–314
Measurement of Air Pollution from Satellites (MAPS), 31, 230, 232, 239, 242, 243, 248
Measurement of Solar Constant (SOLCON), 62, 206, 207, 215, 218
Memorandum of Understanding, 1, 20, 135, 136, 274

Menzies, Robert, 46, 57
Merbold, Ulf, 55, 57–60, 65, 67, 69, 70, 74, 78, 82, 135–137, 139, 291, 294, 399
Mercuric Iodide Crystal Growth (MICG), 112, 119, 130
Messerschmid, Ernst, 137–139, 143
Messerschmitt-Bölkow-Blohm (MBB), 11, 13, 20, 21, 26, 89, 90
Metric Camera, 61, 201
Microabrasion Foil Experiment (MFE), 49
Microgravity Measurement Assembly (MMA), 157
Microgravity Science Laboratory-1 (MSL-1), 339–348, 350–352, 399
Microgravity Vestibular Investigations (MVI), 296, 297
Microwave Remote Sensing Experiment (MRSE), 61, 239
Miller, Tim, 217, 223
Millimetre-Wave Atmospheric Sounder (MAS), 205, 215, 218, 219
Mir space station, 391
Mission Control, 32–34, 46, 67, 68, 70, 71, 80, 84, 127, 128, 131, 142, 149, 156, 179, 180, 219, 224, 231, 234, 247, 255, 258, 282, 284, 285, 313, 328, 334, 342, 343, 349, 360, 363, 367, 382, 384
Mission-Peculiar Equipment Support Structure (MPESS), 110, 112, 114, 140, 141, 199, 200, 203, 205, 230, 235, 239, 252, 263, 275, 301
Mission specialists, 46, 47, 55–58, 86, 90, 96, 116, 137, 139, 149, 151, 152, 163, 166, 167, 182, 200, 208, 211, 227, 236, 242, 243, 249, 261, 262, 267, 276, 283, 291, 303, 315, 323, 331, 339, 353, 371, 379
Modular Optoelectronic Multispectral Scanner (MOMS), 154, 156
Mohri, Mamoru, 254, 255, 262, 266, 267, 269, 271, 272
Money, Ken, 291
Moore, Jesse, 92
Morrison, Dennis, 46
Morrison, James, 43, 44
Mount Pinatubo eruption, 126, 205, 246
Mukai, Chiaki, 267, 315, 316, 318, 379
Mullane, Mike, 117, 128, 293
Musgrave, Story, 46, 47, 83, 85–87, 96, 97, 103, 107
Myers, Dale, 12, 14, 16

N

Nagel, Steve, 53, 137–139, 142, 146, 147, 149, 150, 152, 153, 155, 156, 159–163
National Aeronautics and Space Act, 5
National Aeronautics and Space Administration (NASA), 2, 27, 55, 88, 111, 136, 166, 201, 228, 261, 273, 290, 353
National Space Development Agency of Japan (NASDA), 262, 263, 266, 267, 290, 315, 316, 380
Navigation Experiment (NAVEX), 141
Neurolab, 378–390, 398
Nicollier, Claude, 46, 57–59, 67, 139, 200, 273, 276–279, 281, 283, 284, 286
Night-Day Optical Survey of Lightning (NOSL), 31
Nixon, Richard, 1, 7–9, 11–13
Nordsieck, Ken, 167, 169, 179–181, 183

North American Rockwell, 3, 4, 11
North Atlantic Treaty Organisation (NATO), 7, 12
Nygren, Rick, 53, 391

O

Oberpfaffenhofen, 142, 143, 157, 163
Ocean Color Experiment (OCE), 31
Ochoa, Ellen, 216, 220, 221
Ockels, Wubbo, 45, 57–59, 67, 68, 137, 138, 141, 146
O'Connor, Bryan, 68, 120, 123, 357–360, 363–365, 367, 369
Office of Space and Terrestrial Applications-1 (OSTA-1), 28–34, 229, 230, 235
Office of Space and Terrestrial Applications-3 (OSTA-3), 227, 229, 230, 233, 235
Office of Space Science-1 (OSS-1), 48–51, 53, 94, 95, 104, 196
Operations and Checkout Building, 19, 31, 37, 55, 68, 71, 74, 137, 183, 228, 254, 347, 348
Orbital Manoeuvring System (OMS), 77, 84, 85
Orbiter Processing Facility (OPF), 28, 31, 69, 347
Organic Crystal Growth Facility (OCGF), 298
Oswald, Steve, 167, 182–184, 187, 216, 218, 219, 291, 294
Overmyer, Bob, 109, 117, 119, 120, 123, 125, 127, 132

P

Paine, Tom, 8–10
Palaora, Hans, 15
Panitz, Hans-Joachim, 142
Parazynski, Scott, 220–223
Parise, Ron, 167–169, 174, 177, 181–185
Parker, Bob, 45, 46, 55, 56, 60, 67–70, 74, 76, 78, 82, 119, 165, 167–169, 174, 177, 179, 181
Patchett, Bruce, 96
Pawelczyk, Jim, 379, 382, 385–387
Payload commanders, 56, 149–151, 161, 182, 184, 203, 206, 208, 216, 220, 236, 237, 241–243, 267, 276, 283, 293, 303–305, 310, 315, 316, 319, 323, 331, 339, 370, 371, 373, 378–380, 385, 392, 395
Payload Operations Control Center (POCC), 67–69, 71, 93, 100, 122, 128, 131, 142, 157, 173, 307
Payload specialists, 45–47, 55–59, 67, 88, 90, 95, 96, 100, 119–121, 124, 128, 137, 139, 143, 148, 152, 166, 167, 169, 171, 177, 179, 181, 182, 188, 194, 200, 201, 204, 206, 208–210, 227, 242, 262, 266, 267, 278, 283, 291, 300, 303, 304, 307, 315, 316, 324, 328, 331, 339, 344, 353, 355–357, 359, 371, 372, 379
Payload Specialist Selection Group (PSSG), 57
personalities, 208–211, 358
Peterson, Paul, 96
Pfeiffer, Robert, 35
Phenomena Induced by Charged Particle Beams (PICPAB), 61, 62
Phillips, Bob, 353, 355–357, 372
Physics of Hard Spheres (PHaSE), 351
Pinhole/Occulter Facility (P/OF), 194, 195
Plant Generic Bioprocessing Apparatus (PGBA), 345, 351
Plant Growth Unit (PGU), 51

Plasma Diagnostics Package (PDP), 50, 51, 53, 94, 104–106, 196
Pollack, Herman, 13
Power Extension Package (PEP), 301
Prahl, Joe, 303, 304
Precourt, Charlie, 152, 155, 156, 160, 161, 392–394, 398
Principal investigators, 44, 46, 47, 56, 67, 95, 96, 118, 119, 129, 139, 169, 195, 214, 252, 262, 265, 267, 302, 309, 310, 328, 332, 339, 344
Prinz, Dianne, 96, 100, 188
Protein Crystal Growth (PCG), 263, 265, 268, 269, 298, 305, 307, 308, 326, 396
Prozesskamer, 139, 140
Puddy, Don, 33, 236, 293
Puppi, Giampietro, 283
Puz, Craig, 300, 301

Q

Quistgaard, Erik, 58, 81

R

Readdy, Bill, 291–294, 299
Reagan, Ronald, 43, 81
Redundant Set Launch Sequencer (RSLS), 244
Rees, Eberhard, 15
Regenerative Carbon Dioxide Removal System (RCRS), 303, 382
Reinartz, Stanley, 42
Remote Manipulator System (RMS), 28, 32, 48, 51, 104, 105, 219, 222, 231–234, 279, 301
Research and Applications Module (RAM)/Sortie Can, 4, 10, 12–18, 25

Research Animal Holding Facility (RAHF), 113, 115, 116, 122–124, 132, 133, 354, 361, 362, 373
Research on Electrodynamic Tether Effects (RETE), 274, 287
Research on Orbital Plasma Electrodynamics (ROPE), 274, 287
Reyes, Ernie, 36
Rice, Gene, 29, 63
Richards, Dick, 84, 85, 167, 168, 174, 249, 251, 303, 304, 307, 310, 312–314
Ride, Sally, 32, 203, 227, 229, 231–233, 305
Robotic Technology Experiment (ROTEX), 155, 156
Rockwell International, 3, 64, 229, 302, 303, 322
Rogers Commission, 195, 200, 202, 203, 355, 359
Rogers, William, 12, 17, 18, 202
Rominger, Kent, 324
Ronney, Paul, 339, 344
Ross, Jerry, 149–153, 155, 156, 160–162
Rotter, Hank, 65
Runco, Mario, 314

S

Sacco, Al, 303, 304, 307, 310, 324–329
Sahm, Peter, 142
Sawin, Charles, 46, 47
Schlegel, Hans, 152, 155, 156, 160, 162
Schneider, Howard, 377
Schwinghamer, Bob, 175
Scully-Power, Paul, 227

Searfoss, Rick, 372, 379, 382, 387, 388
Sebesta, Lorenza, 9, 11–13, 20, 26, 35
Seddon, Rhea, 235, 262, 267, 353–367, 369–378, 380, 387, 388, 390
Shaw, Brewster, 55, 58, 65, 66, 69, 70, 73, 74, 76, 78–82, 117
Shaw, Chuck, 283, 287
Shayler, Dave, 47, 64, 102
shifts and shift handovers, 70, 138, 161
Shriver, Loren, 277–279, 281, 282, 284
Shuttle Electrodynamic Tether System (SETS), 274
Shuttle High Energy Astrophysics Laboratory (SHEAL), 171, 172
Shuttle Imaging Radar-A (SIR-A), 29, 30, 33, 226, 230, 238
Shuttle Imaging Radar-B (SIR-B), 226–235, 238
Shuttle Infrared Telescope Facility (SIRTF), 190–192
Shuttle Multispectral Infrared Radiometer (SMIRR), 29, 30
Shuttle Radar Laboratory (SRL), 235–248, 252–254
Shuttle Radar Topography Mission (SRTM), 252, 254–259
Shuttle Solar Backscatter Ultraviolet (SSBUV), 207, 214, 218–220
Shuttle-Spacelab Induced Atmosphere (SIA), 50
Sieck, Bob, 160, 369
Simon, George, 96
simulators, 102, 146, 168, 169, 202, 210, 278, 286, 357
sleep stations/bunks, 70, 103, 124, 138, 161, 237, 241, 270, 286, 310, 317, 324, 356
Slow Rotating Centrifuge Microscope (NIZEMI), 318

Small Helium-Cooled Infrared Telescope (IRT), 92, 94, 103, 104, 195
Smith, Mike, 200
Smith, Steve, 243
Snyder, Bob, 290, 322
Sokolowski, Robert, 263
Solar Flare X-ray Polarimeter (SFXP), 50, 53
Solar Optical Telescope (SOT), 192–194
Solar Optical Universal Polarimeter (SOUP), 93, 95, 96, 98–100, 106, 187, 189, 190
Solar Spectrum (SOLSPEC), 62, 206, 207, 215, 218, 220
Solar Ultraviolet Spectral Irradiance Monitor (SUSIM), 50, 93, 96, 100, 207, 215, 218
Solid Rocket Boosters (SRBs), 2, 3, 31, 71–73, 75, 76, 83, 86, 109, 165, 176, 216, 217, 235, 243, 306
Solovyov, Anatoli, 393, 397, 398
South Atlantic Anomaly, 176, 184
Souza, Kenneth, 265
Space Acceleration Measurement System (SAMS), 270
Spaceborne Imaging Radar-C (SIR-C), 238, 239, 241, 246, 247, 252–254
Space Experiments with Particle Accelerators (SEPAC), 61, 69, 81, 196, 202, 205, 207, 213–215
Spacelab
 'barter' plans for, 43
 costs of, 40, 389
 duplication concerns of, 38
 Follow-On Production (FOP) and second Spacelab, 41
 funding of, 58

Index 429

module
    'core' segment, 62, 199
    end-cones, 61, 63, 78, 129, 300
    experiment racks, 300
    'experiment' segment, 199
    Scientific Airlock (SAL), 10, 62, 111, 127
    Scientific Window Adapter Assembly (SWAA), 10, 63, 114, 201
naming of, 267
ownership of, 26, 137
pallet, 25, 28, 30, 31, 39, 48, 51, 88, 159, 167, 170, 171, 178, 179, 184, 188, 194, 195, 202, 227, 228, 234, 245, 249–253, 275, 279, 398
as partnership, 36
Phase A studies, 17
Phase B studies, 17–20
Phase C/D, 21, 22, 26
Shuttle costs, 40
tunnel of, 6, 15, 41, 57, 63, 64, 80, 115, 140, 146, 147, 151, 266, 270, 271, 294, 298, 301, 306, 310, 324, 347, 393
Spacelab-1, 23, 28, 39, 40, 43, 54–62, 65–73, 76–82, 87, 95, 110, 111, 117–119, 127, 135–137, 139, 166, 167, 193, 199–201, 205–207, 228, 239, 262, 278, 376, 389, 399
Spacelab-2, 78, 82, 83, 86–104, 106, 111, 117, 171, 187–189, 195, 196
Spacelab-3, 25, 88, 90, 111–117, 119, 120, 123–127, 129–132, 147, 205, 215, 266, 298, 303, 324, 325, 328, 361, 389
Spacelab-4, 123, 133, 353, 354, 361, 365

Spacelab-D1, 41, 136–145, 147, 149, 150, 152–154, 157, 270, 390, 395
Spacelab-D2, 142, 147, 150–163, 293, 331, 389, 392
Spacelab-D3, 158, 159
Spacelab-D4, 159
Spacelab-J, 262–271, 294, 316, 390, 392
Spacelab Life Sciences-1 (SLS-1), 353–364, 366–368, 370–374, 376, 377, 389, 390
Spacelab Life Sciences-2 (SLS-2), 332, 354–356, 362, 370–377, 379, 380, 382, 388, 390
Spacelab-Mir, 389, 393–398
Spacelab Mission Development (SMD) tests, 46
Spacelab Mission Operations Control Facility, 173
Spacelab Only modifications, 69
Spacelab Ultraviolet Optical Telescope (SUOT), 165
Space Plasma Laboratory (SPL), 195, 196
Space Radar Laboratory (SRL), 236, 238, 242, 248, 252–254
Space Radar Laboratory-1 (SRL-1), 235–241, 243, 246, 247, 252
Space Radar Laboratory-2 (SRL-2), 235–237, 242, 243, 245, 246, 248, 252
Space Shuttle
    Atlantis, 3, 197, 200–202, 209, 211–215, 220, 221, 223, 225, 236, 268, 275, 276, 278–280, 282, 292, 389, 391–394, 396–398
    Challenger, 3, 41, 72, 83, 109, 137, 165, 199, 227, 263, 273, 290, 355

Space Shuttle (*cont.*)
  Columbia, 3, 27, 55, 128, 140, 167, 199, 226, 263, 283, 290, 353
  Discovery, 3, 4, 120, 182, 216–219, 235, 236, 249–251, 268, 292–294, 299
  Endeavour, 3, 4, 150, 183–185, 187, 216, 237, 240, 241, 243–247, 252, 254–259, 261, 266, 268–272, 302, 323, 339
  Enterprise, 3, 252
  flight deck, 2, 6, 33, 46, 48, 70, 74, 78, 79, 83–85, 87, 90, 93, 95, 100, 102–104, 117, 123, 132, 147, 148, 160, 175, 177, 185, 212, 213, 219, 223, 231, 238, 241, 246, 250, 270, 272, 280, 284, 312, 363, 364, 368, 369
  middeck, 2, 6, 15, 28, 41, 46, 48, 52, 64, 69, 70, 74, 87, 94, 103, 115, 124, 132, 138, 147, 148, 157, 161–163, 184, 212, 213, 219, 231, 237, 241, 243, 250, 265, 269, 270, 272, 292, 294, 299, 307, 309, 310, 312, 317, 318, 324–326, 329, 330, 336, 346, 350, 351, 362, 363, 367, 381, 387, 393, 394, 396
  payload bay, v, 2, 6, 10, 14, 15, 28–30, 41, 48–51, 63, 70, 91, 92, 99, 104, 106, 110, 112, 149, 151, 171, 177, 179, 184, 191, 194, 199, 202, 205, 207, 214, 218–220, 223, 235, 239, 241, 251–253, 259, 275, 276, 278, 279, 282, 286, 294, 299, 302, 340, 342, 347, 348, 358, 363, 389, 390, 397, 398
  payload bay doors, 11, 32, 53, 64, 93, 178, 217, 219, 242, 258, 299, 324, 331, 332, 367, 368

space sickness, 48, 65, 99, 114, 121, 132, 245, 266, 267, 317, 325, 338, 360, 366, 376
Space Station Freedom, 172, 266, 305
Space Task Group (STG), 2, 8–10
Space tug, 10–12
Spartan, 217–219, 221, 249, 250
Springer, Bob, 200, 201
squirrel monkeys, 110, 113, 114, 116–118, 121–123, 354
Starbird, 300, 301
Starlab, 293, 300, 301, 389
Steimle, Hans-Ulrich, 142, 143
Stewart, Bob, 201
Still, Susan, 166, 340, 343, 349, 350, 352, 373
Stoewer, Heinz, 34, 35
Stoltenberg, Gerhard, 7
Stone, Randy, 367
Stonesifer, John, 301
Strategic Defense Initiative Organization (SDIO), 300, 301
Strauss, Franz Josef, 26, 243
Strekalov, Gennadi, 393, 394, 396
Strong, Keith, 96
Structure of Flame Balls at Low-Lewis Number (SOFBALL), 342, 344, 349
STS-1, 27, 28
STS-2, 27–29, 31, 32, 34, 48, 54, 66, 226, 229, 239
STS-3, 48–52, 54, 93, 94
STS-9, 15, 55, 62, 68–74, 77–79, 81, 82, 114, 139
STS-35, 173–175, 177–180
STS-40, 302, 355, 357–360, 362, 366
STS-41G, 227–231, 233, 234, 239
STS-42, 290, 294, 298
STS-45, 204, 208, 209, 211, 213–216, 218, 275
STS-46, 276–279, 281–286
STS-47, 261–263, 268–272, 290
STS-50, v, 303, 304, 306, 307, 312

STS-51B, 109–112, 118, 125, 126, 132
STS-51F, 83–85, 87, 95–97, 99, 101, 102, 104, 106, 189, 192
STS-51H, 200, 201
STS-51K, 139
STS-55, 149, 150, 153–155, 158–160, 163, 301
STS-56, 204, 216, 217, 219–221
STS-58, 204, 322, 369, 372, 377
STS-59, 237, 239–244, 247, 248, 355
STS-61A, v, 137, 140, 142, 144, 145, 148
STS-61D, 353–356
STS-61E, 165, 167–169
STS-61K, 200–202
STS-64, 249–251
STS-65, 314, 315, 317–319, 321, 322
STS-66, 204, 220–223
STS-67, 182–184, 186, 187, 204, 249, 322, 339
STS-68, 243–247
STS-71, v, vi, 391, 392, 395, 398
STS-73, 301, 323, 325, 328, 329
STS-75, 236, 283, 284, 287
STS-78, 204, 331, 332, 334, 337–339
STS-83, 339–342, 344–347, 349, 351
STS-90, 379–381, 387, 388
STS-94, 204, 347, 349–352
STS-99, 252, 254, 255, 257–260
Sullivan, Kathy, 203, 205, 206, 208–215, 227, 229, 231, 232, 234
Sunlab, 187–190, 389
Superfluid Helium Experiment (SFHE), 94
Supernova 1987A, 170, 175, 179
Surface Tension Driven Convection Experiment (STDCE), 311, 312, 326, 327
Szalai, Kenneth, 287, 288

T

Tandberg-Hanssen, Einar, 194
Tanner, Joe, 220, 221
Tarbell, Ted, 189, 190
Taylor, Michael, 46
Terrile, Richard, 57
Tethered Satellite System-1 (TSS-1), 275, 276, 278, 279, 282
Tethered Satellite System-1R (TSS-1R), 283–285, 287
Tether Optical Phenomena (TOP), 274, 278
Thagard, Norm, 116, 117, 119–125, 291, 293, 294, 297, 298, 392, 394, 395, 398, 399
Theoretical and Experimental Study of Beam Plasma Physics (TEBPP), 196
Thermal Canister Experiment (TCE), 49, 53
Thiele, Gerhard, 152, 153, 252, 254, 255, 257, 258
Thirsk, Bob, 331, 335–338, 399
Thomas, Don, 315, 316, 321, 339, 340, 343–346, 349, 350
Thornton, Bill, 47, 56, 66, 116, 117, 119–123, 132, 166
Thornton, Kathy, 323–328, 330
Torr, Marsha, 199, 201
Torso Rotation Experiment (TRE), 335, 337
Tracking and Data Relay Satellites (TDRS), 71, 72, 142, 173, 231, 233
Trafton, Wil, 288
Treetops (Honeywell simulation), 194, 195
Trella, Massimo, 19, 34
Trinh, Gene, 119, 303, 304, 310, 312, 314
Truly, Dick, 27, 28, 31–33
TRW, 25, 26, 72, 196
Two-Axis Pointing System (TAPS), 171, 180, 196

U

Ubar, 225, 226, 233
Ultraviolet Imaging Telescope (UIT), 169, 170, 173, 179, 184, 185, 187
Unique Support Structure (USS), 92, 149, 153, 154, 156, 157, 159
untouchables, 117–120
Urban, Eugene, 95, 105, 106
Urbani, Luca, 331
Urine Monitoring Investigation (UMI), 114
Urine Monitoring System (UMS), 114, 156, 157
U.S. Microgravity Laboratory-1 (USML-1), v, vi, 293, 301–305, 307–313, 322, 324–326, 328, 329, 392, 395
U.S. Microgravity Laboratory-2 (USML-2), 301, 303, 322–329, 331, 340, 399
Utsman, Tom, 217

V

Valery, Nicholas, 14
Vandenberg Air Force Base, 235, 300
van den Berg, Lodewijk, 119–121, 124, 130–132
Vangen, Scott, 182
Vanhooser, Teresa, 342
Vapour Crystal Growth System (VCGS), 112, 119, 130, 298
Vehicle Assembly Building (VAB), 31, 71, 73, 160, 174, 245
Vehicle Charging and Potential (VCAP), 51, 94, 105, 106, 196
Verification Flight Tests, 55–107
Very Wide Field Camera (VWFC), 62, 111, 114, 125–127
Vestibular Sled, 139–141, 145
VFW-Fokker/ERNO, 20–22, 25, 26, 88, 89
Visual and Vestibular Integration System (VVIS), 387
Visuo-Motor Co-ordination Facility (VCF), 386
Voss, Janice, 254–256, 339, 340, 342, 349, 352
Vruels, Fred, 15

W

Walter, Ulrich, 152, 155, 156, 160, 162
Walz, Carl, 315–317, 321, 399
Wang, Taylor, 119, 120, 124, 127–129, 132, 328
Waring, Stephen, 15, 16
Wavefront Control Experiment (WCE), 300
Waves in Space Plasma (WISP), 196
Weaver, Lee, 44, 46
Werkstofflabor, 139, 140, 144, 154, 155
Werner, Michael, 191
Wetherbee, Jim, 277
Whitaker, Ann, 44, 57
White Sands, 52–54, 233, 255
Whitney, Robert, 373, 374
Wide-Angle Michelson Doppler Imaging Interferometer (WAMDII), 196, 197
Wide Field Camera (WFC), 166, 170
Wilcutt, Terry, 243–245, 247
Williams, Bill, 47, 353, 356
Williams, Dave, 379, 382, 384–387

Wisconsin Ultraviolet
 Photopolarimeter Experiment
 (WUPPE), 169, 170, 172,
 177–179, 184–187, 221
Wisoff, Jeff, 243, 245–247
Witteborn, Fred, 191
Wolf, Dave, 287, 371, 374–377,
 394, 399

X-Band Synthetic Aperture Radar
 (X-SAR), 225–260
X-Ray Telescope (XRT), 92, 93, 195

Yardley, John, 42
Young, John, 27, 55, 58, 64, 65, 69,
 70, 73–76, 79–82
Young, Larry, 371, 372
Young, Lou, 191
Youskauskas, John, 58, 93, 124,
 137, 210

Zeolite Crystal Growth (ZCG),
 324, 325